Benchmark Papers
in Soil Science

Series Editor: Charles W. Finkl, Jnr.
Nova University

Volume

1 SOIL CLASSIFICATION / *Charles W. Finkl, Jnr.*

Benchmark Papers
in Soil Science / 1

A BENCHMARK® Books Series

SOIL CLASSIFICATION

Edited by

CHARLES W. FINKL, JNR.

Nova University

Hutchinson Ross Publishing Company

Stroudsburg, Pennsylvania

Copyright © 1982 by **Hutchinson Ross Publishing Company**
Benchmark Papers in Soil Science, Volume 1
Library of Congress Catalog Card Number: 81-6214
ISBN: 0-87933-399-5

84 83 82 1 2 3 4 5
Manufactured in the United States of America.

LIBRARY OF CONGRESS CATALOGING IN PUBLICATION DATA
Main entry under title:
Soil classification.
   (Benchmark papers in soil science; v. 1)
   Includes bibliographical references and indexes.
   1. Soils—Classification.   I. Finkl, Charles W., 1941-   II. Series.
S592.16.S64   631.4′4   81-6214
ISBN 0-87933-399-5   AACR2

Distributed by: world wide by Academic Press,
VAN NOSTRAND Harcourt Brace Jovanovich,
REINHOLD (U.K.)
MOLLY MILLARS LANE
WOKINGHAM, BERKS.
RG11 2PY, ENGLAND

CONTENTS

Contents

PART III: GENERAL CLASSIFICATION EFFORTS

PART IV: LIMITED OR REGIONAL APPROACHES

SERIES EDITOR'S FOREWORD

The Benchmark Papers in the Soil Science series attempt to provide cogent summaries of the field by reproducing classical and modern papers, ones that provide keys to understanding of critical turning points in the development of the discipline. Scientific literature today is so vast and widely dispersed, especially in a multifaceted discipline like soil science, that much valuable information becomes ignored by default. Many pioneering works are now coveted by libraries, and retrieval from the archives is not easy. In fact, many important papers published in the emphemeral literature are no longer available to serious or committed researchers through interlibrary loan. Other professionals devoted to teaching or burdened with administrative duties must be hard pressed to keep up with comprehensive arrays of technical literatures spread through scores of journals. Most of us can, at best, skim only a few select journals to make copies of tables of contents, abstracts and summaries, and reviews in order to remain abreast of specialized and often limited aspects of the robust field of soil science as a whole.

Benchmark Papers in Soil Science, developed as a practical solution to this problem, reprints key papers and investigative landmarks that relate to a common theme. The papers are reproduced in facsimile, either in their entirety or in significant part, so readers can follow major original events in the field, not peruse paraphrased or abbreviated versions of others. Some foreign works have been especially translated for use in the series. Occasionally short, foreign language articles are reproduced from French or German journals.

Essays by the volume editor provide running commentaries that introduce readers to highlights in the field, provide critical evaluation of the significance of the various papers, and discuss the development of selected topics or subject areas. It is hoped that the volume editor's comments will ease the transition for the seasoned investigator who wishes to step into a new field of research as well as provide students and professors with a compact working library of most important scientific advances in soil science.

Areas of specialization in soil science are divided by the International Society of Soil Science into seven divisions or "commissions." The first six commissions cover soil physics, chemistry, mineralogy, biology, fertility, and technology. Because the scope of the field is so great, we concentrate initially on topics traditionally devoted to the seventh commission: soil morphology, genesis, classification, and geography. The series thus begins with volumes dealing with the major soils of the world: their recognition, characteristics, formation, distribution, and classification. Other volumes concentrate on topics in agronomy, soil-plant relationships, soil engineering topics, or melds of pure science with soil systems. Benchmark Papers in Soil Science plow deeply through the field, picking significant but timely topics on an eclectic basis.

Each volume in the series is edited by a specialist or authority in the area covered by the book. The volume editor's efforts reflect a concerted worldwide search, review, selection, and distillation of the primary literature contained in journals and monographs and in industrial and governmental reports. Individual volumes thus represent an information-selection and repackaging program of value to libraries, students, and professionals.

Benchmark books contain a preface, introduction, and highlight commentaries by the volume editor. Many volumes contain rare papers that are hard to locate and obtain, as well as landmark papers published in English for the first time. All volumes contain author citation and subject indexes of the contained papers, usually twenty to fifty key papers in a given subject area.

This inaugural volume introduces the Benchmark Papers in Soil Science series. The topic of soil classification sets the stage for detailed considerations of specific soils to follow in subsequent volumes. The topic has a long history and invites discussion of controversial issues that remain unresolved today. Soil, a complex three-phase natural system, poses unique problems to taxonomists wishing to classify evolving natural bodies that retain vestiges of the past in their parent materials or from former phases of soil development. Important turning points in the history of soil classification are grouped here for the convenience of readers.

CHARLES W. FINKL, JNR.

PREFACE

Classifying things is characteristic of the human mind; it permeates all sorts of scientific endeavor and is essential to organized knowledge. Classification has long occupied a central position in all branches of natural science, including studies of soil. Soil science, a relatively new discipline compared to other fields such as botany, biology, zoology, and mineralogy, lacks an internationally accepted taxonomic system. Many factors account for the wide variety of classification schemes, but such proliferation is perhaps due in part to the unique properties of soils, which exist as multi-dimensional systems in space and time. Different concepts of soil have also led to developments of the specialized classifications employed by pedologists, edaphologists, geologists, and engineers.

In this benchmark volume the evolution of classificatory efforts is traced from early single-purpose classifications to today's modern multi-categorical taxonomic systems. It is thus, as much as anything, a statement of the present state-of-the-art as soil scientists aim for the ultimate goal of an ideal, universal soil classification system. Due to economy of space, only a few papers from a great literature on soil classification could be reproduced here. Although necessity dictated an eclectic approach, it is hoped that this collection of benchmark papers documents a fair selection of critical breakthroughs that clarify questions of definition, nomenclature, and classification.

This volume would not have been possible without the help and encouragement of many others. For the useful comments and generous suggestions of my colleagues, a word of thanks is due: G. Aubert (France), H. de Bakker (The Netherlands), O. W. Bidwell (USA), T. D. Biswas (India), R. Dudal (Italy), F. D. Hole (USA), Ichiro Kanno (Japan), E. G. Knox (USA), W. M. McArthur (Australia), J. A. McKeague (Canada), C. N. MacVicar (South Africa), Masanori Mitsuchi (Japan), Takeshi Matsui (Japan), E. Mückenhausen (Federal Republic of Germany), K. H. Northcote (Australia), H. Späth (Federal Republic of Germany), J. C. F. Tedrow (USA), F. C. Ugolini (USA), E. P. Whiteside (USA), H. F. Winterkorn (USA), D. H. Yaalon (Israel).

A special word of thanks goes to Roy W. Simonson (formerly director of Soil Classification and Correlation, U.S. Soil Conservation Service) for his interest, encouragement, and helpful suggestions. His review of the manuscript is also much appreciated.

CHARLES W. FINKL, JNR.

CONTENTS BY AUTHOR

SOIL CLASSIFICATION

INTRODUCTION

Due to the wide scope of the subject matter, the papers collected here are organized into groups of related topics that reflect important developments in the field. Papers in Parts I, II, and III are ordered chronologically to emphasize the evolution of basic principles and the appearance of pivotal concepts. Those in Parts IV, V, and VI are examples of different approaches to soil classification and correlation. A bibliography follows the single paper in Part VI.

PIONEERS OF SOIL CLASSIFICATION

Men have tilled, drained, and irrigated the soil for at least six millenia. Although these enterprises remain basic to civilization, systematic study of agriculture effectively began in the first half of the nineteenth century when enquiries into the nature of soils were almost exclusively chemical. The development of soil taxonomies, the concern of this volume, came still later. Some forty centuries earlier, however, the Chinese classified soils according to productivity as a basis for tax assessment (Ping-Hua Lee, 1921). Details of what may have been done in long past centuries are uncertain and all known attempts at soil classification are part of recorded history.

Even though there was significant progress in the late 1800s and early 1900s, soil classification as an organized effort is essentially a twentieth-century phenomenon. Soil classification schemes were then, as they are today, tied to concepts of soil formation and knowledge of soil characteristics. Classifications thus represent the current status of pedological thought and as such knowledge accrues, taxonomic schemes are updated, revised, or replaced. The historical development of soil classification must also be considered in relation to concepts of soil as they differ from time to time. Primary or initial thrusts in the field of soil classification took place in Russia, Western Europe, and

North America. Systems in those regions were, in effect, statements of prevalent pedological concepts of the time.

The early schemes produced by Russian pedologists were, for the most part, simple environmental classifications that noted differences in soils on a continental scale. Descriptions and classifications were based mainly on the character of A horizons. It was not until well into the twentieth century that the full A-B-C-horizon designations were adopted in profile descriptions. In fact, improved understanding of soil development came through the work of a Dane, P. E. Müller, who showed in the late 1880s that B horizons of podzols were genetically related to the rest of the profile. Prior to his work, B horizons were simply regarded as transitional between A and C horizons. The classification schemes developed by Dokuchaev (1879, 1893) and Sibirtzev (1895), pioneers in the Russian school of pedology, were based on the character of surface soils and emphasized the role of climate as a soil-forming factor. Both workers regarded the freely draining soils of the main climatic zones as examples of normal, zonal soils. Local exceptions, due to the dominance of some other soil-forming factor, were treated as transitional or intrazonal soils. Immature or poorly developed soils found in any environment were considered abnormal, because they could not be explained according to climatic input. These early efforts established the geographic-environmental approach to soil classification and were important enough contributions at the time. By 1914, Glinka had introduced a refinement making a distinction between soils developed to maturity mainly by external, so-called ektodynamorphic factors such as climate and those whose development was inhibited by internal (endodynamorphic) conditions such as parent materials. In his later classification, Glinka (1924) recognized five major pedogenetic processes that were termed lateritic, podzolic, chernozemic, solonetzic, and swampy. This factorial classification, based on interactions of multiple soil-forming factors, marked an important change in the evolution of Russian soil classifications. Russian workers still cling to modified versions of the geographic, genetic, and soil-process frameworks adopted in pioneering stages.

Scholars in Western Europe generally learned of early Russian classification efforts through the small book by Glinka (1914). A decade later, pedologists in North America gained access to results of the independent efforts of their Russian contemporaries via Marbut's English translation of Glinka's work. Best known for his work on cemented B horizons, Müller (1887) also classified surface-soil organic horizons under beech woodland into mor, moder, and mull. Kubiëna (1953) later recognized similar types of organic materials in his natural classification of soils. In addition to significant contributions in the area of soil chemistry and profile dynamics of "Mitteleuropäische Braunerde" (Brauner Waldboden, Brown Forest) soils in Germany and Austria, Ramann (1918) also developed a classification scheme based on the interaction of soil-forming factors much as some of his Russian contemporaries had done. In Britain, Robinson (1932) presented a general classifcation scheme that appeared to be a spinoff from the Marbut (1927) classifcation in the United States. This was still a genetic classification based on the division of soil-forming environments. Other European efforts zeroed in on classification of specific groups of soils elsewhere, for example, tropical soils (Harrassowitz, 1930; Joachim, 1935; Mohr. 1938).

The origins of U.S. soil-classification systems go back to the middle 1800s, having been initiated about the same time but independent of pedological efforts in Russia. Hilgard (1860, 1892, 1906), Shaler (1891), and Whitney (1909) pioneered research into soil-forming processes and were responsible for many pedological ideas of the time. Whitney's predecessors were among the first to publish results of soil surveys but he became only incidentally involved with this activity. Soil classification in the United States began with a geological bias but gradually moved toward consideration of pedological systems keyed to the concept of independent natural soil bodies with their own genesis and morphology. Marbut's contributions are of particular interest because he brought new ideas to the United States and offered many of his own. Sometimes referred to as the founder of the American school of pedology, Marbut (1927) presented a scheme of soil classification, summarizing much of his own work, at the First International Congress of Soil Science in Washington, D.C. This was the first effort in the United States to devise a comprehensive system of soil classification. Marbut's classification was subsequently replaced in the American soil survey program with the adoption of the scheme presented by Baldwin, Kellogg, and Thorp (Paper 12). The new system retained Marbut's six category names with the addition of order and suborder for the top two but dropped the names pedalfer and pedocal as classes in the highest category. The system remained largely intact during a quarter century of use.

LOGIC AND FUNDAMENTAL PREMISES

Basic principles and units important to soil classification are collected in Part II. Papers 4 through 7 draw attention to the general principles and logical bases of soil classification, to changes in diagnostic criteria, and to different attitudes and contrasting paradigms of soil classification. Early approaches were based largely on suppositions regarding the affects of parent materials, native vegetation, or climate on soil. Still other approaches, based on genetic factors of soil formation or selected attributes of soils, have been tried with varying degrees of success. Ecological principles are, for example, occasionally still favored as a basis for soil classification (for example, Duchaufour, 1978) as are numerical methods (for example, Bidwell and Hole, 1964; Rayner, 1966; Moore, et al., 1972; Webster, 1977). Classification of soils according to parameters that have some practical significance include, for example, those based on the nature of the exchange complex (Gedroits, 1927), duricrusts (Dury, 1969), physical and chemical ripening properties and processes (Pons and Zonneveld, 1965), and salinity (Ivanova and Rozanov, 1939; Bazilevich and Pankova, 1968, 1969), in addition to the technical classifications of engineers (for example, Terzaghi, 1927; Casagrande, 1947; Asphalt Institute Staff, 1969; Portland Cement Association Staff, Paper 25). Experience has shown, however, that the most workable and practical classification of soils as natural bodies are based on general soil characteristics that can be observed or measured (Simonson, 1962), not on external causative factors that are based on inference. The important principle of using soil characteristics themselves and not possible explanations for their occurrence was stressed long ago by Coffey (Paper 2) and Marbut (Paper 4).

The principal elements of soil classification have been studied by numerous workers (for example, Coffey, Paper 2; Marbut, Paper 4; Cline, Paper 5; Robinson, 1932, 1949; Stephens, 1950; Rode, 1957; Manil, 1959; Muir, 1962; Smith, 1965 a and b; Mulcahy and Humphries, 1967; MacVicar, 1969; Schelling, 1970) who concluded that the development of any soil-classification system requires a purpose or specific objective, a definition of what is meant by "soil," delineation of the soil universe to be comprehended, and a statement of the criteria or soil characteristics used in the system. The process of soil classification thus usually involves: (1) identification of the range in characteristics within the population; (2) creation of classes according to similarity among individuals; and (3) prediction of individual behavior from knowledge of class characteristics. Failure to address any one of several systematic conditions often results in schemes that cannot withstand the test of practical use.

Some important advances in the recognition and definition of basic soil entities, as they relate to classification, are grouped in Part II under Units of Classification. Recognition of the soil body and soil individual, and later of the pedon and polypedon, was especially important to classification efforts. There has always existed a fundamental perplexity as to how to proceed with a proper scientific classification of soils. Unlike minerals, plants, or animals, soils occur as integrated (coupled)systems composed of solid, liquid, and gaseous phases. Because soils exist as dynamic, interconnected systems in a complex continuum of change, their classification is more difficult and contentious than are those of most other natural populations. Arbitrary judgments are thus inevitable in the creation of classes due to ranges of properties within soil populations. Many attempts to classify soils are of necessity based on assumptions that the units being classified are discrete and definable. Even if the soil units for classification are units of convenience, subjective and artifical (Cruickshank, 1972), they function as a means of organization, communication, and argument.

GENERAL SYSTEMS FOR CLASSIFYING SOILS OF THE WORLD

There are far more classifications based on the soils of a particular region than general efforts to classify soils of the world. The examples in Part III represent two U.S. efforts (Paper 12 and 14) and one Russian attempt (Paper 13) at a general system of soil classification. Until fairly recently the problem existed as to whether knowledge of soil was adequate for the elaboration of an efficient and tenable world classification. Many early classifiers concentrated on local knowledge inadvertently avoiding pitfalls associated with tackling the development of a comprehensive world system. Even though there are many regions that today remain unmapped, even on a reconnaissance level, the inventory of world soils has increased greatly since World War II, giving soil taxonomists a reasonable grasp of the global pedological data base.

The system adpoted by the national soil survey program in the United States in 1938, as proposed by Baldwin, Kellogg, and

Thorp (Paper 12), received wide international attention and was adopted by numerous other countries with few modifications. After some decades of use, the system with later modifications (Riecken and Smith, 1949; Thorp and Smith, 1949) and minor abridgements was replaced in 1975 by the new *Soil Taxonomy*, the official system of the U.S. Department of Agriculture. Recognition of this system by foreign soil surveys (see discussion in Cline, 1980, pp. 198–201) must be regarded as a measure of international acceptance. Notable features of this most recent effort include attention given to profile morphology and evaluation of diagnostic horizons. Such emphasis by U.S. pedologists may encourage others to adopt horizon characteristics and sequences as criteria for classification.

Many features of the U.S. system, including its distinctive terminology, are incorporated into legends of world soil maps prepared by FAO/UNESCO (1974), which group soils into twenty-six "world classes" on the basis of profile characteristics, but it is not a classification system per se.

Other notable attempts at general classifications include those of Russian pedologists. Carrying genetic overtones and inferrences as to causes of soil properties, the schemes of Ivanova (1956) and Kovda et al., (1967) are regarded as somewhat outmoded by many pedologists in other regions. Although the systems are employed in the preparation of world maps for Russian scientists, they are not widely applied on an international basis.

REGIONAL CLASSIFICATIONS

The plethora of regional classifications emphasizes the need for a universally accepted international system. Though the numbers of systems employed by national soil surveys are legion, the nine papers in Part IV represent approaches used in France (Paper 15), England and Wales (Paper 16), Canada (Paper 17), Brazil (Paper 18), The Netherlands (Paper 19), Japan (Paper 20), Federal Republic of Germany (Paper 21), and Australia (Paper 22). Paper 23 reviews approaches to classification in polar environments. Only a few of many examples could be reprinted, but perusal of the selections here should inculcate an appreciation for the wide variety of regional approaches. The official national systems are of interest because they point out different philosophies of soil classification, special needs, and emphasize a diverse range of terminologies. All aspects may eventually prove useful in the development of a truly universal soil classification, an ideal yet to be attained. Even if an international system is agreed on, it is likely that limited approaches will remain in use for a long time to come, because soils, unlike many other natural features, are considered a prime national heritage.

TECHNICAL AND SPECIAL APPROACHES

Technical or special purpose classifications combine units into groups that are meaningful to selected users. The purpose of many restricted classifications is to provide reliable information for land-management or treatment programs. Such schemes tend to be monocategorical and based on practical *ad hoc* criteria. Procedures inherent in such systems limit the number of soil properties that are observed and soils tend to be classified as mere assemblages of natural objects or materials (Bartelli, 1978). Engineering systems, for example, are quantitative and based on selected criteria such as particle-size grading, plasticity and compressibility, liquid and plastic limits, and shrink-swell behavior. The so-called Unified Soil Classification System, proposed by Casagrande in 1947, groups soils according to their suitability as subgrades for roads and airfields. This system became a standard for the American Society for Testing and Materials in 1969. The American Association of State Highway and Transportation Officials (AASHTO) groups soils on the basis of load-carrying capacities and road-service characteristics. The Federal Aviation Agency (FAA) system, also based on mechanical analysis criterion for coarse-textured soils, incorporates liquid- and plastic-limit tests for fine-grained soils. Both the AASHTO and FAA systems are described in detail in the *Soils Manual* of the Asphalt Institute (1969). The soil classification system of the Portland Cement Association (Paper 25) is based on the influence of soil properties important to the design, construction, and performance of concrete, soil-cement, and other types of pavement.

Numerical taxonomy became feasible with the advent of the electronic computer. In a numerical method, soil profiles are sorted into classes by similarity-grouping techniques. Large numbers of soil properties are used but each is given equal weight in most computations. Advantages of soil classification by computer include data storage and recall and continuity of effort. The resulting classes contain maximum information but, unfortunately, they may not correspond with any natural soil units. Alas! The loss of information due to single measure of similarity and the subjective selection and measurement of soil characters are distinct disadvantages of the method. The use of factor analysis in the numerical taxonomy of soils overcomes many of the objections lodged against previous methods. Arkley's effort (Paper 24) represents one such improved methodology.

REFERENCES

Asphalt Institute Staff, 1969, *Soils Manual for Design of Asphalt Pavement Structure*, Asphalt Institute, Houston, Texas, 53 p.

Bartelli, L. J., 1978, Technical Classification for Soil Survey Interpretation, *Adv. Agron.* **30**:247–289.

Bidwell, W., and F. D. Hole, 1964, Numerical Taxonomy and Soil Classification, *Soil Sci.* **97**:58–62.

Bazilevich, N. I., and E. I. Pankova, 1968, Tentative Classification of Soils According to Salinity, *Pochvovedenie* **11**:3–16.

Bazilevich, N. I., and E. I. Pankova, 1969, Classification of Soils According to Their Chemistry and Degree of Salinization, *Agrokem. Talajtam.* **18**:219–226.

Casagrande, A., 1947, Classification and Identification of Soils, *Am. Soc. Civil Eng. Proc.* **73**:783–810.

Cline, M. G., 1980, Experience with Soil Taxonomy of the United States, *Adv. Agron.* **33**:193–226.

Cruickshank, J. G., 1972, *Soil Geography*, David and Charles, Newton Abbot, 256 p.

Dokuchaev, V. V., 1879, Short Historical Description and Critical Analysis of the More Important Existing Soil Classifications, *Trav. Soc. Nat. (St. Petersburg)* **10**:64–67.

Dokuchaev, V. V., 1893, *The Russian Steppes and the Study of the Soil in Russia, its Past and Present*, J. W. Crawford, trans., Dept. of Agriculture, Ministry of Crown Domains for the World's Columbian Exposition, St. Petersburg, 61 p.

Duchaufour, P., 1978, *Ecological Atlas of Soils of the World*, G. R. Mehuys et al., trans., Masson Publishing USA, New York, 178 p.

Dury, G. H., 1969, Rational Descriptive Classification of Duricrusts, *Earth Sci. J.* **3**:77–86.

FAO/UNESCO, 1974, *Soil Map of the World, Vol. I, Legend*, UNESCO, Paris, pp. 1–41.

Gedroits, K. K., 1927, *Genetic Soil Classification Based on the Absorptive Soil Complex and Absorbed Cations*, Nossov Agriculture Experiment Station, Agrochemical Div. Issue 47. Translated from Russian by the Israel Program for Scientific Translations, Jerusalem, 1966, 70 p.

Glinka, K. D., 1914, The Great Soil Groups of the World and Their Development, C. F. Marbut, trans., 1927, Edwards Bros., Ann Arbor, Mich., 235 p.

Glinka, K. D., 1924, Differents typès d'apres lesquels se forment les sols et la classification de ces derniers, *Comité Internat. Pédologie IV, Comm. No. 20.*

Glinka, K. D., 1931, *Treatise on Soil Science*, A. Gourevich, trans., Israel Program for Scientific Translations, Jerusalem, 1963, 674 p.

Harrassowitz, H., 1930, Boden der tropischen Rezionen, *Blanck's Handbuch der Bodenlehre* **3**:362–436.

Hilgard, E. W., 1860, *Report on the Geology and Agriculture of the State of Mississippi*, Barksdale, Jackson, Miss., 391 p.

Hilgard, E. W., 1892, A Report on the Relations of Soil to Climate, *U.S. Dept. Agric. Weather Bull. No. 3*, Washington, D.C., 59 p.

Hilgard, E. W., 1906, *Soils: Their Formation, Properties, Composition, and Relation to Climate and Plant Growth in the Humid and Arid Regions*, Macmillan, New York, 593 p.

Ivanova, E. N., 1956, Classification of the Soils of the Northern Parts of the European USSR, *Pochvovedenie* **1**:70–78.

Ivanova, E. N., and A. N. Rozanov, 1939, Classification of Salt Affected Soils, *Pochvovedenie* **2**:44–52 (English summary).

Joachim, A. W. R., 1935, Studies on Ceylon Soils. II. General Characteristics of Ceylon Soils, Some Tropical Soil Groups of the Island, and a Tentative Scheme of Classification, *Trop. Agricst. (Ceylon)* **84**:254–274.

Kovda, V. A., Y. V. Zobova, and V. V. Rozanov, 1967, Classification of the World's Soils, *Soviet Soil Sci.* **7**:851–863.

Kubiëna, W. L., 1953, *Bestimmungsbuch und Systematik der Böden Europas*, Enke, Stuttgart, 392 p.

MacVicar, C. N., 1969, A Basis for the Classification of Soil *J. Soil Sci.* **20**:141–152.

Manil, G., 1959, General Considerations on the Problem of Soil Classification, *J. Soil Sci.* **10**:5–13.

Marbut, C. F., 1927, A Scheme of Soil Classification, *1st Inter. Congr. Soil Sci. Trans.* **4**:1–31.

Mohr, E. C. J., 1938, Soils of Equatorial Regions with Special Reference to the Netherlands East Indies, R. L. Pendleton, trans., Edwards Bros., Ann Arbor, Mich., 756 p.

Moore, A. W., J. S. Russell, and W. T. Ward, 1972, Numerical Analysis of Soils: A Comparison of Three Soil Profile Models with Field Classification, *J. Soil Sci.* **23**:193–209.

Muir, J. W., 1962, The General Principles of Classification with Reference to Soils, *J. Soil Sci.* **13**:22–30.

Mulcahy, M. J., and A. W. Humphries, 1967, Soil Classification, Soil Surveys and Land Use, *Soils & Fertilizers* **30**:1–8.

Müller, P. E., 1887, *Studien über dei naturalichen Humusformen und deren Einwirkung auf Vegetation und Boden*, Springer Verlag, Berlin, 324 p.

Ping-Hua Lee, M., 1921, The Economic History of China with Special Reference to Agriculture, *Columbia University Studies in History, Economics, and Public Law* **99**:1–461.

Pons, L. J., and I. S. Zonneveld, 1965, *Soil Ripening and Soil Classification*, International Institute for Land Reclamation and Improvement, Wageningen, The Netherlands, pp. 74–80.

Ramann, E., 1918. *The Evolution and Classification of Soils, 1928*, C. L. Whittles, trans., Heffner, Cambridge, England, 619 p.

Rayner, J. H., 1966, Classification of Soils by Numerical Methods, *J. Soil Sci.* **17**:79–92.

Riecken, F. F., and G. D. Smith, 1949, Lower Categories of Soil Classification: Family, Series, Type, and Phase, *Soil Sci.* **67**:107–115.

Robinson, G. W., 1932, *Soils, Their Origin, Constitution and Classification*, Murby, London, 573 p.

Robinson, G. W., 1949, Some Considerations on Soil Classification, *J. Soil Sci.* **1**:150–155.

Rode, A. A., 1957, Problems of Organization of Work on Nomenclature, Systematics, and Classification of Soils, *Pedology* **9**:89–95.

Schelling, J., 1970, Soil Genesis, Soil Classification and Soil Survey, *Geoderma* **4**:165–193.

Shaler, N. S., 1891, The Origin and Nature of Soils, *U.S. Geol. Surv. Ann. Rep. 1890–91*, pp. 213–345.

Sibirtzev, N. M., 1895, Genetic Classification of Soils, *Novo-Aleksandr. Agric. Inst. Zap.* **9**:1–23.

Simonson, R. W., 1962, Soil Classification in the United States, *Science* **137**:1027–1034.

Smith, G. D., 1965a, La place de la Pedogenese dans le systeme comprehensif propose de classification des sols, *Pedologie* Spec. Ser. **3**:137–164.

Smith, G. D., 1965b, Lectures on Soil Classification, *Pedologie* Spec. Ser. **3**:3–134.

Stephens, C. G., 1950, The Elements of Soil Classification, *J. Dept. Agric. S. Aust.* **54**:141–142.

Terzaghi, C., 1927, Soil Classification for Foundation Purposes, *1st Intern. Congr. Soil Sci. Trans.* **4**:127–157.

Thorp, J., and G. D. Smith, 1949, Higher Categories of Soil Classification: Order, Suborder, and Great Soil Groups, *Soil Sci.* **67**:117–126.

Webster, R., 1977, *Quantitative and Numerical Methods in Soil Classification and Survey*, Oxford, University Press, London, 264 p.

Whitney, M., 1909, Soils of the United States, *U.S. Dept. Agric. Soils Bull. No. 55*, Washington, D.C., 239 p.

Part I

EARLY CLASSIFICATION EFFORTS

Editor's Comments
on Papers 1, 2, and 3

1   SIBIRTZEV
Russian Soil Investigations

2   COFFEY
Excerpt from *A Study of the Soils of the United States*

3   JOEL
Changing Viewpoints and Methods in Soil Classification

The papers in Part I are examples of initial attempts to classify soils as discrete natural entities. Compared to the standards of today, these schemes may appear rather primitive and naive. However, they do represent notable and independent efforts of the time in that they reflect an appreciation of soil as a natural body, recognize an orderly distribution of soils about the earth's surface, and value different criteria in classification. A century of progress has witnessed many attempts at soil classification and despite numerous conceptual advances, many early observations are still valid. The arguments for different approaches to classification follow various forms of persuasion and largely depend on the point of view. In any case, much of what follows in subsequent editorial comments is related to what went on in the beginning as part of early efforts.

In spite of Sibirtzev's modest claims that he was the most celebrated of Dokuchaev's students, one must acknowledge his foremost contribution (Paper 1) as a definitive statement of early Russian classificatory efforts. Of the points emphasized in the paper, several emerge as salient, long-lasting concepts that remain essential to Russian concepts of soil, especially in their approach to classification. Dokuchaev's recognition of soil as an independent natural body was undoubtedly a most important contribution, because this idea still remains central to concepts of soil in a pedological context. Recognition of an orderly global array of soils soon led to the notion that soil-distribution patterns could be broadly correlated with a geography of other natural phenomena such as rock type, climate, and vegetation. Sibirtzev recognized soil zones of belts, as he called them, by which the continents could be subdivided. The pedosphere was, according to his classification, divided into seven major "soil zones," namely:

1. Soils of the warm humid tropics and subtropics (the laterites and lateritic soils);
2. Eolian (dust) soils of the dry continental regions;
3. Soils of dry steppes mainly formed on argillaceous and arenaceous rocks (the Chestnut soils);
4. Soils of herbaceous steppes and prairies in temperate climatic zones (the Chernozem soils);
5. Soils of wooded steppes and deciduous forsts (the Gray Forest soils);
6. Sod (peat) soils and podzols developed under mixed forests of cold temperate regions and that were ordinarily accompanied by concretions (the Sod soils and Podzols); and
7. Soils formed on clays and clayey sands in tundra permafrost regions (the Tundra soils).

These categories, conceived only as an initial ideal scheme, comprised the so-called *zonal soils,* the principal mature or complete soils of the world. They were the "normal" soils of Dokuchaev. Soils due to the dominance of some other factor besides climate and occurring with no zonal regularity belonged to the second class of *intrazonal soils.* Examples included alkali lands, humus-calcareous soils, and marsh soils. Immature, poorly developed, or incomplete soils occurring in any zone formed the third genetic class of abnormal or *azonal soils.* The grouping of soils into three great classes is thus largely based on inferred associations of soil zones with climatic regions, the origin of organic matter occurring in the upper part of the soil profile, and, to a lesser extent, on the impacts of dynamic pedogenetic processes and effects of parent materials. The second part of Paper 1 (pp. 807–818) describes the chief soils of Russia in terms of their characteristics and distribution. This latter point is of interest because modern pedologists probably owe more to their Russian colleagues for their attention to details of profile morphology and inventory of soil descriptions than to their environmental or genetic classification schemes.

Pedologists have been aware, for nearly a century now, of the need for more thorough understanding of soils and also of the value of a perfect classification. In Paper 2, Coffey eloquently reviews the principal bases of classification identifying the following: (1) geology, (2) physical properties, (3) chemical properties, (4) vegetation, (5) processes of soil formation, and (6) combinations of these approaches. Indicating that classifications formulated around any one set of factors alone are unsatisfactory, he calls for a taxonomy of soils based on differences in the soil itself. This was indeed a perceptive observation for the time. Some sixty years later his foresight was vindicated when the U.S. Department of Agriculture adopted a system based on diagnostic criteria (see Editor's Comments and Paper 14 in Part III).

At this juncture it seems appropriate to clarify some points of possible confusion in the naming of taxa in different systems. In the early Russian schemes, as well as in modern ones, the basic unit of classification is the soil type. Russian use of the term at a high level of classification for broad groups of soil has precedence in time over American use of the term for lowest-level taxa. Introduced by Whitney (1909) before contact was made with Russian pedologists, the term was widely used in the United States for narrowly defined taxa. Dropped as a formal category in the current U.S. and Canadian systems, the term now applies to one kind of phase and to a way of naming mapping units. The Russian soil type is roughly equivalent to the great group category in *Soil Taxonomy*.

Joel, basing his comments on Canadian experience (Paper 3), echoes Coffey's plea for an effective classification of soils. He reinforces the premise that soils, like other natural objects, should be systematized according to their fundamental characteristics and not on external modifying factors such as climate, vegetation, or parent material. Soil taxonomies, he claims, must focus on the evolved product, not the factors that determine the course of evolution. Joel's division of the world's soils into two great groups corresponds, more or less, to the so-called timber and prairie soils of the Marbut scheme (see Editor's Comments on Paper 4 in Part II) proposed some years earlier.

REFERENCE

Whitney, M., 1909, Soils of the United States. *U.S. Dept. Agric. Soils Bull. No. 55*, U.S.D.A., Washington, D.C., 239p.

1

Reprinted from *U.S. Dept. Agric. Exper. Sta. Record*, **12**:704–712, 807–818 (1901)

RUSSIAN SOIL INVESTIGATIONS.[1]

N. Sibirtzev

The systematic study of Russian soils may be said to have begun 22 years ago, when Prof. V. V. Dokouchayev, commissioned by the ancient and renowned "Imperial Free Economic Society," took up the study of the Russian chernozem. Energetic and deeply devoted to his work, he soon gathered around himself a number of gifted and enthusiastic young scientists, who at once attacked various questions relating to soils. Professor Dokouchayev succeeded in organizing in various parts of Russia special soil investigations at the expense of Government institutions and private persons. To Professor Dokouchayev belongs the honor of founding a new school in soil investigations, a school which views the soil as an independent natural body. Among the pupils of Professor Dokouchayev the most celebrated is N. Sibirtzev, the author of the classification of soils, described in this article.[2]

The second place in importance among the Russian soil investigators unquestionably belongs to P. Kostichev, an independent worker. An eminent chemist and agriculturist, he contributed much to the knowledge of soils by his skillful and accurate analyses of soils.

The following statement regarding the classification of soils and the characteristics of Russian soils are taken from the reports of Sibirtzev referred to above.

GENETIC CLASSIFICATION OF SOILS.

The conception of a soil as a natural body having a definite genesis and a distinct nature of its own has led to attempts to create a natural classification of soils on a scientific basis. To this important branch of the study of soils the Russian investigators have made important original contributions, and established new principles for the systematic study of soil formations, not confounding the latter with either rock species or with the cultivated horizon of the ground. The

---

[1] Translated and condensed from the original articles of N. Sibirtzev, by Dr. Peter Fireman.

[2] Genetic classification of soils, N. Sibirtzev (Zapiski Novo-Alexand. Inst. Selsk. Khoz. Lyesov. Memoirs of the Instit. of Agric. and Forest. at Novo-Alexandria, Government of Lublin, IX (1895), pt. 2, pp. 1–23). Brief survey of the chief soil types of Russia (Ibid., XI (1898), pt. 3, pp. 1–40). The latter memoir was presented in French (Étude des sols de la Russie) to the Seventh International Geological Congress at St. Petersburg, in August, 1897.

scientists of Western Europe have been at a disadvantage regarding this question, since they have for the most part had to deal with soils only slightly developed, shallow or easily washed away, mixed with various geological deposits, and at the same time strongly altered by cultivation. In reality the "tilled layer of the soil" of Western Europe, under the influence of the intensive and deep cultivation there practiced, is an artificial mixture of natural soil and of the underlying primitive rock. This accounts for the geo-petrographic and the physico-chemical classification of soils in vogue among the scientists of Western Europe.

America, with its virgin soils, vast, frequently still untilled plains, prairies, forests, deserts, and barren alkali lands, and clearly defined climatic, physico-geographical and geo-botanical zones offers an excellent field for the study of natural soils. As a matter of fact the American investigators of the soil, especially Hilgard, have already come to a clear recognition of the soil as an independent formation and have established natural soil types.

The study of soils which has been so industriously carried on in Russia the last 20 to 30 years under the leadership of Dokouchayev and Kostichev has for its starting point the idea of the soil as a natural body which occupies an independent place in the series of formations of the earth's crust.

According to the definition of Professor Dokouchayev, under the term "soil" must be understood the surface horizons of the rocks, more or less altered under the simultaneous influence of water, air, and various organisms, living as well as dead. In other words, the soil is the superficial horizon of rocks in which the general processes and phenomena of weathering, transportation of particles, etc., combine with the biological processes and phenomena due to the influence of plants, animals, and micro-organisms. Weathering of rocks which takes place independently of the action of organisms yields products which must be considered as rocks, and the study of such products belongs to petrography. These products may replace and may be converted by cultivation and fertilizing into artificial soils, but must be distinguished from natural soils. However, such soils are of rare occurrence. It is well known that many organisms, such as nitrifying bacteria, lichens, alpine plants, etc., play an important part even in the first stages of the disintegration of the massive and sedimentary rocks. On the other hand, in the class of natural soils should not be included the mechanical deposits of dead organisms or their excretions (peat beds, guano, and the like) and those derived from rocks of organic origin.

As a superficial geo-biological formation of the earth's crust, the soil differs from the parent rock from which it is derived in composition, complexity of the dynamic factors, and external morphological peculiarities. Natural soils vary with (1) the petrographic type of the parent rock; (2) the nature and intensity of the processes of disintegration, in connection with the local climatic and topographic conditions; (3) the quantity and quality of that complexity of organisms which participate in the formation of the soil and incorporate their remains in it; (4) the nature of the changes to which these remains are subjected in the soil, under the local climatic conditions and physico-chemical properties of the soil medium; (5) the mechanical displacement of the particles of the soil, provided this displacement does not destroy the fundamental properties of the soil, its geo-biological character, and does not remove the soil from the parent rock;[1] and (6) the duration of the processes of soil formation.

All these may be termed genetic conditions of formation of natural soils. Such existing types of natural soils always correspond to a definite combination of the soil-forming factors. The parent rocks, the organisms (with their subsequent transformations), and the physico-geographical conditions of the country, including climate (humidity, temperature), recent geo-physical history and relief, are the chief agents of soil formation. The correlation among these factors may assume various forms, a certain connection or parallelism being observed either among all or only a part of them. Thus, the composition and distribution of the ancient sedimentary and crystalline rocks do not, of course, depend on those conditions (even climatic) of the country to which the formation of the existing soils is subject. But the weathering of the rocks and, in general, all the physical and chemical processes which take place in the soil are influenced by climatic conditions. Climates in which wet and dry seasons alternate produce laterites, the climatic conditions governing the biological processes which result in the formation of lateritic soils. Eolian loess and pulverulent rocks which resemble it are characteristic of continental regions with a dry climate. From this point of view the nature of a given soil type presents, in a certain measure, a function of the climate.

The soils of a given territory are also influenced by the life activities and the dead remains of plant and other organisms. The soils influence the development and the life activity of these organisms and their decomposition after death. On the other hand, the character of the plant growth, for example, plays not only a direct, but an intermediate rôle in the formation of the soil. The relief of the soil has an important influence in determining the drainage, temperature, etc. And lastly, the relative duration of the soil-forming processes which have gone on since the removal of the glacial or water cover, as for example, the successive changes which have taken place in the climate, the encroachments of the forests upon the prairies, the spread of marshes, the drying up of the soil, etc., must in their turn influence the character

---

[1] Otherwise the soil is converted into alluvium, diluvium, etc., or, in general, into mechanical deposits of secondary formation.

of the soils. The knowledge of the laws and the forms of these influences make it possible to obtain from the study of soils a basis for the reconstitution of the recent past of the country and for sketching its recent geo-physical history. The essential factors determining the characteristics of natural soils are as follows:

(1) The conditions and the factors of the origin of the given soil type (the material and the organic agents); (2) the morphological properties of the soil, *i. e.*, its color, depth, constitution,[1] structure, transition into the parent rock, etc.; (3) the physical, chemical, and chemico-biological properties; (4) the modification with the type; and (5) the geographical and topographical distribution.

The natural classification of soils can be elaborated, taking the genetic principle as a starting point. In establishing the chief groups of soils the existing types of formation of soils in nature must be recognized, the homogeneous or similar combinations of soil-forming agents (such as climate, parent rocks, organisms, relief of country, etc.) must be formulated. As is well known, the weathering of the rocks alone, provided it takes place under similar physico-geographical conditions, may efface to a considerable degree the differences which exist among the rocks, and may give alluvial products of fine earth more closely resembling one another than the original rocks; this similarity is more manifest when the biological factors also tend to produce a uniform result. We can, consequently, establish an ensemble of natural conditions which will produce as a result soils, say, of the chernozem group. A characteristic feature of these soils is the peculiar accumulation of humus under the sod. Wherever analogous conditions prevail soils of the chernozem type are formed. Similarly, we know the climatic conditions which favor atmospheric-eolian weathering, the pulverization of the soft rocks, and where these conditions obtain eolian dust soils result. The soils of these groups in their principal features are the natural resultant of the physico-geographical type of the given continental region or zone. The soil of the different zones will, of course, not be uniform, but will exhibit similarity to the extent to which their content of fine earth and humus reflect the analogous influences of a definite and constant combination of geo-physical factors of soil formation.

In this way the first class of zonal soils is determined. In the proc-

---

[1] A vertical section of a soil always shows two, three, or even more horizons, detailed descriptions of which are given in Russian works on soils. Of these horizons the most remarkable are: (1) The upper horizon, the most uniformly and strongly colored by humus; (2) the lower horizon, distinguished from the upper by its structure and color and gradually merging into the subsoil; and (3) the subsoil or parent rock preserving its fundamental petrographic features. The first two horizons taken together give the depth of the surface soil. Sometimes in these horizons sub-horizons can be distinguished, with peculiar differences in composition, structure, and tint (alkali soils, forest soils, etc.).

esses of their formation general ectodynamic and special biological phenomena manifest themselves in accordance with the physico-geographical types of continental zones. Such are the following types of soils:[1]

(1) Lateritic soils. These are the soils of the tropical and subtropical regions with alternating wet and dry seasons.

(2) Atmospheric-eolian soils. Formed of the dust rocks in the central regions of the different continents under arid conditions.

(3) Soils of dry steppes or steppes deserts. Being formed of argillaceous and arenaceous primitive rocks, they are chestnut and fawn-colored.

(4) Chernozem soils. These occur in connection with the grass steppes or prairies of the temperate or warm-temperate regions. They develop best from argillaceous rocks.

(5) Soils of wooded steppes and deciduous forests (gray soils), resembling chernozem soils, but differing from them in the conditions of their origin, and in their morphological and other properties.

(6) Sod soils and podzol soils[2] which are peculiar to the temperate-frigid zone. They are typically developed under mixed woods and bushes and are ordinarily accompanied by concretions.

(7) Tundra soils. These are formed from the clays and argillaceous sands of the tundras, in a cold climate with a very long winter. They are characterized, to a greater or less degree, by being perpetually frozen (the subsoil waters are in a solid state).

The groups of soils named represent the soil zones or belts into which the surface of the continents may be divided.

The lateritic soils belong to the coastal zone of tropical and subtropical continental regions which is broken and cut up by seas. After them follow toward the north and south, in the order indicated in the above enumeration, the regions with the other soil types. In the zone of the continental plateaus and the inclosed or partly inclosed plains of the northern hemisphere—in central and southwestern Asia (China Persia, Arabia, Turkestan), in the Caspian region, in northern Africa, and in the western and southwestern States of North America are found the atmospheric-eolian soils and the soils of the steppes deserts. In the southern hemisphere are corresponding zone soils covering central Australia, inland sections of southern Africa (the country of the Hottentots, the region to the south of the sources of the Zambezi

---

[1] Only the best known types are mentioned here, use being made of the results of studies of the natural soils of Russia, Western Europe, of some regions of Central and Southern Asia, of America, etc., partly of Australia, and Africa.

[2] Podzol soils are unproductive soils consisting mainly of very fine sand, but containing more organic matter than their color would indicate. They resemble ashes in appearance, hence the name "podzol," which indicates this resemblance. They correspond nearly with the Bleisand of Germany.

River), and Argentina. In the open grass plains, such as the Hungarian, Russian, and Siberian steppes or American prairies in the northern hemisphere, and the eastern provinces of Argentina (Entrerios, Corrientes, Buenos-Ayres) in the southern hemisphere, occur the soils of the chernozem group. In Asia, Europe, and North America between the chernozem and tundra soils those of the fifth and sixth group are situated. In the southern hemisphere there is no such complete grouping of soils as in the northern. This is due to a different configuration of the southern continents.

The system of soil zones enumerated above is only an ideal general scheme. In reality no one of these zonal types of soils embraces the continental surface of the globe in a continuous belt. All of them extend in interrupted bands and spots, now expanding enormously in breadth, now becoming narrow, now intermixing with one another at their boundaries, now forming circumscribed areas separated by greater or smaller distances from the principal zones. The reason for this is found in the effect of local orographic, geological, and climatic peculiarities, which interfere with the development or cause a displacement of certain soils.[1]

The division into types distributed in zones or belts does not begin to exhaust the whole variety of natural soils. As stated above, among the soil-forming factors there are some which may individualize themselves by diverging from the concordant action of the other factors. Thus, for example, a particular composition of the parent rock may retain its influence on the soil and thus impart special features which are not proper to the dominant zonal type; a similar effect may be caused by the local saturation of soils with water, due to the configuration of the surface. Humus soils of this second class may be called intrazonal or semizonal. They are dispersed among the main zones in circumscribed areas and spots, occurring chiefly, although not exclusively, in connection with some of the zones. Certain types of the intrazonal soils are met with in those zones whose general conditions favor the most or interfere the least with the action of the individualizing factor.

There are undoubtedly very many types of intrazonal soils. We shall mention the following as examples: (1) Alkali soils, which form when the parent rock contains soluble salts and the drainage is poor. Since the salt contents of the rock may depend on causes purely geological, having no direct connection with the other soil-forming factors, there is, generally speaking, no zonal regularity to be observed in the distribution of alkali soils. However, they occur mostly in the arid regions

---

[1] In Russia, *e. g.*, the soils of the steppes deserts extend to the south and southeast of the chernozem, and in North America to the west and southwest (in conformity with the increasing aridity of the climate). It may be added that vertical zones may also be observed on broad slopes and plateaus which appear in a measure as local repetitions of the horizontal zones in an analogous order.

of Europe, Asia, America, Africa, and Australia, *i. e.*, in the second, third, and part of the fourth zones. (2) Humus-calcareous soils. Humus-containing soils are formed from calcareous rocks (limestone, marble, chalk, etc.) accumulating much humus in consequence of the rapid leaching out of the calcium and magnesian carbonates and the retarded decomposition of the organic remains in the feebly alkaline medium. (3) Marshy soils. Under this term are understood soils which owe their origin to the influence of stagnant waters (water-logged soils) dispersed over the surface of continents, wherever the relief and the hydro-geological conditions favor their formation. They occur most frequently in temperate and frigid zones, although some-times found in the arid zone. They are formed (a) in a medium of fresh water (sour meadows, the marshes of the lowlands), or (b) in sec-tions which are or have been subject to inundations by the sea or by the waters of estuaries (sea marshes, salt marshes, delta marshes, etc.). The different stages in the formation of the swamps, the diverse com-position of the organisms, the character of the aqueous medium, the drying up of the marshy basin due to various causes, give to the soils of this type a great variety.

Lastly, there are many natural soils which are composed of the unaltered parent rock (when forming *in situ*) to the almost complete exclusion of fine earth and humus, or which are formed by a mixed process (1) by the mechanical deposition of particles, mineral as well as organic (alluvium); and (2) by the periodic action on the alluvial deposits of the special factors which form humus soils. The soils of this nature stand, so to speak, on the border line between soils proper and rocks, in one case merging into soils, in another approaching rocks. They form the third class of incomplete or azonal soils; they are met with everywhere. When they are formed *in situ*, outside of alluvial depressions and valleys, they can be divided into two large groups, (1) crude soils and (2) skeleton soils. By crude soils are meant those in which there is a considerable quantity of clay-like particles (clays, silt, and fine sand), but in which the horizon of vegetable humus is not clearly defined. Every humus soil passes downward into a crude soil, but the term is applied here only to those soils which are wholly or almost wholly crude. The name skeleton soil is applied to those in which granular and sandy, gravelly, or pebbly elements, or in general, the skeleton mechanical elements which take the place of the humus and fine earth, entirely predominate.

Among the conditions which conduce to the formation of crude and skeleton soils are the following:

(1) Unalterability or difficult alterability of the parent rock or of the rocky components of the soil (sand, rock fragments, pebbles, com-pact sedimentary rocks, etc.).

(2) The washing off of the humus horizon by the snow and rain waters (crude soils on hills and slopes).

(3) The short duration of the processes of soil formation (undeveloped soils on comparatively recently uncovered or deposited rocks).

(4) The interference with the soil-forming processes by unfavorable climatic influences (especially in deserts and arctic regions).

The fundamental feature of alluvial soils is their formation with the aid of mechanical transportation and deposition of particles by water. Such are the soils of the river valleys. Alluviums, however, must not be confused with alluvial soils. The former are purely mechanical deposits of varying depth—geological formations—while an alluvial soil is the horizon of this deposit which has been subjected to the action of the general dynamic agents of weathering and to the influence of organisms.

To sum up the above considerations, natural soils may be divided into the following genetic classes and types:

Class I.—Zonal soils, complete.

　　Type 1. Lateritic.

　　　　2. Atmospheric eolian.

　　　　3. Soils of the steppe, deserts or dry steppes.

　　　　4. Chernozem.

　　　　5. Soils of wooded steppes and gray forest soils.

　　　　6. Sod soils and podzol soils.

　　　　7. Soils of the tundras.

Class II.—Intrazonal soils.

　　Type 1. Alkali lands.

　　　　2. Humus-calcareous soils.

　　　　3. Marshy soils, etc.

Class III.—Incomplete or azonal soils.

　　Soils formed *in situ.*

　　　　(*a*) Crude ⎫
　　　　(*b*) Skeleton ⎬ of various groups.

　　Alluvial soils (of different types).

In nature transitional forms are found among the soils of the various genetic types. These transition types may result (1) from the fact that the soil-forming agents (*e. g.*, the climatic conditions) do not change suddenly, but more or less gradually, and thus can produce intermediate results; or (2) from the changes which take place in the soils themselves in the course of their formation and development. Soils may pass through various phases and forms of development in correspondence with the external influences which act upon them. Thus, some alkali soils, losing little by little their salts by leaching, are converted into soils of dry steppes or even into chernozem. Alluvial soils, having passed out of the sphere of river inundations, approach the local zonal types. If a locality, for one reason or another, loses its

22

drainage, the soils may become swampy, and, *vice versa*, marshy soils, by drying and drainage, lose their characteristic peculiarities and approach other local types. If, during the period of formation of chernozem, the steppe or prairie is encroached upon by forests, the latter change the structure and composition of the soil in the direction of soils of wooded steppes and forest soils, etc.

The genetic types of soils are large categories which include many subtypes, groups, and subgroups. A detailed classification of soils may be based on two kinds of facts: (1) On the degree or force and on the variation of those dynamic processes which impart to the soil the fundamental features of the given genetic type. Thus, for example, there exist conditions which lead to the formation of chernozem soils, but these conditions may vary, may deviate from a certain mean, and, in consequence of these fluctuations, from one and the same or a similar parent rock there may result unlike chernozems with a different content and quality of humus. (2) On the changes in the composition and structure of the soils in connection with the composition and structure of the parent rocks. The subdivisions of this category are based upon (*a*) the physical properties of the soils, *i. e.*, their skeleton and fine earth; (*b*) the chemical and chemico-petrographic peculiarities of the soils. Chernozem, for instance, may be argillaceous, subargillaceous, subarenaceous, marly, phosphoritic, etc. The division of the genetic types and subtypes of soils into groups and subgroups, a division based on the mechanical, physical, and chemical properties of the soil mass, connects the system here described with the common soil classifications of the German and Russian authors (Mayer, Schübler, Knop, Senft, Ramann, Feska, Kostichev, and others). It is believed that a soil classification such as that described above, which is based on the quantitative contents in the soil of skeleton and fine earth and on the particular character of these two constituents (mechanico-physical groups and subgroups), is more general than the commonly accepted system. Following these subdivisions, or, more properly, within them, are the chemical subdivisions based on (1) the chemico-petrographic composition of the soil skeleton, (2) the composition of the siliceous substances of the fine earth of the soil (the chemical nature of the soil clay, of the zeolitic compounds, etc.), and (3) the oxids and salts containing no SiO_2, their quantity and nature (carbonates of alkaline earths, of alkalis, ferrous and ferric oxid, phosphates, sulphates, their solubility in water, etc.).

In the previous article the system of soil classification adopted by Dokouchayev and his collaborators was explained. This classification was in brief as follows: (1) Zonal soils, including lateritic soils, eolian or loess soils, soils of the dry steppes, chernozem, gray forest soils, sod and podzol soils, and tundra soils; (2) intrazonal soils, including alkali, humus-calcareous and marsh or swamp soils; (3) incomplete or azonal soils, including crude and skeleton soils, and alluvial soils. The following article discusses the characteristics of these various types of soils as they occur in Russia.

BRIEF SURVEY OF THE CHIEF SOIL TYPES OF RUSSIA.

ZONAL SOILS.

Russia, being a country of temperate and cold climates, has no lateritic soils.

Loess soils.—Loess or eolian soils occur in the hot, windy, dry climates of Turkestan and the trans-Caspian region, alternating with sandy and alkali soils. The loess soils are yellowish, bright orange, or straw colored. The percentage of humus does not exceed 2.5, and is usually less than 1. About one-half of the soil particles are less than 0.01 mm. in diameter. The other half is usually a mixture in which particles ranging in diameter from 0.01 to 0.05 mm. predominate. In a grayish loess soil from the vicinity of Tashkend there was found fine sand 65 per cent, ferric oxid 3.6, alumina 10, calcium carbonate 7 to 15, potash 2.8, and phosphoric acid 0.28 per cent. The amount of zeolites present ranged from 15 to 20 per cent and more. Loess or eolian dust soils are widely distributed, *e. g.*, not only in the Aral-Caspian basin, but in China, northwestern India, Arabia, Africa, and the drier portions of North America.

Soils of the dry steppes.—In European and Asiatic Russia, between the loess and chernozem, are found the brown and chestnut soils of the dry steppes. The area occupied by these soils in European Russia includes the vast regions between the Ural River and the lower Volga (with the exception of the sandy soils) and between the lower Volga and the district of Manitch, extending also into the steppes of Crimea

24

and over the coast of the Black Sea. In Asiatic Russia these soils
cover parts of Uralsk, Turgai, Akmollinsk, and Semipalatinsk. The
annual rainfall of this soil zone varies from 30 to 40 cm., one-third of
which occurs during the three summer months. The natural vegetation consists mainly of drought-resisting grasses and other plants
which dry up early in the season and are driven about over the steppes
by the winds. The predominating parent rocks of these soils are
brownish, greenish gray, and reddish Post-Tertiary clays, compact,
frequently marly, and containing gypsum and soluble salts in some
instances. In other cases they are loess-like or sandy. Rock fragments,
pebbles, etc., are also found in the soils. The conditions in this soil
zone are not favorable to rapid weathering. Light brown or brown-
gray soils, poorer in humus, occupy the southern or more strictly
desert portion of the belt. The chestnut soils, richer in humus, and
merging into the chernozem, are found in the northern portion. The
upper horizon of the first class of steppe soils is not more than 1 ft.
in depth and gradually merges into the subsoil. The humus content
is variable, but averages about 2 per cent. The humus is very slightly
soluble in water except when alkaline salts are present. The richness
of the humus in nitrogen is a characteristic feature of these soils. In
a sample of steppe soil which contained only 1 per cent of humus there
was found 0.12 per cent of nitrogen, equivalent to 12 per cent of nitrogen in humus. A similar observation has been made by Hilgard
regarding the nitrogen content of the humus in soils of the arid region
of America (E. S. R., 6, p. 197). The amount of zeolites found varied
from 8 to 12 per cent. The amount of matter soluble in cold 1 per
cent hydrochloric acid, excluding the carbonates, was 1½ to 2 per cent.
The upper horizon of the chestnut soils is from 1 to 1½ ft. deep. These
soils contain on the average from 3 to 4 per cent of humus, the amount
sometimes being as high as 5 per cent. From 2 to 3 per cent of the
soil is soluble in cold 1 per cent hydrochloric acid. A bulk analysis
of subsoil from this zone showed silica 68.2 per cent, alumina 11.56
per cent, iron oxid 3.56 per cent, lime 4.63 per cent, magnesia 1.92
per cent, potash 1.98 per cent, soda 1.36 per cent, carbon dioxid 3.74
per cent, and phosphoric acid 0.15 per cent. Similar soils are found
in California, Colorado, New Mexico, and other parts of the arid
region of the United States.

Chernozem.—The southern third of European Russia is preeminently a region of chernozem. The area occupied by it reaches
approximately 216,000,000 to 270,000,000 acres. The chernozem
zone extends from the southwestern boundaries of Russia, over the
basins of the Dnieper, Don, and part of the Volga, to the southern
half of the Ural Mountains. It also extends beyond the Ural River
and into Asiatic Russia, although it does not form a continuous belt
over the mountainous region of eastern Siberia. All of the chernozem
soils of Russia are found between 44 and 57° north latitude.

The chernozem territory is an undulating plain with occasionally extensive elevations and furrowed by ravines and river valleys. There is no doubt that in prehistoric times it was flatter and more uniform than at present. The climate is preeminently continental, but with less pronounced characteristics than in the zone of the dry steppes. The annual rainfall fluctuates between 40 and 50 cm., 30 cm. occurring during the period of plant growth. Agriculture suffers occasionally from droughts and from high winds which are sometimes intensely cold and at other seasons hot and dry. It is believed that at an early period of the history of the steppes, when their surface was more uniform and retained the cover of dead vegetation, the moisture conditions of the soil during winter were better than they are at present, although it is not likely that there was ever an excess of water. The chernozem zone of southern Russia has never been an uninterrupted swamp, as has been maintained by some scientists who believe the chernozem to be derived from the decomposition of peat. It was a prairie with a luxuriant growth of grass. Its natural plant cover consisted mainly of thick tall grasses interspersed here and there with bushes and shrubs. There were originally no forests except on the sandy strips and in the river valleys. The investigations of Ruprecht, Middendorf, Krasnov, Tanfilyev, Korzhinski, and other geobotanists have explained the complex character of the vegetation of these steppes meadows.[1]

Chernozem is as a rule formed by the admixture of humus with loess, but it is also sometimes derived from other parent rocks. In general, it may be stated that calcareous formations which yield fine particles on weathering are more favorable than other rocks to the formation of chernozem. In addition to this, there must be a particular combination of topography, vegetation, climatic conditions, etc., favorable to the accumulation of humus in the soil. Chernozem is usually black, the shade varying in intensity and passing sometimes into chocolate and cinnamon. Its average depth is about a meter, but this varies, the sandy chernozems being generally deeper than the clayey. The structure of the uncultivated soil is granular, the aggregates being from 2 to 4 mm. in diameter. As the soil merges into the subsoil this structure disappears and the soil becomes more compact and irregular in color, gradually assuming a brown color as it merges into the parent

---

[1] The list of plants growing on these steppes includes *Adonis vernalis*, *A. wolgensis*, *Pæonia tenuifolia*, *Lavatera thuringiaca*, *Linum perenne*, *L. flavum*, *Medicago falcata*, *Aster amellus*, *Trifolium* spp., *Oxytropis pilosa*, *Onobrychis sativa*, *Vicia tenuifolia* (and others), *Centaurea marschalliana*, *C. rathenica*, *Scorzonera purpurea*, *Hieracium virosum*, *Campanula sibirica*, *Echium rubrum*, *Lychnis chalcedonica*, *Thymus marschallianus*, *Salvia pratensis*, *S. nutans* (and others), *Nepeta nuda*, *Phlomis tuberosa*, *Ajuga genevensis*, *Euphorbia procera*, *Asparagus officinalis*, *Poa pratensis*, *Festuca ovina*, *Stipa pennata*, *S. capillata*, and others.

rock. The percentage of humus is quite variable, but in general declines quite uniformly from the center toward each edge of the chernozem zone. This variation is so uniform that it has been utilized by Dokouchayev in the establishment of so-called isohumic bands.

On the basis of humus content the chernozem may be divided into four genetic subtypes: (1) The humus or rich chernozem of the eastern central belt, which contains more than 10 per cent of humus; (2) the medium or ordinary chernozem, which occupies the larger part of this soil zone and contains 6 to 10 per cent of humus; (3) the southern chocolate-colored chernozem, which merges into the chestnut soils of the dry steppes, containing 4 to 6 per cent of humus; and (4) the northern cinnamon-colored chernozem of central Russia, which occurs in strips and spots, alternating with forest and light loess soils, and which contains 3 to 6 per cent of humus.

The chernozems also show wide variations in the composition of their mineral constituents, being clayey, sandy, calcareous, peaty, alkaline, etc., according to the sources from which they are derived or the conditions of their formation. The humus is but slightly soluble in water. The total nitrogen content varies from 0.2 to 0.7 per cent in the soil or from 5 to 8 per cent in humus. The clay content varies from 20 to 40 per cent, zeolites from 15 to 35 per cent. Cold 1 per cent hydrochloric acid dissolves from 3 to 5 per cent of matter from the soil, excluding carbonates. The absorptive power varies from 20 to 43 per cent. The silicates of chernozem have undergone a high degree of weathering and decomposition. Thus, of the 2 to 2.4 per cent of potash, from one-fifth to one-half dissolves in 10 per cent hydrochloric acid. Of the 8 to 10 per cent of alumina from one-half to four-fifths dissolves in the same reagent. The phosphoric acid varies from 0.12 to 0.3 per cent. In the upper horizon of the soil the carbonates, mainly calcium carbonate, do not usually exceed 1 to 3 per cent, but in chernozems derived from limestones the carbonates sometimes reach 10 to 15 per cent. The sandy portion of the chernozem is very fine, consisting of quartz, with an admixture of mica, feldspar, and other silicates. According to Kostichev, the mineral portion of chernozem, excluding the carbonates, is very similar in composition to the loess from which it is derived, there being a slight increase of phosphoric acid, due to the accumulation of humus. In the foothills of the southern Ural Mountains there occurs a variety of chernozem which contains as much as 2 per cent of phosphoric acid.

It may be said in general that the chemical properties of chernozem are more favorable than the physical. The particles are as a rule too fine, from 60 to 80 per cent of the particles being ordinarily less than 0.05 mm. in diameter, and the proportion of silt (particles less than 0.01 mm. in diameter) sometimes reaches 58 per cent. Particles larger than 0.5 mm. in diameter are either entirely absent or present in very small quantities.   As long as the chernozem preserves its natural granular structure the high percentage of fine particles has comparatively little influence upon its relation to water, but in cultivation under the climatic conditions prevailing in the steppes of southern Russia these soils to a large extent lose this structure and consequently present the properties of fine porosity, high capacity for absorbing and retaining water, and low permeability.   With irregular rainfall followed by droughts the moisture of the surface soil has been observed to decrease to 6 per cent (one-seventh of its water capacity), and the soil dries and hardens, resulting occasionally in serious failures of crops.

The chernozem of Siberia has not been very fully studied.   Analysis shows that it contains from 5 to 11 per cent of humus and from 0.28 to 0.6 per cent of nitrogen.   In the clayey types there is from 15 to 25 per cent of zeolites, 7 to 10.5 per cent of alumina soluble in sulphuric acid, and 0.16 to 0.28 per cent of phosphoric acid.   The soils of the Amur prairies are generally richer in humus than the ordinary chernozems of Russia.   Soils of the chernozem type are found alternating with alkali lands and sandy soils in Banat and in the plains of eastern Hungary, which are separated by the Carpathian Mountains from the steppes of southern Russia.

The chernozem zone also embraces a considerable part of the United States.   The soils of the humid prairies in Wisconsin, Minnesota, Iowa, Missouri, and other States are quite similar in character to the chernozems of the Amur region.   In States such as the Dakotas, Montana, Nebraska, Kansas, and Arkansas, where the rainfall is deficient, the soils are similar to the ordinary and the chocolate colored chernozems of the steppes of southern Russia.   In the more strictly arid States, such as Arizona, southern California, etc., are found analogues of the chestnut and light brown soils of Russia.

It is of interest to note that there is a southern chernozem zone represented by the soils of the pampas of Argentina.   Especially fine examples of this type of soil are found in the Province of Entrerios.

Gray forest soils.—Under this name are included the soils of the wooded steppes, adjoining the chernozem or even penetrating far into the region of chernozem, but which have been modified by forest vegetation.   They merge by a gradual transition into chernozem on the one hand and peaty soils or podzols on the other.   They extend in a narrow, rather regular, not always continuous belt across central Russia from the governments of Lublin and Volinsk on the west to the basin of Kama and Viatka on the east.   In the chernozem zone they are found usually along the rivers and valleys, where the soils are well drained and free from alkali.   The observations of soil experts and geobotanists show concordantly that fine grained soils, which possess a great capacity for humidity and a low degree of permeability, and those which contain a large amount of soluble salts are unfavorable to forest growth, particularly if the soils receive a limited supply of moisture; but that as soon as these conditions are corrected and the forest vegetation has gained a foothold in the steppe on the slope of some ravine, it is at once in condition to protect itself against the unfavorable climatic and soil influences. It gathers the snow, moderates the winds. lowers the range of the temperature, prepares for itself the soil necessary for its growth, and advances little by little into the neighboring steppe. The different stages in this process of transformation of chernozem may be observed in progress under natural conditions and may be duplicated under artificial conditions. Prof. Kostichev filled a cylindrical vessel with chernozem, covered it with a layer of leaves, and maintained it in a moist condition. In three years the chernozem was transformed into a gray soil with $2\frac{1}{2}$ per cent of humus.

The upper horizon of these soils in virgin condition is $1\frac{1}{2}$ to 3 dcm. in depth, gray, gray-cinnamon, or dark gray in color and almost structureless. The lower horizon, 3 to 4 dcm. and more in depth, is ashgray, sometimes friable, but more frequently of a crumby structure. It consists of brown-gray rounded or polyhedral aggregates mixed with fine quartz and siliceous flour. An admixture of humus gives to this powder an ash-gray color. Lower down the aggregates become larger, the amount of the ash-gray powder decreases, and the horizon. gradually assuming a brown color, merges into the subsoil.

The parent rocks (subsoils) of the forest lands are usually weathered morainic clays, diluvial clays (sometimes loess-like), leached loess, and ancient sedimentary rocks—clays, marls, etc.—also weathered and leached.

The content of humus fluctuates in the upper horizon between 3 and 6 per cent; in the lower horizon it rapidly falls to 2 and even 1 per cent. The solubility of the humus in water is greater than in the case of the chernozem. The total amount of nitrogen varies from 0.01 to 0.16 per cent (4 to 5 per cent of the humus). The amount of zeolites does not exceed 20 per cent, frequently falling as low as 16 or 12 per cent. The total amount of mineral substances decomposed by 1 per cent cold hydrochloric acid is ordinarily about one-half that found in chernozem. The potash varies from 1 to 2.4 per cent, lime from 0.4 to 1 per cent, and phosphoric acid from 0.1 to 0.14 per cent. As high as 0.28 per cent of calcium carbonate has been observed. The soils are much less soluble in 10 per cent hydrochloric acid than chernozem.

The ash-colored powder of the lower horizon is considered to be a product of the action of humus acids upon the silicates, causing the separation of a part of the silica in pulverulent form.

The mechanical composition of these soils is variable. In the forest subclays of the Nijni Novgorod, Orlov, or Poltava governments the amount of particles less than 0.01 mm. (20 to 25–32 per cent) was to that of the larger ones (80 to 75–68 per cent) as 1:4, 1:3, 1:2. The general absence of structure of the upper horizon contributes to its pulverization in plowing, resulting in an increased capacity for humidity and decreased permeability.

The subclays of the wooded steppes occupy in all respects an intermediate position between the chernozem and the "forests" subclays proper, approaching first one then the other in character. By a study of the distribution of the forest subclays and the subclays of the wooded steppes in the territory of the chernozems Dokouchayev was able to determine the areas which have been in the past occupied by forests, but which are now under cultivation. Tanfilyev has lately prepared a map of the prehistoric steppes of European Russia. Wooded steppes and true forest soils extend into Siberia. Soils identical with or very closely resembling them are also found on the plains of western Europe, namely, in Galicia, Hungary, and in central Germany. There is little doubt that this type of soils occurs on the American continent where the prairies begin to be replaced by forests.

Sod and podzol soils.—The Russian term "podzol" very nearly corresponds with the German "Bleisand" (lead sand), with this difference, however, that the term is applied not only to sandy but also to more sticky, clayey soils if they have been affected to a marked degree by chemical leaching processes under the influence of the solvent action of humus acids. In the regions where podzol soils occur the climatic and other conditions are especially favorable to the decomposition and leaching of the soil constituents by the solvent and reducing action of the humus.

The upper horizon of the podzol is light gray or gray, frequently with a light cinnamon tint, and 1 to 1½ dcm. in depth. It has no marked structure, and its coherence varies with the content of clay, sand, and humus.

The underlying horizon is much lighter, sometimes almost white, sometimes with a yellowish or pale-blue tint. This is the podzol proper. It presents a mass of fine particles, flour-like in a dry state, sticky in a wet state, very rich in silica. The thickness (depth) of the podzol layer varies from a few centimeters to over 4 decimeters. The subsoil or the parent rock is most frequently red-brown sandy morainic clay with pockets of podzol, or argillaceous sand, but the subsoils may also be pebbly clays, feebly coherent and friable sands, clay or loamy yielding rocks, or even loess-like deposits.

When the second horizon is near the surface the whole soil is called podzol; when it is not individualized, indistinct, or entirely absent, a sod or peat soil results. Between the first and the second there exist in nature gradual transitions, as can be seen in northern Russia on every cultivated field and under every forest.

Concretions are ordinarily found in podzol soils in the form of bullet-like grains, small veins, or continuous layers in the lower part of the second horizon or at the border between the latter and the parent rock.

The soils of this group occupy not less than two-fifths of the area of European Russia, the greater part of Poland being included. At the north they extend as far as Archangel and penetrate in strips and circumscribed areas into the borderland of the tundra soils. At the south they comprise parts of the governments of Perm, Kazan, Nijni Novgorod, Vladimir, Riazan, Kaluga, Oryol, Chernigov, Volyn, and Lublin, where they intermix with the forest subclays and the chernozem. Typical podzols are found especially in the governments of Mogilyov, Smolensk, Vitebsk, Tver, Novgorod, Pskov, and St. Petersburg.

In podzol soils which were once covered with woods and are now cultivated the content of humus is not large, varying from a few tenths of 1 per cent to 2 or 3 per cent, rarely more.[1] In the lower horizon the amount of humus rapidly falls to 0.1 to 0.3 per cent. The nitrogen fluctuates between 0.1 and 0.15 per cent in the upper horizon. The solubility of the humus is remarkably high. From soil of the upper horizon water extracts from one forty-eighth to one-twentieth of the total humus and of the lower horizon from one twenty-seventh to one-tenth. Nitric acid is often found in these extracts.

The soils contain on an average 95 to 97 per cent of mineral matter, of which 80 per cent and more is silica. The amount of zeolites usually does not exceed 10 to 12 per cent, frequently falling much lower (7 to 5 per cent); the amount of substances soluble in 1 per cent cold hydrochloric acid is rarely more than 2 per cent. The total quantity of phosphoric acid varies from 0.05 to 0.08 per cent, but is larger in soils containing a large amount of organic matter. The investigations of Kostichev have proved that in this case it is present mainly in combination with the humus. The absorptive capacity does not in general exceed 12 to 13 per cent.

The podzol soils vary widely, according to the nature of the parent rock. The composition of samples of three different horizons of a podzol soil from the Novgorod Government is given in the following table:

Composition of a podzol from the Novgorod Government.

	Humus.	Lime.	Magnesia.	Alumina.	Iron oxid.	Phosphoric acid.	Silica.
	Per cent.	Per cent.	Per cent.	Per cent.	Per cent.	Per cent.	Per cent.
Upper horizon	2.8	1.172	0.378	7.032	1.84	0.085	81.02
Lower horizon (podzol proper)	.3	.790	.240	4.790	.67	.050	90.70
Subsoil		1.030	.340	7.210	1.62	Undetermined.	84.50

[1] If the upper horizon is turf-like it contains sometimes up to 15 per cent and more of partly decayed organic matter.

If the second horizon is near to the surface or the whole soil is transformed into podzol the land is, of course, very poor. In sandy soils the second horizon contains much less of the alkalis, lime, magnesia, iron oxid, alumina, and phosphoric acid than is found in clayey soils of this class.

The relation of sand to fine earth in podzol soils varies from 5:1 to 7:1. The capacity for water is only from one-half to two-thirds that of chernozem, while its permeability is 2 to 6 times as great. On the better class of podzol soils, when well provided with moisture, the crops, although not large, are more uniform and constant than on the chernozem, especially if well fertilized. In the true silty podzol, however, there is frequently more than 70 per cent of fine earth in the form of quartz dust. It absorbs moisture with avidity and retains it for a long time, turning into a sticky dough-like mass. On drying it breaks up into dust or hardens and forms crusts. This is one of the worst and most unproductive soils, both on account of its poverty in fertilizing constituents and of its unfavorable physical properties.

Soils of the podzol type are found in Siberia, northern Germany, France (the landes), Holland, Denmark, and Scandinavia, and North America (mainly in the British possessions).

Tundra soils.—The soils of the arctic tundra of European Russia and Siberia may be classified as rocky, turfy, clayey, and sandy. The level surface and the treeless condition of the tundra of the basins of the Petchora, Obi, and Yenisei Rivers impart to it a steppe-like appearance. The vegetation consists of lichens, mosses, Arctostaphylos, Andromeda, Empetrum, *Rubus chamæmorus*, Vaccinium, Carex, etc. *Betula nana* and the polar dwarf willows appear as almost the only representatives of bushes. The humus is crude and accumulates only in the surface horizon of the clayey or sandy soil, to a depth of 3 to 5 cm.; everywhere can be seen denuded places, surrounded by mosses or lichens. The temperature fluctuations are striking. The summer is very short; even in July the temperature falls at night to $+ 3°$ C., and at the end of the month even to $-2°$; in August it snows, and soon the long winter, with its icy winds, begins. The perpetually frozen layer begins in the clayey tundra at a depth of 0.7 to 1 meter and in the sandy at a depth of about $1\frac{1}{2}$ meters. The turfy tundras are characteristic mounds of turf, frozen inside, which are 15 to 20 meters in length and 4 meters in height. The forest penetrates into the tundra from the south, along the river banks, where the perpetually frozen horizon is deeper than in other places.

INTRAZONAL SOILS.

Alkali soils.—Alkali lands are found in the southern part of European Russia, in southwestern Siberia, in the Transcaspian region, and in Turkestan. In the territory of the chernozem they occur in spots, usually on the gently sloping southern declivities or on the slight depressions of the steppes. Sometimes these areas occupy dozens of square kilometers and contain saline lakes, but more frequently they are scattered over the steppes in small spots. In a vertical section of a chernozem alkali soil there are seen: (a) The upper horizon, black, dark gray, dark brown, or gray, sometimes homogeneous, sometimes pervaded by a whitish dust; from 1 to 3 dcm. deep; (b) a light gray or whitish horizon, 1 to 3 dcm. deep (sometimes almost absent), merging into (c) a brownish or yellowish compact and sticky clay.

On the surface of the alkali soil, especially after a rain, appear efflorescences or crusts consisting of whitish siliceous powder and minute saline crystals. The content of humus in the upper horizon is in general much less than in the adjacent chernozem, but sometimes reaches 8 per cent and more. The water extracts are colored light cinnamon or light cherry from the alkaline humates in solution. The solubility of the humus reaches one-seventieth in the upper horizon and one twenty-fifth in the second horizon (b), *i. e.*, it is twice or three times as great as in the chernozems. This is due to the greater humidity of the alkali soils and brings them into close relation with the soils of the podzol type. The whitish color of the lower horizon and the siliceous dust of the efflorescences and crusts is due to the same cause. Of the mineral salts soluble in water in the alkali soils of the chernozem zone there occur sodium carbonate, sodium sulphate, sodium chlorid, calcium sulphate, magnesium sulphate, and calcium bicarbonate. Many alkali soils are marly. The total amount of salts extracted by water varies according to Kostichev and others from 0.5 per cent to 5 per cent and more. With regard to physical characteristics, the alkali soils of the chernozem territory are distinguished by becoming very compact and hard upon drying.

The alkali soils of the dry steppes and of Turkestan are mostly yellowish and brownish in color, like the zonal soils which surround them, but dark colored alkali soils are also met with. The white incrustations consist of sodium sulphate, sodium chlorid, magnesium sulphate, calcium sulphate, and carbonates. Extensive alkali deserts without any cultivation whatever occur, as well as saline mud flats.

In general the alkali lands of European and Asiatic Russia bear a close resemblance to those of Hungary, India, Arabia, the western States of North America, Argentina, Australia, and other level and dry regions.

Humus-calcareous soils.—The soils which are formed from limestones and marls are frequently skeleton soils and contain little humus, especially if distributed over steep river banks and along ravines, but from the same parent rocks—soft limestones, chalk, and chalky marl—originate gray and dark gray soils, sometimes very rich in humus. In the southern part of Poland they attract especial attention, being in

33

marked contrast with the surrounding light gray podzol soils. They are known under the local terms of "rendzina" or "borowina." The upper horizon of the rendzina is most frequently gray, without a cinnamon tint, not rarely spotted with white undecomposed chalk; lower down the color becomes lighter and the soil gradually merges into the marly, sticky clay which is mixed with chalky gravel. Still lower lies the white parent rock—chalk or limestone. The content of humus varies from 3 to 5 per cent and more; its solubility from one one-hundredth to one one-hundred-and-thirtieth. The amount of calcium carbonate varies from 3 to 17 per cent and more. The clayey character of the mineral matter renders the soil sticky in wet weather and hard in a drought. However, lighter sandy rendzinas also occur.

Marsh or swamp soils.—Soils of this type extend largely throughout the whole northern half of Russia, but are of little economic importance. In the basin of the Pripet River they occupy more than 2,000 square kilometers. Throughout the podzol soil areas spots and strips of grassy marsh soils are formed under the influence of excessive stagnating water. The vegetation consists of species of Carex, Scirpus, Phragmites, Acorus, Menyanthes, Parnassia, Nasturtium, Ranunculus, Butomus, Sagittaria, etc. The roots of these plants penetrating into the slimy mineral rock oversaturated with water, give humus which slowly oxidizes and which accumulates in large amounts (4 to 20 per cent). The borders of the marshes are frequently cultivated and are known as "black earth" in contrast to the adjacent light sod and podzol soils.

The thickness of the dark-colored horizon varies from 2 to 8 and more decimeters. The solubility of the humus of the soil as a whole is not great ($\frac{1}{800}$ to $\frac{1}{270}$), but rapidly increases with the depth in the soil, being one-tenth at a depth of 1 meter. The abundance of moisture which dissolves humus acids favors the decoloration and leaching of the lower horizons of the soil, making them very similar to the podzols. Under the marshes are frequently found white, light gray, or bluish, and grayish-white slime, either clayey or sandy. The total quantity of nitrogen in the upper horizon varies from 0.3 to 4 per cent.

In the mineral part of the soil the proportion of the clay and sand is variable. Brown veins and concretions of limonite, vivianite, iron sulphid, etc., are usually present. A considerable amount of carbonate and sulphate of calcium are also characteristic of many marshy soils which contain animal remains (shells of mollusks, etc.).

INCOMPLETE OR AZONAL SOILS.

To this class belong the crude and skeleton soils originating from compact, pebbly, conglomerate, and sandy rocks, and morainic and alluvial soils, which are more or less widely distributed throughout

Russia. The Russian rivers, with the exception of some which flow through mountainous regions, overflow regularly in the spring. The alluviums which they deposit consist of sands, clays, and sandy or marly clays, containing some limonite, peat, vivianite, etc.

The prairie vegetation which springs up after the water has receded results in an accumulation in the upper horizon (soil proper) of varying quantities of humus.

The petrographic character of these soils approaches that of the soils from which the alluvium is derived—in northern Russia the podzols, in southern the chernozems.

2

Reprinted from pages 23–38 of *Bureau of Soils Bulletin No. 85,* U.S. Department of Agriculture, 1912, 114 p.

A STUDY OF THE SOILS OF THE UNITED STATES

G. N. Coffey

[*Editor's Note:* In the original, material precedes this excerpt.]

CLASSIFICATION OF SOILS.

NEED·OF CLASSIFICATION.

From the foregoing discussion it is evident that many influences and factors have been instrumental in giving to the soil its present properties and characteristics. Numerous kinds of rocks or soil material, subjected to the action of so many agencies and processes, with an ever-changing degree of intensity, have resulted in the formation of many varieties or types of soil, related to each other in various ways.

In the study of soils, as well as any other subject which has to do with a large number of individuals, classification is of the utmost importance in order that the various and complex relations may be shown as far as practicable. A complete and perfect classification will give an epitome of our knowledge in regard to any subject. Such, however, can not be made until a thorough understanding of the properties and characteristics of the objects to be classified is had—a condition far from being attained in our present imperfect knowledge of soils. Much valuable work, however, has been done and the knowledge thus obtained is being used as an aid to further studies. No more fundamental work than the proper classification and correlation of the soils of the country confronts soil investigators to-day, for this information is essential to the final solution of some of the most important questions with which they have to deal. Natural phenomena are always difficult to classify, because the various individuals merge into others by almost or entirely imperceptible gradations. No sharp lines of division exist, and such as are drawn must be more or less arbitrary. This fact is as truly applicable to soils as to any other great group of natural objects.

The need of arranging soils so as to show their relation to one another has long been recognized and some system of classification has been employed whenever the study of the soils of any extensive area has been undertaken. Different investigators, looking at the problem from different viewpoints, have used different bases of classification, each of which presents some important relation. One of the principal difficulties with most systems has been that they are founded upon one group of factors alone, while others of equal or even greater magnitude have been entirely neglected. A satisfactory system must be such that all of the important relations will be indicated, else it will not answer the needs of an enlightened agriculture.

PRINCIPAL BASES OF CLASSIFICATION.

In order to secure as full and clear an understanding of this subject as possible it is well to consider the principal bases, or systems, which have been used in classifying soils. These may be grouped under the following heads: (1) Geological, including (a) the age, (b) the kind of rock, or lithological, and (c) the agency of rock formation, or dynamic; (2) physical, including (a) texture, (b) structure, and (c) color; (3) chemical; (4) vegetation, especially native, or ecological; (5) processes of soil formation, or climatic; and (6) a combination of all these.

GEOLOGICAL.

The geological basis of classification has been used very extensively in many countries and by many investigators. Most of the earlier soil work in this country consisted largely in descriptions of soils of the different geological formations. In France[1] and Japan,[2] especially at the present time, the main soil divisions are based upon the geological formations which are then divided according to physical

[1] See Rapport sur les Cartes Agronomiques par M. Adolphe Carnot, Ministère de l'Agriculture, Bulletin, 1893.

[2] See various maps published by the agronomic section of the Imperial Geological Office of Japan.

characteristics. The geological basis of soil classification has usually been approached from the viewpoint of the age of the material, its mineralogical character or the agency of its formation.

Age of rock.—As most of the earlier work upon soils was done by geologists, and as age is the principal factor upon which geological divisions are based, the classifying of soils according to the geological formation from which they have been derived naturally resulted. Peters,[1] for example, in his work upon the soils of Kentucky, divided them into Silurian soils, Subcarboniferous soils, Carboniferous soils, Quaternary soils, Recent Soils, etc.

The same idea of age is involved in the separation of glacial soils according to the different advances of the ice, as has been done by Hopkins[2] in Illinois and Stevenson[3] in Iowa. The principal reason for this separation is the greater amount of leaching to which the earlier glaciations have been subjected. This does not depend, however, entirely upon the length of time that the material has been exposed. The topography of the country, the permeability of the material, as well as the climatic conditions, have much to do with this also. The geologists themselves are often uncertain as to the time and number of advances of the ice, so that it is not wise to follow these separations very far; too many other factors have relatively so much greater influence upon the nature of the soil.

In general, it might be said that, so far as age has actually produced differences in the soil, it should be given due weight in classification, but when such is not the case it should be entirely ignored.

Kind of rock, or lithological.—As the soil consists very largely of mineral matter formed from the breaking down of the rocks, its nature must necessarily depend to a large extent upon the kind of rock from which this material has come. This has led to the classification of soils according to the kind of rock from which they have been derived. Thus we have granite soils, shale soils, limestone soils, etc., expressive of the relation existing between the rock and the derivative soil.

If the same kind of rock always gave the same kind of soil, and the same kind of soil was always derived from the same kind of rock, this classification might answer all practical requirements. Such, however, is not the case, as was shown in the discussion of the relation of soils to the underlying formation, and while this classification embodies essential features, it is incomplete and therefore in itself unsatisfactory.

Agency of deposition.—Probably the most frequent basis of classification is what is commonly termed "origin," although in this con-

[1] Geological Survey of Kentucky, 1885, vol. 2, p. 160.
[2] The Fertility in Illinois Soils, by Cyril G. Hopkins and James H. Pettitt, Bul. 123, Illinois Experiment Station. See especially soil map and pp. 193–194 and 257.
[3] The Principal Soil Areas of Iowa, by W. H. Stevenson, Bul. 82, Iowa Experiment Station.

nection the term is used with a somewhat restricted meaning In this system the soils are usually divided into residual or sedentary and transported, according to whether they are derived from the breaking down of the underlying rock or from material which has been transported since it was broken down. The latter are further divided into (1) colluvial, (2) alluvial, (3) æolian, and (4) glacial, according to the agency of transportation.

While the above division has been used by many investigators, scarcely any two agree as to the exact line of separation. Merrill,[1] for example, uses colluvial deposits to " include those heterogeneous aggregates of rock detritus commonly designated as talus and cliff débris," and states that they are " comparatively limited in their extent." He practically confines them to gravitational deposits. Hilgard,[2] on the other hand, says that " when the soil mass formed by weathering has been removed from the original site to such a degree as to cause it to intermingle with the materials of other rocks or layers, as is usually the case on hillsides and in undulating uplands generally, as a result of rolling or sliding down, washing of rains, sweeping of wind, etc., the mixed soil, which will usually be found to contain angular fragments of various rocks and is destitute of any definite structure is designated as a colluvial one." The latter definition is thus seen to be much more inclusive than the former.

Most investigators are agreed that sedentary or residual soils consist of those which are formed in place, but there is a difference of opinion as to what constitutes formed in place. The more common usage is to confine residual soils to those derived from hardened rocks, while those formed from unconsolidated deposits are called transported, and these terms are used in this sense in this report. However, the soils derived from the unconsolidated sandy-clays and clays of the Atlantic Coastal Plains are as truly formed in place as those from older consolidated rocks. In a strict sense, only the most recent alluvial soils are transported soils; practically all others are derived from the underlying material, although this may have been previously transported by water, wind, or ice.

Differences in the agencies of transportation of soil material necessarily determine to a certain extent its character and, consequently, the nature of the resultant soil, but differences in the physical, mineralogical, and chemical composition of the material transported, together with changes produced by variations in the soil-forming processes, may so overshadow these, in some cases at least, as to render them almost if not entirely negligible.

This classification takes into consideration only one phase of the origin of soils, the geological, while the biological and climatological

[1] Rocks, Rock-weathering, and Soils, by Geo. P. Merrill, p. 319.
[2] Soils, by E. W. Hilgard, p. 12.

are omitted. In short, it fails to show some of the most important physical, chemical, and biological relations between soils, relations of the utmost significance to the agriculturist, in whose interest any system of soil classification is primarily made.

PHYSICAL PROPERTIES.

That soils differ in physical properties and that these differences have an influence upon the growth of plants has long been recognized by practical agriculturists as well as scientific soil investigators, but it is only within comparatively recent years that the importance of these properties has been appreciated. The work of Whitney, King, and Hilgard has done much to bring about this result. The three most important physical properties to be considered in soil classification are texture, structure, and color.

Texture.—The term "texture" is employed somewhat loosely in soil literature. As used in this paper it has reference to the relative proportion of the different grades or sizes of particles, while structure refers to the arrangement of these particles. The necessity of distinguishing between the two is at once evident. The former remains practically constant, while the latter is continually changing. Every time a field is plowed the structure is changed, and every rain that falls afterwards makes additional changes in the arrangement of the particles. As texture is a property that remains practically constant, it can be used as a basis of classification, but only those differences in structure which are due to the inherent character of the material can be considered.

According to texture, soils are divided into a number of classes, as sand, sandy loam, loam, clay loam, and clay. The number of classes will depend upon conditions. Whenever the difference in texture is sufficient materially to influence plant growth, a separation, if practicable, should be made, although the fineness to which these divisions shall be carried depends upon the amount of detail desired. The five classes just given are the ones most commonly recognized, but in detailed work three or four times this number have to be made. In addition, the presence of stones or gravel necessitates the introduction of stony and gravelly members of the different classes also. There is need of large textural groupings, with subdivision of each group, in order that this classification may be adapted to any amount of detail to which one may wish to go.

The diagram given below shows the distribution of the material as obtained by averaging the mechanical analyses of a large number of soils in the 12 most important soil classes. It does not necessarily represent the most typical distribution.[1]

---

[1] See Physical Principles of Soil Classification, by George N. Coffey, Proceedings of the American Society of Agronomy, vol. 1, p. 181.

This textural classification is undoubtedly of the very greatest importance, and any system of which it does not form a part can not be satisfactory. On the other hand, the fact that in this system texture is the only factor considered and the further fact that it has

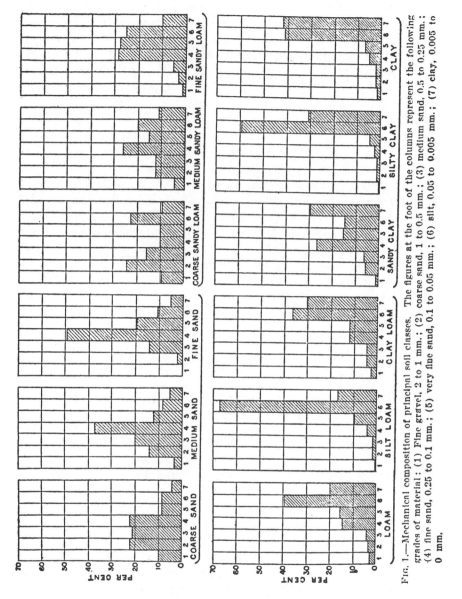

Fig. 1.—Mechanical composition of principal soil classes. The figures at the foot of the columns represent the following grades of material: (1) Fine gravel, 2 to 1 mm.; (2) coarse sand, 1 to 0.5 mm.; (3) medium sand, 0.5 to 0.25 mm.; (4) fine sand, 0.25 to 0.1 mm.; (5) very fine sand, 0.1 to 0.05 mm.; (6) silt, 0.05 to 0.005 mm.; (7) clay, 0.005 to 0 mm.

been found necessary to recognize as many as 50 to 75 types having the same texture, shows that it can not in itself answer all the requirements. The importance of considering other differences besides texture can not be too strongly emphasized.

Structure.—Structure, or the arrangement of the particles, is another very important property of the soil, although one which is really very little understood. Porous or impervious, loose or compact are terms used to express differences in structure. While some of the differences in structure may be so artificial and transitory, like those induced by plowing, that they can not enter into the question of classification, still there are others which seem to be inherent in the nature of the mineral matter and no amount of manipulation can entirely eliminate them. The particles in some soils fit very closely together and good tilth is maintained with difficulty, while in others the arrangement is loose and cultivation easy. Numerous observations prove that subsoils, which shrink and swell a great deal as the result of a change in moisture conditions, are nearly always impervious. The cause of this is probably the presence of a large amount of " colloidal clay." Such differences in structure are easily discernable in the field, and while further study is needed to explain the cause, or causes, they should certainly be recognized in detailed soil mapping.

Color.—Color is one of the most obvious physical properties of the soil, one used by practical farmers as an index to its character from time immemorial. A black color has come to be almost synonymous with productiveness. " Red soils," " gray soils," " brown soils," "white soils," etc., are terms in very common use. In itself color may be of very little importance, but as an indicator of physical and chemical conditions it is of the greatest moment. The practical agriculturist will no more class together soils markedly dissimilar in color than the ethnologist would consider a white and a black man as belonging to the same race.

In judging soils by their color one must always consider to what this color is due, else entirely erroneous conclusions may be reached. In a humid region, for example, a white color is indicative of a soil poor in lime and " available " phosphoric acid as well as of low agricultural value. In an arid region this appearance is often due to the presence of large amounts of lime, and such soils are usually very productive when supplied with moisture.

While color is properly classed as a physical characteristic of the soil, probably its greatest significance is as an indicator of chemical differences. By its use it is often possible to detect differences which the present methods of chemical analysis are unable to explain. When its importance in chemical analysis is considered one can readily understand how valuable a factor it may be in judging of differences in the chemical condition of the soil constituents. In analytical chemistry certain colors are indicative of certain elements or compounds. The same is doubtless true of soils, although the conditions here are so complicated that the assumption may often be impossible of proof.

CHEMICAL PROPERTIES.

A great many chemical analyses of soils have been made and classifications according to the percentage of the different elements of plant food found in them have been undertaken. The mineral theory of plant nutrition advanced by Liebig gave the greatest stimulus to this line of soil investigation.

That differences in chemical composition have a marked influence upon soil fertility is not to be questioned, but no method of soil analysis has ever been devised by means of which the fertility of the soil can be determined with certainty a priori. There are so many factors which have an influence upon the growth of plants besides the mere amount of plant food present that the determination of this alone can never be used as a safe and sure guide to its agricultural possibilities. Even if a method could be devised, only a very small part of the soil in a field can be analyzed, and it is therefore necessary to use other means of distinguishing differences, else there is no way of knowing whether the surrounding soils are similar in character, and unless this is true the chemical analysis would be of practically no value. Chemical analyses may be used in the laboratory as a means of studying the causes of differences in soils, but a practical classification must be based upon those obvious differences which can be determined in the field. For example, one of the most important chemical differences in soils relates to their calcareous or noncalcareous nature. Several laboratory methods have been devised for determining this question, none of which are entirely satisfactory, and Hilgard[1] states that "the decisive feature in this matter must evidently be the native vegetation, which expresses the nature of the land much more clearly and authoritatively than any arbitrary definition or nomenclature can possibly claim to do. A soil must be considered as being calcareous whenever it naturally supports the vegetation (calciphile) characteristic of calcareous soils." It is possible, however, for a soil to be calcareous in the deeper subsoil and noncalcareous near the surface, so that it would support a deep-rooted calciphile and a shallow-rooted calcifuge vegetation. The writer doubts whether a soil should be considered as truly calcareous unless it has a sufficient amount of lime carbonate to give the dark tint to the humus, provided moisture conditions are favorable.

It is nearly always possible for a soil expert to detect, by means of color, origin, native vegetation, or crop growth, the existence of important chemical differences in soils, although he may not be able with our present methods to explain the exact cause or nature of the differences. If they can not be detected by these means they are not usually of sufficient importance to justify giving them consideration in soil classification.

---

[1] Soils, by E. W. Hilgard, p. 496.

NATIVE VEGETATION.

The most important function of the soil is the support of plant life. The kind of plant life which is found is dependent more or less upon the character of the soil, and this fact has been made use of by some investigators as a basis of classification. One of the most ardent advocates of this basis is Hilgard, who used it in the Tenth Census as the principal foundation for dividing the soils of the Cotton States into broad groups and regions. In his recent book on "Soils" the same author has devoted the entire fourth part to a discussion of "Soils and Native Vegetation." Peters used it to some extent in his work upon the soils of Arkansas and Kentucky, and other investigators have given it more or less attention. Hilgard laid especial emphasis upon the native vegetation. The same principle, however, has been applied in classifying soils according to the crops grown upon them. Thus we have corn soils, cotton soils, wheat soils, pineapple soils, etc. The character of the crop grown, however, depends too much upon economic conditions to use this as a basis of soil classification. A study of the adaptation of soils to different crops is very profitable, and the Germans especially have done much work along this line, but it is more a classification of land than of soil.

There is no doubt that a change in native vegetation is usually indicative of a change in soils, but there are striking exceptions. In many instances where there is a marked difference in the natural timber growth it is possible to find differences in the soils to explain it, but in other instances markedly different soils will show the same character of timber growth. The post and black-jack oaks seem to thrive best upon soils subject to drought. This condition may be found in both deep sands and impervious clays and these trees are found upon both kinds of soil.

It is also a well-known fact that where one kind of timber growth is removed trees of a different genus often take possession of the land. It is a matter of common observation that when fields originally covered with deciduous trees, like the oak, are abandoned, there almost invariably springs up a growth of pine, and this is so common that the tree has received the name of "old-field pine."

A fundamental essential of scientific classification is that it must be based upon inherent and invariable properties of the materials classified. Classifying soils according to native vegetation is therefore going at the matter backwards; it is putting the effect before the cause. It is better to determine the cause of the change in the vegetative covering, and, if found to be due to a variation in the soil, base the classification upon this; otherwise the destruction of the vegetation will destroy the foundation of the system. Vegetation should, therefore, be used as an indicator of agricultural value and an aid in rather than a basis of classification.

CLIMATIC.

That soils formed under arid and humid conditions are essentially and strikingly different in character is clearly recognized by all investigators who have made any study of the subject. As has already been pointed out, this is primarily due to differences in the processes of weathering resulting from the markedly dissimilar climatic conditions, especially moisture, although temperature also has an influence. This contrast in the soils of different climates has led to their classification upon this basis, or, more strictly speaking, upon the differences which have resulted from a dissimilarity in climatic conditions.

While American and other investigators have generally recognized the influence of climatic factors in soil formation, Russian workers, especially Dokouchayev, who is credited with founding a new school of soil investigation, have laid greatest emphasis upon the importance of climate in soil classification. With them it is given first place, because all processes of soil formation or weathering vary more or less with the climate, and therefore soils formed under markedly dissimilar climatic conditions will be very unlike in character. The climatic classification coincides to a certain extent with the chemical and geological classification. According to this classification, the world has been divided into the following " soil zones: "[1]

(1) Laterite soils, or soils of the warm, humid Tropics and Subtropics.

(2) Eolian dust soils, formed in continental regions of dry climate.

(3) Soils of the dry or desert steppes.

(4) Chernozems, formed in connection with herbaceous steppes and prairies of the temperate or cold temperate zones.

(5) Soils of the wooded steppes and of the deciduous forests, resembling the chernozems, but distinguished by conditions of origin.

(6) Peat soils and Podzols, properly soils of a cold temperature. They are typically developed under the mixed forests and are ordinarily accompanied by " ortsteins."

(7) Soils of the tundras, formed upon the clay and sandy tundras.

COMBINATION.

Since all of these viewpoints represent more or less important relations between soils, it is evident that any system based upon one set of factors alone is not entirely satisfactory, and attempts have therefore been made to construct a system by combining one or more of these.

In France, for example, a commission was appointed in 1893 to report upon all matters connected with the making of agricultural maps. This commission[2] recommended that geological maps be

---

[1] See Etude des sols de la Russie, by N. Sibirtzev, Seventh International Geological Congress, St. Petersburg, 1897. Also Russian Soil Investigations, Expt. Sta., Record, Vol. XII, pp. 704–712 and 807–888.

[2] Rapport sur les Cartes Agronomiques, par M. Adolphe Carnot, Ministère de l'Agriculture, Bulletin, 1893.

used as the basis, and that the variation in sand, clay, lime, and humus of the geological formation be shown by the initial letter of the proper ingredient.

Pagnoul[1] classified soils according to the amount of humus, lime, sand, and clay. They are divided (1) into ordinary soils, having less than 50 per cent of humus, and (2) humus soils, having more than 50 per cent of humus. Ordinary soils are subdivided into calcareous, clayey, and sandy, and a combination of these gives further subdivisions.

In Germany the Prussian Geological Survey has constructed many "Geologisch-agronomischen" maps. In the general description of these it is stated that "the explanation accompanying each colored map contains a geological and an agricultural part. In the latter the weathering processes and the nature of the several soils are described. These are divided into groups according to the chief soil components, and are loamy, sandy, clayey, humaceous, calcareous, etc.

The classification used upon the Japanese maps is based upon geology, which is shown in colors, while the physical characteristics— sand, loam, clay, etc.—are indicated by hatching over the colors. By means of symbols the soils are divided into those "rich in humus" and those "moderately rich in humus." The Japanese classification is therefore much like that of the French and Germans.

As already pointed out, the Russians make their primary classification upon the origin of the soil—not so much the geological as the climatic and organic origin. Taking the origin as the starting point, the classification is elaborated and further subdivisions made according to differences in texture or other physical and chemical properties.

The system of classification used by the Bureau of Soils of the United States Department of Agriculture is founded upon those differences which are obvious in a field examination. In one sense it is therefore largely physical, but many of these differences are due to geological, chemical, or biological causes, and it is therefore proper to say that these factors are also given a place. In fact, all features which have any obvious influence upon plant growth are taken into consideration.

From a study of the soils in various parts of the United States, it has been found possible to divide the country into a number of provinces, each of which contains soils peculiar to itself. This is due to a similarity in the geological processes by which the soil material in these provinces has been formed. Thirteen such provinces have thus far been recognized.[2]

---

[1] Les Terres Arables du Pas-de-Calais, par A. Pagnoul.
[2] See "Soils of the United States,' by Milton Whitney, Bul. 55, Bureau of Soils. U. S. Dept. Agr., p. 27.

Owing to variation in the source of the material or in the processes of weathering, soils may differ in character and appearance, although formed by similar processes. The soils of a given province are therefore further divided into series, which consist of soils very similar in all characteristics except texture. Soils are made up of particles of various sizes or grades, and the relative proportions of the different grades determine the texture. Different divisions of soils based upon texture have thus been established, and the soil series is subdivided into different classes. A combination of the series and class characteristics constitutes a type, which is the unit of soil classification. An example will suffice to illustrate the system:

Province _____ Piedmont Plateau.
Series _____ Cecil.
Class _____ Clay.
Type _____ Cecil clay.

IDEAL.

The ideal classification would be one in which the individual types are grouped in a number of divisions, each larger grouping representing more and more distant relationships in the soils. Thus in one of the major divisions there would occur no type which closely resembles any soil in any of the other divisions. The working out of such a system in detail would require a complete knowledge of the soil and of the relative influence upon plant growth of the numerous factors which constitute soil differences. Such knowledge is far from realization, if not indeed absolutely unattainable.

In making such a classification it would be necessary to recognize inherent differences in the soil itself as the fundamental idea; to consider it as a natural body having a definite genesis and distinct nature of its own and occupying an independent position in the formations constituting the surface of the earth. The various constituents which make up this bio-geological formation would have to be studied and their relative influence upon ecological relations determined. The major divisions should then represent differences in the most influential of these constituents and further subdivisions made upon those of less consequence.

The task is made more difficult by the fact that the relative importance of the various differences may change with a change in any of the other interdependable factors or conditions. In regions of dry farming, for example, texture is of very great importance because of its pronounced influence upon soil moisture, which is the determining factor in successful agriculture in the semiarid regions. In an irrigated district, on the other hand, the moisture supply is under human control, and texture, therefore, loses its prominent position.

Likewise differences which have a decided effect upon the growth of one plant may have little significance if some other crop be grown. Alfalfa will not do well upon any soil that does not contain a considerable amount of lime, but wheat or corn will grow upon both calcareous and noncalcareous soils, although there is strong evidence to show that the same variety will not produce equally well on both. On account of these and other difficulties, it seems that it will never be possible to construct an ideal system of classification, although it is well to have it in mind and to make our classification correspond as nearly as possible to the ideal one.

GENETIC.

In the first part of this paper it was shown that all differences in soils are due to variations either in the processes of derivation or in the character of the material from which they have been formed through weathering; in other words, to origin in the broad sense of that term. A classification based upon differences due to these two sets of factors may, therefore, be termed genetic. The question naturally arises as to which of these shall constitute the basis for the larger groupings.

In considering the constituents of the soils the great importance of humus, lime, and clay was pointed out, and in view of their preeminent influence in determining the relation of the soil to crops, it is evident that the larger division should represent, as far as possible, differences in the proportion of these constituents. The study of the soils of a large area like that of the United States shows that the percentage of humus and lime, and to a less extent of clay, depends very largely upon the amount of leaching which has taken place, and, therefore, the processes of derivation rather than the character of the material should be used as the basis for the larger divisions. As these constituents vary with the climate, especially the amount of precipitation, the major divisions, while representing variations in the most important constituents, will correspond rather closely to climatic regions. For convenience, therefore, we might say that the primary divisions are based upon climate, although strictly speaking they are based upon differences in the soil which are due to variations in the processes of formation resulting from dissimilar climatic conditions.

Emphasis, however, should be laid upon the fact that differences in climate are of importance in soil classification only in so far as they have produced variations in the nature of the soil. Other factors may come in and intensify or diminish the action of the processes sufficiently to make it impossible to draw the line of division strictly upon the amount of precipitation. In Alabama and Mississippi, for example, the very calcareous and only slightly consolidated nature of the Selma chalk (Cretaceous) has caused the formation of Dark-colored Prairie soils in a region where the rainfall is approximately twice the average of that in the great prairie belt. In the limestone valleys of the Appalachian region the rainfall is no greater than in Alabama, but the hard nature of the rock makes necessary a long lapse of time for its degeneration, and the lime, which is essential in the formation of the dark prairie soils, is leached out. These Prairie soils in Alabama and Mississippi should be grouped with other Prairie soils, though they lie in a region of much greater rainfall.

Differences in topography and in the nature, especially the texture, of the material will also have an influence upon the amount of leaching that will result from a given rainfall. As these determine almost entirely the amount of moisture which remains in or upon the soil, they bring about marked variations in its character. Constant saturation, for example, will cause the accumulation of organic matter and result in the formation of Black Swamp soils or, in extreme cases, of muck and peat.

While these variations in the processes of soil formation thus give rise to differences in the percentage of lime and organic matter, they also affect other essential, although less important, soil constituents. The best illustration is the large amount of soluble salts found in connection with lime in the arid regions. A grouping of the soils upon differences due to the soil-forming processes, or climatic origin, will not only represent marked contrasts in the important soil constituents—lime and humus—but will also coincide with variations in many other ingredients. As the climatic conditions are generally fairly uniform over extensive areas the major divisions will usually embrace large stretches of country and represent such profound dissimilarities in both soil and climate, which constitute the two parts of a plant's environment, that the variety of crop suited to the soils of one division need not be expected to do well in another.

Upon the above factors or differences the soils of the United States can be divided into a number of divisions, each division being distinguished by soils having certain more or less well-defined characteristics. As the climatic conditions change gradually rather than abruptly from one section of the country to another, the soils of one region will likewise merge into those of an adjoining region, with usually no sharp line of division. In fact, near the boundaries, or where the differences are due to drainage, there may be small areas of one kind surrounded by another. The following divisions may be recognized in the United States: (1) Arid or unleached soils—low in humus, (2) Dark-colored Prairie or semileached soils rich in humus, (3) Light-colored Timbered or leached soils low in humus, although containing considerable organic matter, (4) Dark-colored Swamp or leached soils high in organic matter, and (5) Organic or muck and peat soils. Of these divisions the last two are of local development, and, therefore, of much less importance than the other three.[1]

While extremely important soil differences, as represented by these major divisions, have resulted from dissimilarity in the process of weathering or climatic origin, other differences due to variations in the character of the material or geological origin, occur within the above division (except peat and muck) so that further subdivisions of these can be made from this viewpoint.

In studying this phase of the subject it will be found that variations here are likewise due to differences in the agency of deposition and in the nature of the material deposited. The first is the basis of subdivision most commonly used in classifying soils, and while a difference in the agency of deposition does not necessarily imply an unlikeness of the material, or vice versa, such is usually the case. A knowledge, therefore, of the agency concerned in the formation of the soil material will aid in reaching a proper understanding of the true nature and properties of the soil. Where these agencies have had little appreciable effect upon the soil-forming material too much stress should not be laid upon them.

According to differences in the agencies or method of formation of the rocks, subdivisions of the major soil groups can be made. The following are suggested: (1) Soils from crystalline rocks formed through the agency of heat and metamorphism; (2) soils from sandstones and shales or consolidated material deposited by physical agencies; (3) soils from limestones formed by organic agencies or as chemical precipitates; (4) soils from ice-laid material (glacial); (5) soils from unconsolidated water-laid material (sedimentary); (6) soils from wind-laid material (æolian); (7) soils from gravity-laid material (colluvial); and (8) alluvial soils representing recent stream material deposited as actual soil.

Not all of the rocks formed by the same agencies or processes consist of identical material, and therefore variation in the soil derived from them is the result. In some cases the difference may be physical, in others mineralogical, and in still others chemical. We may thus have the same material overlying gravel or clay; may have alluvial soils washed from limestones or granite; may have glacial material composed of ground-up shale or limestone; or there may be sedimentary deposits rich in iron, or some other elements, while other deposits will contain these elements in only small amounts. For this reason further subdivision of the soils, even though derived from

---

[1] Two other divisions might possibly have been added, "alkali" soils, representing an excessive accumulation of soluble salts and semiarid soils or slightly leached surface soils containing considerable humus but having unleached subsoils. As the former can be so readily changed by leaching out the injurious salts it seems best to consider the "alkali" soils as a variation of the arid, while the semiarid constitute a transition from the Arid to the Dark-colored Prairie soils.

rock formed by the same agencies, becomes necessary.   These differences are usually more easily detected by color than otherwise, and the division into series might be considered as based more upon this property than any other, although other factors are, in some cases, used also. The soils of a series are similar in all other characteristics except texture. Owing to the fact that very nearly similar soil material may be deposited by different agencies there may be a closer resemblance in two series formed from rocks of different origin than between two derived from rock of similar origin. For this and other reasons a classification based upon difference in the soil itself would be more desirable and some time will doubtless be made.

It has already been pointed out that texture is the most important physical property of the soil. According to differences in this property the series are divided into classes, or into sand, loam, clay, or some intermediate class. In order to bring out the importance of this property of the soil, the texture is made a part of the type name.

There are undoubtedly a large number of soil series, and it is possible in this paper to describe only some of the more representative. In some cases no division of the larger groups into series was attempted.

[*Editor's Note:* In the original, material follows this excerpt.]

3

Reprinted from *Sci. Agric.* **6**:225–232 (1926)

Changing Viewpoints and Methods in Soil Classification.

A. H. JOEL

My principle reasons for choosing this theme for presentation are as follows:

First, to direct your attention to new terms and methods and modified points of view arising from the findings of a number of comparatively recent investigations in soil science. Much of the subject matter is available in papers read and published at various soils conferences. However, most of it is not readily accessible. and I am therefore presuming that it is not common knowledge to most agronomists.

Second, to outline some of the broader relations and more important characteristics of the major groups of our soils as determined by field and laboratory investigations of the Saskatchewan Soil Survey. These will be considered principally from the standpoint of the newer viewpoints which I purpose to discuss.

Much of the information of the general plan of soil classification is taken from papers of Dr. C. F. Marbut, Chief of the Division of Soil Survey of the U. S. Bureau of Soils, and from various papers presented at conferences of the American Soils Survey Association. For assistance in gathering Saskatchewan data I am indebted to my colleagues, Prof. F. H. Edmunds and Mr. J. Mitchell.

The Need of An Effective System of Soil Classification

One of the fundamental needs of any natural science—in fact, one of its first requirements—is an effective system of classification of the bodies which it purposes to study. Such a scheme must not only be convenient and logical, but also very comprehensive.

Botany, Zoology and many other natural sciences have long had such systems. Soils Science, however, has not yet developed a satisfactory classification that could be readily applied. One of the principle reasons for this, I believe, is the great difficulty in determining and evaluating the important characteristics to be used in differentiating soil groups and individual types. This is due to the complicated chemical and physical make-up of such a mass of material as is found in the soil body. Another reason for the lack of a satisfactory classification scheme is, no doubt, the comparative youth of the science.

Soils investigators have recently come to realize strongly the serious handicap of lacking a good classification key, especially when trying to apply the results of investigations in one part of the world to soils in another. Until quite recently they have managed with makeshift soil descriptions involving colorless and often quite meaningless terms such as soil, subsurface and subsoil, timber and prairie soils, humid and arid soils, etc., coupled with conflicting conceptions of colours and other soil characters.

Botanists and agronomists, may, with no great effort, obtain a comprehensive description of a plant if given its scientific name. Soils men lack such a key. We need a Rydberg or a Gray.

The International Society of Soil Science has keenly felt the need of an effective classification in its various conferences, and recognizing the bar to the progress of Soil Science, special international committees were appointed to work on the problem. The valuable contributions of the committees of the Fourth Conference held in Prague, Czechoslovakia, in 1922, and of the Fifth, in Rome, in 1924, represent great strides of progress in the solution of the problem. A very promising foundation scheme has been established and generally agreed upon. Since the Rome conference a great deal of work has been carried on, and will be intensively pursued during the coming year. This is in preparation for the elaborating and building up of the proposed system of soil classification, with the intention of presenting and moulding it at the Sixth International Soils Congress to be held at Washington, D.C., in May, 1927. Field

_____

* Paper given before the Western Canadian Society of Agronomy, Saskatoon, Sask., December, 1925.

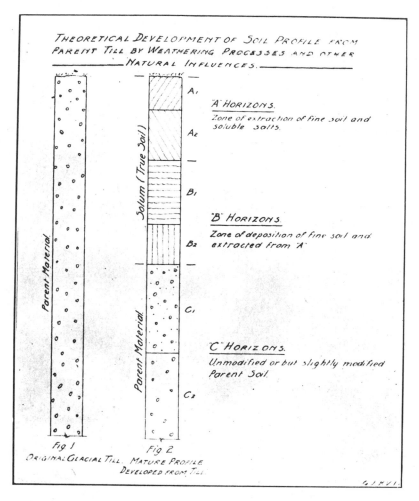

Theoretical Development of Soil Profile from Parent Till by Weathering Processes and other Natural Influences.

"A" Horizons.
Zone of extraction of fine soil and soluble salts.

"B" Horizons.
Zone of deposition of fine soil and extracted from "A"

"C" Horizons.
Unmodified or but slightly modified Parent Soil.

Fig. 1   Fig. 2
Original Glacial Till.   Mature Profile Developed from Till

workers, especially, are giving the problem considerable attention, collecting a great mass of data, numerous samples, pictures, etc. As one immediate outcome of this international effort it is planned that the special committee will present the first reconnoissance soil maps of the world and of the continents at the coming conference. In short, pedologists are, at present, greatly concerned with the study of the fundamental characteristics of soils. We want to become really well acquainted with the character of the natural body with which we are concerned. It seems to be the concensus of opinion that this should take precedence over many other studies previously pursued.

The Inadequacy of Recent Popular Soil Classification Systems

Most natural science classifications are based on the characteristics of the objects or bodies studied. Soil classifications, on the other hand, have been based mainly on external modifying factors such as geological origin and climatic and vegetative influences. It is not my purpose to discuss the reasons for this. However, I do wish to outline briefly the reasons why such bases are illogical and generally unsatisfactory and are consequently giving way to a scheme based primarily on soil characteristics. This complete change is the principal newer viewpoint to which I wish to call your attention.

Classification based on geological origin is one of the first and most popular in use. This

53

was probably a natural result of the close relationship of soils to Geology and of the earlier work done in the latter science. The scheme has been very useful, but, as is true of other methods based on external factors rather than on the inherent characteristics of the soil itself, it has proven inadequate for general application. In studying local areas the scheme worked fairly well, but in studying separate groups many of the relationships of soil features and geological origin failed to hold good.

To illustrate, in Saskatchewan the unsorted boulder clay or glacial till plain deposits are the original or parent materials from which most of our soils developed. Even at widely separated points this material possesses a number of identical characteristics, and in the same local area the unmodified till is usually very much the same wherever found. Furthermore, in these local areas the derived soils are as nearly alike as their "parent" materials or the original till deposits. Consequently, in such restricted areas the geological

scheme works very well. On the other hand, in widely separated areas, soils derived from similar glacial till are often very dissimilar. The geological system usually falls down when applied to separated districts.

Another basis of classification commonly used has been that of climate. It too has been found wanting, but in a different way. It has proven to be a far more effective basis than the geological in making broad groupings, but is a poor criterion for soil classification when used for local separations.

For example, our prairie province soils are in an entirely different class than Ontario soils. In fact, the separation is so marked that we have no fear of duplicating names for the same soils in both places. Furthermore, we know that climatic differences are chiefly responsible for the soil differences, that our prairie province lands are made up of sub-humid and semi-arid soils and that Ontario lands are made up of humid soils. These differences between these two great groups are always well marked as I shall

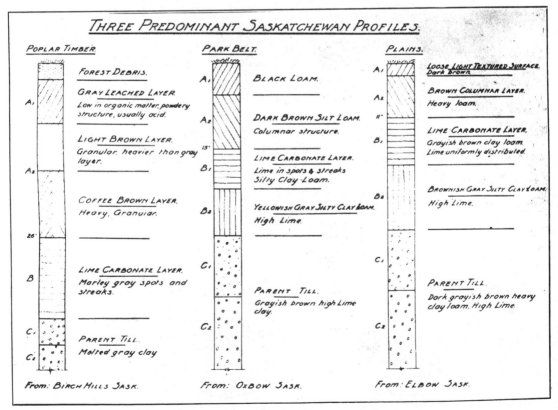

Ged, climate + veg^t.

later illustrate. However, within each area mentioned, there are many variations that cannot be accounted for by climate. This basis, therefore, is of little use as a common basis for separations, in local districts.

Still another basis of classification commonly used is that of natural vegetation. I shall dismiss it with a few general considerations. Suffice it to say that although it, too, has been a useful guide in explaining many soil features, and in separating broad groups, it is too limited in application.

In our field work we have observed some interesting ecological relations. In fact, in many cases the plant and soil associations have helped us considerably in mapping our types. For example, in South-east Saskatchewan, South of Oxbow, in a certain area belts of sand or gravelly soils were distinctly marked out from the loam areas by the absence of "bluffs" (clumps of willow, poplar, etc.). In fact, in the whole district there seems to be a gradual encroachment of this typical Park Belt growth over the medium soil types. In other districts Golden and Black Willows usually mark out poorly drained spots; Rose, Wolf Willow, Snow Berry, Wild Cherry, and others, the light well-drained lands; various salt bushes, the alkali spots, etc.

The most striking general separations are between the prairie and the timber soils. In all parts of the world soils developed under these two conditions present distinctly different characteristics, and many interesting soil features have been explained on this basis. Still, the differences in vegetation will not explain many variations either in the same local area or in separated belts.

The Soil Profile as a Basis for Classification

The various schemes of classification mentioned have all proven inadequate for various reasons. It has therefore been necessary to remould old viewpoints completely, and to make a more direct attack. The basis chosen by the special International Committees appointed for the purpose is that of actual soil characteristics as determined by studies of the soil profile.

By the soil profile is meant the complete vertical soil section to a point well within the unmodified parent or original material from which the natural weathered soil has been developed. This manner of approach deals directly with the soil itself and not with external causative factors or associations. It bases the classification on the evolved product rather than on the various factors which determine the course of the evolution.

Most agronomists are very likely familiar with the soil profile in a general way. I am assuming, however, that they are not so familiar with its details, special terms and application. I hope that this is not taking too much for granted.

A common characteristic of a normal or developed soil profile is the layered or tiered arrangement. The various natural layers, known as horizons, seem to be the final product of the various factors involved in soil development: climatic, vegetative, geological origin, etc. The significant facts are that we herein have a property common to practically all normal soils, yet inherently different in details in different soil types, with these differences representing the results of varying development, as well as variations in fundamental features.

The general character and theoretical development of the most representative type of Saskatchewan profile or soil section is illustrated in the chart on page 226.

Figure 1 in this chart represents a profile of upland glacial till as originally deposited in this general vicinity by the ice sheet supposedly about 25,000 years ago. By far the greater portion of Saskatchewan surface deposits originated in this way. The original soil deposit was very likely a heterogeneous mass of rock material ranging in size from large boulders to colloidal clay. We have reason to believe that this material is still well preserved in an unmodified or very slightly modified condition underneath the weathered portion of the soil column, generally below a depth of about 6 feet.

During the centuries intervening between this deposition and the present, a number of processes have been slowly but steadily changing this heterogeneous mass of material into our present soil profile and its definite horizons. The repeated percolation of water through the column has extracted much of the finer soil particles and soluble salts from the top layers and deposited them in layers

A road cut in the short grass plains area of Saskatchewan. The dark A horizons, the well developed lime layer—B horizon—and the characteristic columnar structure are all clearly shown.

below. The accumulation of organic residues and their subsequent decay has radically changed the layers affected in their physical and chemical properties. Chemical reactions of great number and complication have also left their stamp. As a consequence of these and other modifying influences the general type of soil column illustrated in Fig. 2, has been developed from that shown in Fig. 1. The two roughly represent the genesis of the soil profile.

Four general divisions are usually present in the natural profile of a normally well developed soil. The first, not always considered as a part of the profile itself, is a layer of surface organic residue which I have arbitrarily labeled as O.M. This is typically a very thin layer on prairie soils, but often as thick as one foot on some forest soils. Second is a section beneath this O.M. known as the A horizon, representing the zone of extraction or eluviation. Beneath the A horizon is a zone of accumulation or concentration, the B horizon. Finally, at the base of the profile is the zone of unmodified or but slightly modified original soil material, known as the C horizon. The surface organic matter accumulation with the A and B horizons constitute the true soil. The International Committee on Classification has chosen to call it the "Solum" and the original geological deposit the "parent material".

Most soil profiles show a more detailed division of zones based on differences in colour, structure, texture, etc. The subdivisions most generally found are known as the A_1, and A_2; B_1 and B_2; and C_1 and C_2 horizons, and the term horizon usually applies to these subdivisions rather than to the principal zones.

A normal or mature soil profile is considered to be that profile developed on well-drained, smooth uplands, in virgin soil, and under conditions where the soil has been undisturbed for a long period. In other words, it should represent the well oxidized, well aerated. well-drained and deeply weathered soil column of the region. Such conditions are considered essential to the development of normal agricultural soils.

Immature or imperfect profiles are the abnormally developed or poorly developed types such as are found in recent deposits of streams, on slopes where repeated erosion and deposition destroy the profile, in poorly drained areas or under other conditions where erosion, youth or insufficient or excessive drainage have prevented the normal and complete action of weathering processes over long periods of time. On this basis extremely light and extremely heavy soils are not considered as having normal profiles. The mature profile is taken as the standard of the region and the other types described by comparison to it.

The terms soil, subsurface and subsoil, are practically in discard unless used with definite terms to locate them in the profile.

The principal differentiating characteristics used in soil classification on the basis of the profile are as follows: the number, colour. texture, structure, chemical composition, thickness, relative arrangement, resistance to penetration and consistency of the various horizons and the geology and nature of the parent material.

Great World Groups of Soils Based on Profile Separation

Based on this general scheme the soils of the world have been divided into a number of great groups; and, where extensive work is being done in soil surveys, subdivisions have been made into minor groups. Possibly the final arrangement will be quite similar to Zoological divisions, classes, orders, families, genera, species and varieties.

To illustrate its application I shall briefly describe and illustrate four profiles representing as many world groups of soils. At the same time I shall point out how these major groups may be subdivided into minor groups representing some of our own Soil Series and Types.

Soils of the world may be divided into two great groups on the basis of the presence or absence of a zone of accumulation of lime carbonate and salts of alkali and other alkaline earths. Our soils are in the group in which the zone is present. This feature is peculiar to dry climates, ranging from subhumid to arid. You have no doubt all noticed the layer in road cuts, deep plowing, holes, etc.

This lime layer represents a zone of deposition or accumulation of salts, the greater portion of which are leached out of the A horizons by percolation water. The probable reason for its concentration in this particular zone is that the quantity of percolation water is so limited that the depth of penetration is usually within this zone. The salts simply precipitate out of solution. In humid regions, on the other hand, there is sufficient water to form an almost constant percolation flow right through the soil column. Under such conditions the salts are finally leached away as seepage water, and the zone of concentration does not form.

This zone of salt accumulation may easily be determined by the effervescence reaction of dilute HCl with the lime carbonate, the liberated CO_2 gas forming the bubbles. The layer is usually of a marly gray color. The salts may either be quite uniformly distributed in the layer or concentrated in granules or streaks. Another characteristic feature of most dry land soils is the presence on the surface of a thin layer, often a mere film in thickness, of light textured loose soil of single grain structure. It resembles wind-blown material and may have such an origin.

Still another feature is a characteristic type of structure in the A_2 and B_1 horizons. This is a columnar arrangement, rather poorly shown in the pictures.

The humid climate profiles are characterized not by a zone of salt accumulation but by a zone of accumulation of fine soil particles.

Both salts and fine soil material are removed from the A horizons, but only the soil particles are held back by the filtering action of the B horizons. Two important profile features result from this action. One is the change in texture of the zones of extraction and of accumulation, the loss of fine particles leaving the A horizons lighter textured (sandier), and the addition of fine particles to the zone of accumulation making the B horizons heavier. This probably accounts for the heavy subsoils of most humid areas. The other feature produced by this strong leaching action is the much lower content of lime and many other mineral salts in the Solum of humid soils as compared to "dry" soils. The great prevalence of soil acidity and the absence of alkali conditions are two important practical results.

In humid soils then we have intensive extraction of fine soil from the surface layers and of salts from the whole profile. The A horizon then is light textured and the B heavy textured. The same general differences of colour resulting from variations in organic matter occur as in dry land soils.

Fine soil particles may also be transferred in dry land soils and generally are under our conditions, but to a lesser extent.

Here then we have two broad groups of soils based on fundamental differences in profiles. True, the contrasts are due principally to differences in climate, yet the classification is based not on climate but on profile characteristics alone.

Two other major soil separations may be made on the basis of two distinctly different profile types. These correspond generally, to our so-called timber and prairie soils. The profiles are in strong contrast, especially in the upper layers. We have both here in the west. The real timber profile, however, is quite limited in occurrence. As is true of two profiles just described, a study of the horizonal characteristics serves to explain a number of peculiar soil conditions.

The timber profile developed under northern conditions in Russia—the so-called podsol type—seems usually to possess an organic accumulation of forest debris only slightly decomposed, usually acid in reaction and brown in colour. Just beneath is a layer of light textured, light coloured soil, usually acid in reaction, low in organic matter and badly leached. Below this is a heavier layer, yellowish brown to coffee brown in colour, the zone of accumulation of fine particles, the B horizon. The layer of salt accumulation may or may not occur depending on climatic conditions. In Saskatchewan it practically always does, although at lower depths than in prairie soils.

Many of our poor, so-called burnt-over and white-mud northern bush lands are very likely soils of the profile just described, though not so well developed as the Russian podsols. The surface organic matter soon disappears. The white-mud appearance is merely the badly leached light coloured A horizon. It composes most of the plough-depth surface soil. Its acid reaction, low organic matter and leached condition explain its low productivity and poor physical state. The excessive leaching in this layer is likely due to the presence of organic acids in the percolation water, such acids forming in the layer of leaves, moss and other forest debris on the surface.

The Park Belt lands should not be included in forested soils. They are an entirely different proposition as I shall later point out.

The last major profile which I wish to consider is that of the treeless type. It also may or may not have the lime layer present, depending on climatic conditions. Under our conditions it is practically always present in mature profiles; on the humid corn belt prairie it is not.

In this group there is little or no accumulation of undecomposed surface organic matter, such as was pointed out for forest soils. On the other hand there is an accumulation of well decomposed organic matter in the A horizons and sometimes in lower layers, the quantity usually decreasing with depth. The gradual change from darker to lighter shades of colour from surface to subsurface horizons, the result of differences in organic content, is one of the most characteristic features of prairie profiles.

Saskatchewan Profiles and Their Interpretation

The most prevalent type of Saskatchewan profile combines the prairie and semi-arid features, namely, decomposed organic matter in the top layers, decreasing in quantity with depth; a thin surface layer of light textured loose soil; a definite layer of salt concentration; columnar structure, especially of the A_2 horizon, and a compact heavier horizon beneath the loose surface layer, also generally the A_2 horizon.

A number of soil relations and peculiar soil conditions in Saskatchewan may be at least partially explained and interpreted on the basis of profile characteristics described. For example, the nature of the Park Belt profile strongly indicates that in most cases, at least, this belt was originally prairie and that the tree growth has encroached upon it quite recently in soil history. I am still more convinced of this after interviewing early settlers in a typical park area during this past summer. All agreed that the Park "bluffs" had come in since their arrival and that previous to cultivation the whole area was open prairie. At present much of this district supports quite a dense growth of mixed poplar and willow, with open prairie between the bluffs.

Had the opposite condition been the case—that is, an original forest belt thinned out to present Park conditions,—I dare say that this very fertile black land belt would hardly measure up to its present productive power.

Another conclusion which seems reasonable is that our various prairie profiles are fairly accurate records of climatic conditions in various parts of our province for a long period of years. Under humid conditions the A horizons of upland prairie soils are usually dark brown to black in colour and a definite layer of salt concentration absent. On the other hand, under desert or arid conditions, the surface layer is usually very light brown to grayish brown in colour, with the zone of salt concentration right at the surface. In Saskatchewan, as well as in other areas of similar climate, it has been noted that the gradation in surface colour from light brown to black and the increase in depth of the layer of salt concentration both vary quite consistently with changes in climatic conditions which tend to increase the quantity of effective soil moisture. In short, the surface colours tend to darken and the salt layer to become deeper as the quantity of available soil moisture is increased. The changes agree pretty well with long time rainfall records coupled with evaporation observations.

It seems reasonable to assume from this, that these profile features furnish us not only with valuable bases for soil classification, but also with average long-time meteorological records at least as accurate as many records of geological history as interpreted from rock formations.

Other conditions which may be at least partially interpreted from a study of soil profile features are the acidity of our northern "bush" soils, the presence of hardpan layers, the relative age and amount of leaching of lands, drainage history, etc.

However, the real purpose of soil profile studies at present is not to explain soil conditions so much as to develop a good scheme of classification. The interpretative work is at present secondary to the work of classifying.

Soil Sampling by Profiles

Many changes of methods of studying soils, both in the field and laboratory, have been necessary in order to conform with the profile system of classification. Most of these directly concern the soils man. I shall limit this phase of my discussion to one proposition, namely that of soil sampling. This matter concerns anyone dealing with soils to any appreciable extent.

In brief, in my own humble opinion, I believe that, in general, samples should be taken by profile layers rather than by arbitrary depths. The former method gives both the sampler and examiner a great deal more information of the particular soil dealt with. Furthermore, the results of horizon analysis and of other profile studies may generally be interpreted in terms of arbitrary depths, but similar information on the basis of arbitrary depths is seldom applicable to profile layers. For example, analyses made of the A and B horizon of known thicknesses may be easily calculated to depths of 0 - 6 2/3 and 6 2/3 to 20 inches, the usual depths selected, and recent studies have shown that such calculations usually approximate the results of actual analyses.

As to the actual sampling process a wide bladed pick and a shovel and post hole digger have proven to be far more satisfactory than the soil auger. In fact the auger has lost favour considerably with soil surveyors. It is still used by the field parties, but usually not as the principal sampling tool. The pick, shovel and post hole digger expose the complete soil profile and enable one to take more accurate samples without seriously changing the soil structure or mixing chemically and physically different layers.

Part II

PRINCIPLES AND UNITS OF CLASSIFICATION

Editor's Comments
on Papers 4 Through 7

PRINCIPLES OF CLASSIFICATION

Classification, in the traditional sense of a natural system, involves the incorporation of groups of objects into a logical, hierarchial system where each group occupies a unique position. The classificatory groups, called taxa, are of different grades. Each taxon, except the lowest category, includes one or more taxa of the next lower category. Strictly speaking, classification involves placing things in classificatory groups whereas systematics embraces the general theories underlying this activity. Although proposed long ago, the word "taxonomy" has recently become a vague synonym for classification, systematics, and even nomenclature.

The general principles of classification were worked out in the classical period and subsequently by such notables as Aristotle and Linnaeus. Modifications of so-called Aristotelean natural classifications or refinements of the Linnaean order over the centuries proved especially beneficial to botanists and zoologists. More recent developments as they apply to the biological sciences or classification in general are summarized by Mill (1874), Schenk and McMasters (1956), Cain (1959), Simpson (1961), Crowson (1965,

1970), and in the informative discussions of Blackwelder (1967) and Mayr (1969). The articles reprinted here (Papers 4 through 7) relate specifically to the classification of soils, though there are also many others of merit (for example, Afanasiev, 1927a and b; de Bakker, 1970; Gerasimov, 1952; Hallsworth and Costin, 1950; Kubiëna, 1958; Leeper, 1952, 1956; Muir, 1962; Schelling, 1970; Simonson, 1952; Smith, 1963; Stephens, 1950; Stewart, 1954; Whiteside, 1953).

It may appear to some that soil taxonomists have not yet achieved the classificatory precision of other natural scientists when pedological classifications are compared with the *International Code of Botanical Nomenclature* (Larijouw et al., 1956), the *International Code of Zoological Nomenclature* (1964), or the *International Stratigraphic Guide* (Hedberg, 1976). It is only fair to point out, however, that botanical, zoological, and geological classifications have been developed over considerably longer periods of time than have pedological schemes. Further, in spite of a rich developmental history and a greater familiarity with their subjects, classifiers of other natural objects have their troubles too. Far more, in fact, than one might expect from perusal of the formal codes, rules, and conventions. Croizat (1945), Cronquist (1978), and Levin (1979), for example, point to some persistent and unresolved problems that botanists have debated for several decades. In short, biological scientists are also engaged in a struggle between patching and preserving the old systems on the one hand and constructing new frameworks on the other. The principles of soil classification laid down in the last half century represent important strides in the formulation of a comprehensive soil classification. Such principles are briefly summarized here because of their crucial role in this development.

In 1922 Marbut formulated several principles, or propositions as he referred to them, that he regarded as pertinent to the natural classification of soils. He clearly believed that soil classification was a worthy scientific endeavor and placed it on a par with the classification of other natural objects. With respect to the latter, this was so much the case that he drew a parallel between the pedologists' soil type, series, and group and respective Linnaean species, genus, and family categories. Reinforcing the earlier premise that soils should be classified on the basis of inherent properties, he listed ten criteria for differentiation at a low level where those differentiae would vary narrowly. Modified conceptions of soil, especially further development of the idea that soil is a dynamic natural body, led to modest revisions of Marbut's cirteria. Discussing such changes, Whiteside (Paper 6) points to the introduction of shape, moisture, temperature, age, and biology of the soil body. To the properties of individual horizons were added mineralogical composition, in place of Marbut's vaguely defined "character of soil material," and consistence. These observable and measurable features of soil serve as differentiating criteria in the grouping of soils at different categorical levels in *Soil Taxonomy*.

The revised list of some twenty differentiating criteria, complete as it may seem, might require additional modification for, as Cline points out in Paper 5, pedologists may still be unaware of important relationships. The properties listed by Marbut were intended for differentiation among soils at the level of the soil type. Additional criteria, such as mineralogical composition, are applied in *Soil Taxonomy* at a higher level, but they, like criteria used in lower categories, must not exceed their *ceiling of independence*. This principle states that no cirterion can be used to differentiate that separates like soils in categories below it. Further, all classes of the same category of a single population should be based on the same characteristics in order to enhance visual relationships. Thus, according to the principles of differentiation, as they affect classes, so clearly outlined by Cline, a differentiating characteristic must: (1) be important for the objective or purpose of the classification; (2) be a property of the thing being classified; (3) carry as many accessory properties as possible; and (4) provide homogeneous classes.

The Basinski paper (Paper 7), summarizing the Russian approach to soil classification, stands in marked contrast to American perception of classification principles. The different approach has much to do with Russian concepts of soil where intrinsic soil-forming processes are undetachable from its definition. Russian principles of soil classification, which differ on this basis because the evolutionary-genetic approach is considered the only proper one, represent further development of traditional systems. The definition of soil type, the basic unit of classification, is based on the following criteria as originally established by Sibirtzev: (1) genetic properties of the soil; (2) pedogenetic processes that determine these properties; and (3) pedogenetic factors that define and determine those processes. As Basinski tactfully points out, the grouping of soils should be based on the characters of a given soil body, as was originally stressed by Dokuchaev, but Russian pedologists have not adhered to this principle, seeming to choose instead interpretations of genesis. The bases of genetic

classifications are summarized by Basinski for geographic-environmental, factorial, process, and evolutionary classifications.

REFERENCES

Afanasiev, J. N., 1927a, The Classification Problem in Russian Soil Science, *Russian Pedological Investigations No. V*, Academy of Sciences of the USSR, Leningrad, 51 p.

Afanasiev, J. N., 1927b, Soil Classification Problems in Russia, *1st Intern. Congr. Soil Sci. Trans.* **4**:498–501.

Blackwelder, R. E., 1967, *Taxonomy*, Wiley, New York, 376 p.

Cain, A. J., 1959, Taxonomic Concepts, *Ibis* **101**:302–318.

Croizat, L., 1945, History and Nomenclature of the Higher Units of Classi-fication, *Torrey Botan, Club. Bull.* **72**:52–75.

Cronquist, A., 1978, Once Again, What is a Species?, in *Biosystematics in Agriculture No. 2, Beltsville Symposia in Agricultural Research*, Allanheld, Osmum & Co., Montclair, N.J., pp. 2–30.

Crowson, R. A., 1965, Classification, Statistics and Phylogeny, *Syst. Zool.* **14**:144–148.

Crowson, R. A., 1970, *Classification in Biology*, Aldine, Chicago, 350 pp.

de Bakker, H., 1970, Purposes of Soil Classification. *Geoderma* **4**:195–208.

Gerasimov. I. P., 1952, Scientific Foundations of Soil Taxonomy, *Pedology* **11**:1019–1026.

Hallsworth, E. G., and A. B. Costin, 1950, Soil Classification, *J. Aust. Inst. Agric. Sci.* **16**:84–89.

Hedberg, H. D., ed., 1976, *International Stritigraphic Guide*, Wiley, New York, 200p.

International Code of Zoological Nomenclature, 1964, International Trust for Zooligical Nomenclature, London, 176p.

Kubiëna, W. L., 1958, The Classification of Soils, *J. Soil Sci.* **9**:9–19.

Larijouw, J., chairman, 1956, *International Code of Botanical Nomen-clature*, Kemink en Zoon, Utrecht, 338p.

Leeper, G. W., 1952, On Classifying Soils, *J. Aust. Inst. Agric. Sci.* **18**:77 et seq.

Leeper, G. W., 1956, The Classification of Soils, *Soil Sci.* **7**:59–63.

Levin, D. A., 1979, The Nature of Plant Species, *Science* **204**:381–384.

Mayr, E., 1969, *Principles of Systemic Zoology*, McGraw-Hill, New York, 428p.

Mill, J. S., 1874, *A System of Logic*, vols. I and II, Harper and Row, New York 659p.

Muir, J. W., 1962, The General Principles of Classification with Reference to Soils, *J. Soil Sci*, **13**:22–30.

Schelling, J., 1970, Soil Genesis, Soil Classification and Soil Survey, *Geoderma* **4**:165–193.

Schenk, E. T., and J. H. McMasters, 1956, *Procedures in Taxonomy*, Stanford University Press, Stanford, 199p.

Simonson, R. W., 1952, Lessons from the First Half-Century of Soil Survey. I: Classification of Soils, *Soil Sci.* **74**:249–257.

Simpson, G. G., 1961, *Principles of Animal Taxonomy,* Columbia University Press, New York, 247p.

Smith, G. D., 1963, Objectives and Basic Assumptions of the New Soil Classification System, *Soil Sci.* **96**:6–16.

Stephens, C. G., 1950, The Elements of Soil Classification, *J. Dept. Agric. S. Aust.* **54**:141–142.

Stewart, G. W., 1954, Some Aspects of Soil Taxonomy, *5th Intern Congr. Soil Sci. Trans.* **4**:109–116.

Whiteside, E. P., 1953, Some Relationships Between the Classification of Rocks by Geologists and the Classification of Soils by Soil Scientists, *Soil Sci. Soc. Am. Proc.* **17**:138–142.

4

Reprinted from pages 85–94 of *Life and Work of Curtis F. Marbut,* Artcraft Press, Columbia, Mo., 1951, 134 p.

SOIL CLASSIFICATION

Curtis F. Marbut

Bul. III. American Association of Soil Survey Workers. 1922.

In Soil Science we must not forget that we stand in a different position from that occupied by the specialists of other sciences.

In the older Natural History sciences the fundamental principles were worked out, at least in a broad way, long ago. The characteristics of the subjects concerned have been long known and the laws governing their development are well understood in their broad outlines. The characteristics of plants for a large part of the world were determined a hundred years ago and earlier. The same statement can be made of the characteristics of animals, of geological formations, of the bodies whose study lies in the special field of the astronomer, and many others.

Long before any special attempt was made to use the knowledge of the objects concerned, of plants especially, for practical purposes a large body of knowledge concerning their characteristics had been accumulated. This applies to the conscious attempts to use this knowledge in a technical and rational way, based on the fundamental nature of the objects themselves. It does not apply to the use of plants by primitive man, though even his use was based on his empirical knowledge or what we may call his unconscious scientific knowledge. In other words long before modern technology undertook to realize on a knowledge of the fundamental character of plants, a great deal of the knowledge had been collected.

This is not the case with soils. Man throughout his history has depended upon empirical knowledge and in the main still depends upon it. This is true almost entirely of the farmer and it is also true of many scientific men concerned with the practical study of matters with which the soil is more or less concerned. On what other basis can we explain the almost universal method of stating the results of laboratory or experimental work. For example, consider the several papers presented a year ago at the meeting of the Society of Agronomy on soil liming, where no one seemed to assume that all soils are not essentially alike. There was no suggestion of a recognition of the fact that large areas of soil throughout the world are not in need of lime. We are all familiar with the controversy over the use of rock and acid phosphate in which neither side seems to have considered that the differences of results obtained were due primarily to differences, fundamental differences, between the soils concerned, and that both parties may have been right, each in his own particular environment. There can be no explanation of this failure to recognize the possibilities in each case on any other basis than a lack of the recognition of the fundamental differences in the soils concerned.

No blame however, can be laid on any of these parties for this failure. It lies in the complete lack throughout the world, until less than 40 years ago of a knowledge of the real characteristics of soils, and the lack of this knowledge in this country until less than 20 years ago. How large a proportion of the agronomists of the country now, of the men concerned directly in their work with the soil and therefore with its fundamental characteristics, know how many fundamentally different soil profiles have been developed within the area of the United States? How many of them really know that there are characteristic soil profiles or soil sections in this country, each prevailing over a wide area of country, and each just as characteristic of the region over which it prevails as the character of the native vegetation, the character of the agriculture, or the character of the climate? How many of us who do know that such soil characteristics exist, know the significance of all the features of these soil profiles? I will answer this question by saying that not one of us knows.

Summing up the situation: we do not yet know soil characteristics. We are still engaged in the accumulation of, in fact we must say that we have really just begun to accumulate, the knowledge concerning the characteristics of the objects with which our science or our branch of science deals.

Why this late date for beginning the accumulation of the knowledge needed? It is partly due to the assumption and to the seeming satisfaction we have all found in that assumption, that the soil characteristics, the fundamental nature of any soil is described when we are able to say of it that it is a granite soil, a sandstone soil, a limestone soil, etc. Scientific history will probably record no greater mistake, none with more profound effect in delaying the advent of the period of real investigation, than this one. We have assumed that this statement of the source of the material stated all that was needed to tell the whole story, not realizing that the source of the soil material has about as much relation to the fundamental character of the soil itself as does the source of the brick, or wood or other building material used, have to the type of architecture prevailing in a given region.

This being our situation we are under the necessity, more or less pressing, of accumulating the fundamental knowledge of our subject, of developing the laws of its existence, of creating the whole science of the subject at the same time that we must make some attempt to apply the knowledge and principles thus gained to the solution of the practical problems of the use of the soil in the growing of crops. In this respect let me emphasize by stating again that we are in a position entirely different from that occupied by the specialists of the other sciences. Their fundamental knowledge was accumulated long ago and the great principles developed during the progress of its accumulation.

Although we are under the necessity of carrying a double burden, of following two lines of work, we must not fail to keep in mind at least a clear distinction between the two if we expect to obtain results of permanent value. Confusion of ideas will be just as fatal here as anywhere else. If in confusing them we undertake to merge the two into one operation by limiting the one according to the demands of the other we shall make a fatal mistake and build up a structure that will have to be destroyed by some more rational generation following us. One of our functions is scientific, the other is technical. One consists in the accumulation of knowledge, the other in the application of that knowledge to the solution of certain practical problems. If we make the fatal mistake of limiting the range of the first work, restricting the methods of attack, or narrowing the point of view in accordance with what may seem to be the probability of our practical use of the results that may be obtained we shall fail utterly to perform the task set before us. It seems superfluous to state that the rational handling of the soil for obtaining practical results cannot be performed until at least some knowledge of the actual characteristics of the soil have been accumulated. Yet that has been the course followed until recently.

In order that we may use readily the data we accumulate regarding the soils in order that we may be able to comprehend what we have and determine the gaps in our knowledge, it must be systematized. It must be classified. We must not forget however, that before a classification of objects on anything like a permanent, sound basis can be constructed we must have accumulated at least some knowledge of the characteristics of the objects concerned. This knowledge must itself be accumulated in at least a consistent way. The method of approach in obtaining it is of itself a matter of development. For example, in the first work of the Soil Survey, practically nothing was learned of the soil, in our attempts to study the soil and in our attempt to describe the information we had acquired, except its geology. We did not in reality study the soil at all, we determined but little more than the nature of the parent material and the geological processes by which it was accumulated and assumed that to be the end of the task assigned to us.

It is not necessary here to outline the successive stages through which we have passed from that simple beginning to our present position. It should be said however, that the earlier method of attack was the inevitable one because of the state of knowledge existing at that time, and such changes as have been made in either method or point of view have been brought about solely through the accumulation of knowledge.

Where do we stand at present? What is our method of approach at present? What questions do we ask of the soil? and what information do we expect it to furnish us?

Before attempting to describe our present methods of attack and our present point of view I desire to call your attention to certain propositions that seem to be of fundamental importance in any consideration of the method of approaching the soil or of classifying the knowledge after it has been accumulated.

In the course of development of every science certain fundamental conceptions become accepted as applicable and as a prerequisite to the proper attitude in approaching any problem connected with it. In soil science, at least that branch of it concerned with the study of the soil in the field, the following propositions may be considered fundamental—at least they express the attitude of the Federal Soil Survey at the present time.

1. The true soil considered as a natural body, is the weathered surface horizon of the earth's crust.

2. If the soil be the weathered surface horizon there must be an unweathered portion which we shall designate the soil material, or parent material. We have therefore (1) The soil, and (2) the soil material. In clear thinking these two things must be clearly differentiated. The soil must be thought of as a body distinct from that of the soil material or parent material although the former has been made from the latter. The pile of lumber is entirely distinct from the forest and no one thinks of confusing them and of calling them both one object.

3. The soil is a natural body, developed by natural forces acting through natural processes on natural materials. Its true nature cannot be determined except through a study of the natural or virgin soil. Any effort to arrive at a complete understanding of the soil by studying it only or mainly as it has been modified by man in cultivated fields will involve necessarily a great loss of time and the danger if not almost the certainty of committing gross errors. It is almost as reasonable and practicable to study the soil in cultivated fields alone or mainly, and expect fundamental results from such studies as it would be to study the botany of the world's vegetation by devoting attention to cultivated plants only. The botanist does not pretend, nor does any one urge that he attempt, to study his subject by studying cultivated plants mainly, yet in much of our work in the past the cultivated soil is the soil that has been examined, sampled, and described.

4. Soil classification must be scientific and on a basis comparable with that employed or used in the classification of other natural bodies. No deviation from strict scientific considerations for the sake of the so-called practical use of the soil can safely be permitted. Soil classification can no more deviate from the strict scientific viewpoint than can botanical classification, geological classification, or zoological classification.

5. Every soil, in the course of its development, and as a product of the forces under the influence of which the development takes place, develops a series of horizons or what might be called a soil section or profile. This seems to be universal and to be characteristic of all soils at maturity and later.

In addition to these fundamental propositions there are certain general propositions regarding classifications in general that should be stated here.

1. Classification is merely an arrangement or grouping of objects.

2. The grouping of objects can be made only after the objects have been created or produced or assembled. Merely to assemble them however, will not enable one to classify them, or at least the mere assembling or piling up of objects does not constitute the only preliminary to classification. Characteristics must be determined.

3. Grouping of objects can be effected only on the basis of something or according to something.

4. In classifying men for example we must group them according to age, height, color of skin, features of face, shape of body, country of birth, or residence or some other feature. It is evident that most objects may be grouped on a number of bases.

5. While it is undoubtedly true that objects may be grouped according to a number of characteristics or features, yet the feature or group of features selected in any particular case is not only not a matter of entire indifference but on the other hand is a matter of the greatest importance.

6. I think I may lay down the proposition as an absolute one that the basis of grouping should be the characteristics of the objects grouped. They should be tangible, determinable by a study of the objects themselves and by direct observation and experiment. For example, a classification of insects should be and is made according to the characteristics of the insects—their color, size, number of legs, etc. not according to the things they eat. It is universal practice and recognized as scientifically sound that natural objects are grouped according to their characteristics. I do not now recall a single science whose objects are grouped except for some temporary purposes, and not as a permanent scientific grouping, on any other basis than that of the features or characteristics of the objects. The nomenclature of the several groups does not necessarily describe them but it may do so.

7. A complete grouping consists of or includes major and minor groups and subgroups of the objects running downward in subgrouping through several categories. Any one characteristic may be arbitrarily selected as the basis for the major grouping. For example, mankind may be grouped into those who farm and those who do not farm, or into lawyers, and not lawyers. Each of these may be divided into subgroups according to any other characteristic that may be arbitrarily selected. Such a haphazard way however, is not logical and therefore not scientific.

8. For the major grouping we should select that characteristic that will gather all the objects to be grouped into the smallest number of groups possible so that each group will contain the largest number of objects possible. Thence downward to the lowest series of groups that basis or feature must be selected giving the fewest subgroups and the greatest number of objects in each subgroup.

9. In order therefore to group a number of objects according to their characteristics, according to similarity of characteristics in each group, the first thing to do is to determine the characteristics of the objects. Determine what they are, study each and every object to see what it is like. This is so evident a matter that it seems foolish to refer to it but when soil study began in this country the first grouping had to be made mainly on characteristics other than those of the soil. It is evident that the first groupings that have been made, throughout scientific history, of the objects knowledge of which when accumulated, was called a science, were tentative, made either on the basis of a very few characteristics of the objects concerned and in many other cases based on features that were merely supposed to be those of these objects. In some cases these features were found later to be only remotely connected with the objects under consideration.

10. In determining the characteristics of bodies by direct study, observational or experimental, it is soon found that there are characteristics representing maturity of development and those of immaturity of develop-

ment, and if no discrimination were made on that account we would collect a mass of facts that could not be used. The botanist determines the charactcristic of his plant in bloom or fruit. The zoologist gives greatest attention in his grouping of animals to the characteristics of mature individuals. For purposes of grouping, at least for the broadest groupings, we must determine the characteristics of mature bodies. In those cases where the course of development of individuals is a matter of milleniums or even of thousands of years, the youthful individuals or the immature individuals must be reckoned with but their characteristics would be concerned with the formation of subgroups rather than of major groups. It is true that in biology the embryology of an object is an important factor in its grouping but we must bear in mind that that phase or feature of embryology or of the animal in its embryonic stage to which attention is directed in classification is the course of development in that stage rather than size or shape or any other physical feature.

Returning now to the characteristics on which we must base a soil classification let us see what questions we ask of our soils. Soil characteristics have been referred to in this paper very often and they are doubtless abundant, but what are those of most importance and to which attention must be directed; you will recall that one of the fundamental propositions formulated above stated that all soils at maturity develop a soil profile, and that the features of the soil is expressed in the features of this profile. What are the features of the profile that we seek? In actual practice we have been forced by the weight of facts to determine the following:

1. Number of horizons in the soil profile.
2. Color of the various horizons, with special emphasis on the surface one or two.
3. Texture of the horizons.
4. Structure of the horizons.
5. Relative arrangement of the horizons.
6. Chemical composition of the horizons.
7. Thickness of the horizons.
8. The thickness of the true soil.
9. The character of the soil material.
10. The geology of the soil material.

All of you are familiar with the fact that the ultimate soil unit, the soil type, the species, includes all areas of soil having a uniform profile—uniform in all respects. You are also familiar with the fact that the soil series, or soil genus, includes all areas of soil having profiles that are uniform in all respects except that of the texture of the surface horizon, or the upper two horizons in those cases where the surface horizon is very thin.

Many of you are doubtless not familiar with the fact that soil series may be grouped according to the broad features of the soil profile into groups which may be considered comparable to the family groups in the grouping of animals and plants.

The profile of the mature soil at any given locality is characterized by
certain fundamental features that are characteristic of such soils over
a wide area of country. Throughout such a region all mature soils will
have profiles essentially alike in their broad features, and which will be
wholly unlike the profiles in certain other regions. By accumulating and
analyzing the information regarding soil profiles of mature soils over a
region like the United States and grouping them according to similarity
of general features we are able thus to determine what the common and
therefore important features of the soils of the region are, assuming that
features of wide extent are important, and by correlating these with other
natural features and with natural forces and processes determine their relation to the latter and the causes of their development.

The mature profiles within any given region will not be identical but
they will be comparable, and variations will be confined to unimportant
details. The profiles should have the same number of horizons, the colors
of corresponding horizons should vary but slightly, the arrangement or
relative order of the horizons should be the same in all the structure
of corresponding horizons should be uniform and the depth to the parent
material should be approximately uniform in soils of uniform texture.
They will vary within narrow limits in the thickness of the several horizons
in details of color shade, and in other rather unimportant details.

The area over which a given soil profile prevails in the mature soil
is a soil province in the true sense of that word, though it need not consist
of one continuous area. On the other hand, it may consist of a number
of detached areas.

Associated with each mature soil within a given soil province there are
a varying number of soils in various stages of immature development and
certain others in past mature stages of development. In these the profiles
may vary considerably from that of the standard or mature soil. For example, a soil in association with a given mature soil may be so young as to
have no true soil profile features whatever, being nothing more than soil
material recently accumulated. Freshly laid alluvium may serve as an
illustration. Another soil may have developed a surface profile in part or
completely with parent geological material underlying it at shallow depth.
In still other cases two or more of the upper horizons may have developed.
In all cases, except a few of rare occurrence, the horizons developed at any
one stage assume the general characteristics of the corresponding horizons
of the mature profile. In extreme cases however, the parent rock may be
of such a nature that the surface horizon may be when first developed
entirely unlike the surface horizon of the mature soil and may remain
unlike it through a long period of time. The Houston black clay is a
case in point in which the surface horizon is black and granular in structure entirely unlike the corresponding horizon, in both color and structure,
of the Oktibbeha clay which is thought to be the mature soil corresponding
to the youthful Houston.

The characteristics of the horizons of the mature soil profile change
considerably as a result of past maturity of development. These features

vary somewhat but the most frequently met with examples consist in the presence of a compacted or indurated horizon in the lower part, often at the base of the soil profile. This when well developed tends to retard the movement of water through the soils upward or downward. Meteoric waters held up, even temporarily, above this horizon causes waterlogging to a greater or less degree of the horizon above it, changing its character accordingly. This influence may in extreme cases influence the character of the surface horizon.

Each varying soil profile associated with a given mature profile to whatever stage in development it may be due or whatever variation in character of parent material may have produced its variations constitutes a distinct kind or, as we designate it, series of soils and the whole group of soils consisting of the mature and its related and associated immature soils constitutes what could be designated as a soil family.

The soil profiles of that part of the United States east of the Rocky Mountains may be grouped around some ten mature soil profiles, each differing in one or more important respects from any other one of the ten.

The accompanying outline map shows the general area in which each of these profiles is predominantly characteristic of the mature soil. The Roman numeral in each area refers to the corresponding soil profile description below:

I.   1. Forest debris; rohhumus, trockentorf. As a rule it rests directly on the soil with very thin layer or no layer at all of decomposed organic matter (mull) at the bottom.
     2. Gray horizon, ranging in thickness from a film in very heavy soils to several inches in thickness. The podsolized horizon.
     3. Chocolate brown, coffee brown, rusty brown horizon. The orterde or ortstein horizon. In some cases may be indurated. May range up to 12 inches in thickness.
     4. Partially weathered parent soil material. No sharp line between this and No. 3.

II. 1. Forest debris consisting of leaves at the surface and forest mold or *mull* below usually mixed with dark colored earth due to work of worms and insects.
     2. Dark brown horizon consisting of brown mineral soil material mixed with forest mold. Ranges up to 3 or 4 inches.
     3. Brown horizon, often with yellowish shade, ranges up to 18 inches in the medium textured soils and deeper in the sands.
     4. Reddish brown to yellowish brown horizon heavier than any of the overlying horizons; the reddish brown shade being more common in the lighter textured members.
     5. Parent material usually partially weathered at top.

III. 1. Forest mold with mull horizon as in II but both mold and *mull* usually thinner than in the latter.
     2. Dark gray horizon, consisting of light colored mineral soil material mixed with forest mold.

75

3. Pale yellow to yellow horizon ranging up to about 18 inches in the medium textured types.

4. Brownish yellow horizon, heavier than 3. Extends to depths of several feet.

5. Reddish horizon, mottled throughout or only in lower part.

6. Parent soil material.

IV. 1. Dark brown to black horizon, filled with grass roots but usually little or no accumulation of mold on the surface. Lower part of horizon becomes browner through decrease of organic matter. Ranges up to 15 inches.

2. Brown to yellowish brown horizon which is usually similar to 1 in texture.

3. Brown horizon, heavier than higher horizons. Ranges up to 3 feet or more in thickness.

4. Parent soil material, partially weathered.

V. Similar to No. IV, being less dark in color of the surface horizon and having a reddish brown horizon 3. Depth to the parent material somewhat greater than in IV when other conditions are equal.

VI. 1. Black horizon, usually with no surface accumulation of undecomposed organic matter. Soil filled with grass roots. Ranges in depth up to 15 inches.

2. Yellowish brown to brown horizon ranges up to three feet in thickness.

3. Yellowish brown to greenish brown or greenish yellow horizon filled with streaks and spots of gray lime carbonate. Ranges up to three feet more or less in thickness.

4. Parent soil material.

VII. Somewhat similar to VI, differing from it in the dark chocolate brown soil and the pinkish to reddish color and usually higher carbonate concentration of horizon 3. Horizon 2 is usually reddish also.

VIII. 1. Chestnut brown to dark brown horizon. No accumulation of undecomposed organic matter on the surface. Soil filled with grass roots. Ranges up to about 10 inches in thickness.

2. Yellowish brown horizon, usually a few inches in thickness.

3. Carbonate horizon, usually enough carbonate to color whole horizon gray. Ranges up to 2 feet in thickness.

4. Parent material.

IX. Similar to VIII but redder brown in the surface soil and pinkish or reddish in the subsoil.

X. 1. Brown horizon.

2. Thin lighter brown horizon, often absent.

3. Calcareous zone.

4. Parent material.

There is doubtless a reddish representative of No. X in the southern part of the United States but it has not yet been identified.

The classification here outlined is based on soil character only the character being expressed in the features of the soil profile. The groups of soil series represented here by the soils grouped around each of the fundamental profiles constitutes a soil family. The term soil *province* does not enter into the system as an essential part. The latter term merely designates the general region in which a given soil profile predominates but the soils within this area are differentiated not on the basis of their occurrence within such area but on the basis of soil profile. It is conceivable that an area of one soil family may occur in the midst of a large area of a totally different family. The word province is merely a geographic term and not a term to be used in a soil classification pure and simple.

The nomenclature of soil families will be left open for future discussion.

A little study of the various profiles will show that they may be grouped into two large groups, one of them including all soils having a zone, somewhere in the soil profile, of carbonate accumulation, the other, without such a zone. The latter would include families I, II, III, IV and V. The former all the others. Each of these may be designated a soil *order*.

Other soil families not here described are found in the western parts of the United States.

5

BASIC PRINCIPLES OF SOIL CLASSIFICATION

MARLIN G. CLINE[1]

The purpose of any classification is so to organize our knowledge that the properties of objects may be remembered and their relationships may be understood most easily *for a specific objective*. The process involves formation of classes by grouping the objects on the basis of their common properties. In any system of classification, groups about which the greatest number, most precise, and most important statements can be made *for the objective* serve the purpose best. As the things important for one objective are seldom important for another, a single system will rarely serve two objectives equally well.

CLASSES

The smallest natural body that can be defined as a thing complete in itself is an *individual*. All the individuals of a natural phenomenon, collectively, are a *population*. Plants, animals, or soils, for example, are populations, each consisting of many individuals. The individuals of a population have many common properties, but the variation within a population is so great that man is unable to see similarities and understand relationships among individuals in the disorderly arrangement in which he finds them. He attempts to make the variation that he finds in nature orderly for his convenience by framing classes—by grouping individuals that are alike in selected characteristics. A *class* is a group of individuals, or of other classes, similar in selected properties and distinguished from all other classes of the same population by differences in these properties.

Classes as segments of a population

There is diversity in the degree of difference among classes. Among living things, for example, plants as a class are so unlike animals as a class that one thinks of them as separated by an insurmountable barrier of differences. Amoebae as a class and algae as a class are not so different; yet one is an animal and the other a plant. Similar situations may be found within every population. If one could place all the individuals in a row ranked from highest to lowest value of one property, the series formed would pass by almost imperceptible stages from one extreme of the property to the other. Any two adjacent individuals would be much alike; the two end members would be vastly different. Classes based on that property would be segments of a continuous series. The end individual of one class would be more like the adjacent end member of the next class than like an individual at the other extreme of the same class. Classes of natural objects are not separated by insurmountable barriers; they grade by small steps into other classes.

The modal individual

Within every class is a central core or nucleus to which the individual members are related in varying degrees. The mean, the median, and the mode as used in

[1] Marlin G. Cline is Professor of Soil Science, New York State College of Agriculture, Cornell University, Ithaca.

statistics, for example, are estimates of the central nucleus of a class. A class of natural objects may be considered in terms of a frequency distribution according to value of a selected property. Commonly, within some small increment of value of that property, the frequency of occurrence of individuals is a maximum. This is the modal value of the property which defines the central nucleus of the class—the modal individual. The definition of the modal individual may be based on observed properties of a real individual, or it may be estimated by a statistic representing a hypothetical individual and derived from the observed properties of a sample of the class.

One may visualize a class as a group of individuals tied by bonds of varying strength to a central nucleus. At the center is the modal individual in which the modal properties of the class are typified. In the immediate vicinity are many individuals held by bonds of similarity so strong that no doubt can exist as to their relationship. At the margins of the group, however, are many individuals less strongly held by resemblance but more strongly held by similarity to this modal individual than to that of any other class. A class is a group of individuals bound from within, not circumscribed from without. The test of proper placement of any marginal individual is its relative degree of similarity to the modal individuals of different classes.

Characteristics of classes

So far, variation in only a single property has been considered, but the same concepts and principles apply when two or more characteristics vary at the same time. Their application, however, depends upon an understanding of the different kinds of characteristics of classes.

Some property must be chosen as the basis of grouping. Individuals that are alike in that characteristic are placed in the same group; those that are unlike are placed in different groups. The property chosen as the basis of grouping is called the *differentiating characteristic*.[2] It serves to differentiate among classes; its mean value within each class defines the modal individual of that group. The test of any grouping is the number, precision, and importance of statements that can be made about each class for the objective. If the classes are well formed, a precise statement about the differentiating characteristic for each class should always be possible.

If the basis of grouping is good, the differentiating characteristic should be associated with a number of covarying properties. Texture, for example, is the differentiating characteristic of soil types within a soil series. A number of other properties, such as cation-exchange capacity of the inorganic fraction and water held at various tensions, change as texture changes. A precise statement can be made about these *accessory characteristics* as well as about texture in the soil type. Through accessory characteristics one multiplies the number of statements about each class and increases the significance of the classes formed. A

---

[2] The term is used here in a less restricted sense than that defined by Mill (5). As Mill used the term, it applied only to distinctions used to differentiate between greatly contrasting populations, such as plants and animals.

well-conceived grouping is based upon that differentiating characteristic that (a) is itself important for the objective and (b) carries the greatest possible number of covarying accessory characteristics that are also important for the objective.

Those properties of the individuals of a class that vary independently of the basis of grouping are called *accidental characteristics.* They are not related to the differentiating characteristic, and no statement can be made about them for the class as a whole. Slope, for example, may vary independently of soil texture and is an accidental characteristic in classes based on texture.

Now consider the concept of a class as a group the members of which are ranged about a modal individual when many properties are variables. The modal individual may be defined precisely not only in terms of a value of the differentiating characteristic but also in terms of values of all its accessory characteristics. The class as a whole can be defined precisely in terms of (a) values both of the differentiating and of the accessory characteristics of the modal individual and (b) the respective deviations from those values within the range of the class. These should be the components of precise definition of classes. The

TABLE 1

A simple two-category system of classification of surface soils

CATEGORY	DIFFERENTIATING CHARACTERISTIC	CLASSES			
2	Color value	Light		Dark	
1	Reaction	Acid	Alkaline	Acid	Alkaline

mean values for both differentiating and accessory characteristics derived from a sample of the class define the modal individual. Their standard deviations define the variability of the class. ·The tests of statistical significance are measures of whether the differences between classes are real or only apparent.

Accessory characteristics grade into properties that are purely accidental. Some attributes are almost always accessory but in a few individuals bear no relationship to the differentiating characteristic. The dark color of the A_1 horizon of soils, for example, is usually associated with high organic matter, but in the "Regur" soils of the tropics it is not. Such properties cannot properly be specified in the definition of a class, but they can and should be described.

MULTIPLE CATEGORY SYSTEMS

When a population is so diverse that any single grouping fails to show the relationships desired, the classes formed may be subdivided to show more relationships. This is illustrated in table 1. A *category* in such a system is a series of classes, collectively, formed by differentiation within a population on the basis of a single set of criteria. A category must include all individuals of the population; groups within a category are classes at a defined *level of abstraction.* Classes of category 2 in table 1 consist of groups of classes of category 1. Both include all individuals of the population. The technical meaning of the term *category* in classification is not to be confused with its common use as synonymous with *class.*

MARLIN G. CLINE

Categorical rank and homogeneity of classes

The number of statements about classes in a multiple-category system increases in going from higher to lower categories. In table 1, the classes of category 2 are differentiated on the basis of color values. Statements about either of the two classes are limited to color and associated accessory characteristics. Reaction is purely an accidental characteristic in this category, and no statement can be made about it. This category by itself is a simple single-category classification. In category 1, however, each class of category 2 is subdivided on the basis of reaction. Each of these four classes may be defined in terms of (a) the differentiating *and* accessory characteristics of category 2, *plus* (b) the differentiating property of category 1, *and* all of its accessory characteristics. In any multiple-category system, regardless of the number of categories, the properties that are homogeneous in a given class consist of the accumulated differentiating and accessory characteristics of that category and all categories above it. The greatest number of statements can be made about classes of the lowest category; the least number of statements, about units of the highest category. Categories in which few differentiating and accessory characteristics have accumulated are at a high *level of abstraction* and have high *categorical rank*; those in which many have accumulated are at a low level of abstraction and have low categorical rank. Homogeneity of classes increases with decrease of abstraction and categorical rank.

Dependence of classification systems on state of knowledge

When viewed in the opposite perspective, each successively higher category is a grouping of classes of the preceding lower category. Classes of a lower category are treated as individuals and grouped into classes of a higher category to show relationships among them.

The units of a lower category are not groups formed by applying a single differentiating characteristic to the entire population. They are homogeneous with respect not only to the differentiating property of that category but also to all characteristics used to differentiate in all higher categories. Formation of classes at a low categorical level in their final form, therefore, presupposes knowledge of the population adequate to complete all categories above it. *As the body of knowledge about any phenomenon increases, therefore, attempts to effect a complete natural system of classification must pass through a series of approximations.* The degree of improvement of each over its predecessor is in proportion to the increment of accumulated knowledge and its effects in basic concepts.

NATURAL AND TECHNICAL SYSTEMS

Although there are as many "best" systems of classification as there are objectives for grouping, not all are equally significant in organizing man's knowledge. The lowest category of the natural classification is a prerequisite of all other groupings.

In a natural classification, one classifies in such a way that the name of each class will bring to mind many characteristics and will fix each group mentally in relation to all others. The objective is to show relationships in the greatest number and most important properties. In the process, however, one gathers into classes of the lowest category the multitude of individual objects which are impossible of separate investigation by reason of sheer numbers. It has been shown that these classes are homogeneous with respect to the accumulated differentiating and accessory characteristics of all categories of the system. They are homogeneous within the limits of existing knowledge about the properties of the population and about the significance of differences within it. The natural classification, therefore, performs the extremely important function of organizing, naming, and defining the classes that are the basic units used (a) to identify the sample individuals that are the objects of research, (b) to organize the data of research for discovering relationships within the population, (c) to formulate generalizations about the population from these relationships, and (d) to apply these generalizations to specific cases that have not been studied directly. No other grouping provides such units; it is this fact that sets the natural classification apart from all other groupings.

Given the classes of the lowest category of the natural classification, one can group them for a great variety of technical purposes, as illustrated by Orvedal and Edwards (6). One may group soil classes of the lowest category, for example, to select a sample representative of specified soil conditions for research on phosphorus fixation, to discover genetic relationships, or to apply engineering principles in highway construction. These are *technical groupings*, each for a limited objective, each with a special bias dictated by the objective. Each must be based on those few properties that are most significant for the objective. Although only a few properties need be homogeneous in the classes that are grouped for any one objective, only the classes of the lowest category of the natural classification provide units that are homogeneous with respect to the variety of properties important for the large number of objectives for which technical groupings are made.

PRINCIPLES OF DIFFERENTIATION

As they affect classes

1. *A differentiating characteristic must be important for the objective.*

The statements about each class are confined to (a) the differentiating characteristics, (b) their accessory characteristics, or (c) interpretations of both. A grouping based on a property that is not important for the objective leads to classes about which the statements possible are not the most numerous or the most important for the purpose of grouping. In spite of this fact, attempts to interpret a grouping made for one purpose in terms of another objective are common. The category of soil series, for example, is a grouping purely to show similarities in properties of the soil profile, but one sees repeated attempts to define suitability of an entire soil series for land use. The bases of grouping soils into series do not provide units homogenous in all properties important for land use; the classes may range rather widely in such features as slope, degree of erosion, stoniness, and texture. A technical grouping of soil types and phases on the basis of an interpretation of all factors important in land use is required for this objective.

2. *A differentiating characteristic must be a property of the things classified or a direct interpretation for the objective.*

'If the grouping is a natural classification, the bases must be properties of the things classified. The objective is to see similarities and understand relationships among those properties; to classify on the basis of any other factor is to conceal those similarities and relationships. If soils were to be classified on the basis of factors of soil formation, for example, similarities and relationships among properties would be brought out only to the extent that the cause and effect relationships between factors of soil genesis and soil properties are known. The system formed would conceal relationships as yet unknown. A technical grouping of areas to show known relationships of soil genesis or to discover new ones, however, may well be based on those factors for correlation with properties of natural soil groups. In many technical groupings, various interpretations of characteristics may serve the objective best. A grouping of soils according to lime requirement might be based on an interpretation of the aggregate of known soil characteristics, including such measurable properties as pH and cation-exchange capacity of the various horizons. The grouping might be based on direct experimental evidence if that were available.

3. *The differentiating characteristic should carry as many accessory properties as possible for the objective.*

In the grouping in table 1, color was purposely chosen to differentiate in category 2 as a property that carries few accessory attributes. What can one say about dark-colored surface soils beyond the fact that they are dark, if he considers that such a group would include Chernozem, Solenetz, "Regur," Half-Bog, and varieties in which color was inherited from the parent material? Choice of a differentiating characteristic that carries many covarying properties, like reaction in category 2, increases the number of statements about each class and, therefore, the usefulness of the grouping.

4. *The class interval of a differentiating characteristic must provide classes homogeneous for the objective.*

When differentiation is based upon degree of expression of an attribute, the limiting value of that property between classes of a continuous series may be placed arbitrarily at any point in the series. Not all points within such a series are equally pertinent for a given objective, however, and the establishment of the class interval for the most useful grouping is not arbitrary. Accessory characteristics may not be straight-line functions of the differentiating property, and the best limits of classes for a given objective may be largely determined by them. The graph of percentage base saturation against pH, for example, is a characteristic titration curve. In northeastern United States, many mineral soils are roughly 30 per cent base-saturated at pH 5.0; 80 per cent, at pH 6.0; and 95 per cent, at pH 7.0. Thus a significant break in the curve occurs near pH 6.0; that pH is also associated with important changes in the solubility of iron, aluminum, and phosphorus. Soil classification must rely on the data of observation and experiment for establishment of the significant limiting values of classes. The points of greatest significance may vary not only among different objectives but also among different soils for the same objective.

As they affect relationships among categories

1. *The differentiating characteristic must classify all individuals in any single population.*

Nikiforoff[3] calls this the "principle of wholeness of taxonomic categories." Every category must include all existing individuals of the population; therefore, the differentiating

---

[3] The author is indebted to C. C. Nikiforoff for many of the ideas expressed in this paper, particularly for the original expression of the principles of "wholeness of taxonomic categories" and "ceiling of independence of differentiating characteristics."

characteristic of each category must apply to all individuals, or some will remain unclassified. Violation of this principle was a serious error in Marbut's Classification of soils in 1935 (4); his "family groups," which were azonal and intrazonal soils, were not classified by the criteria used in higher categories. This does not mean that a system of classification must immediately provide a pigeonhole for every individual in every category. It must provide for expansion in the number of classes based on a given differentiating property to accommodate new individuals as they are discovered. (In many cases one class may be defined as the zero degree of the differentiating characteristic.)

2. *Greatly different "kingdoms" require different differentiating characteristics at the same level of abstraction.*

This principle rests upon a concept of degrees of difference among groups of things at different levels of abstraction. All living things might be considered a unit separate and distinct from all things that do not possess life. Each is a segment of the universe and might conceivably be included in one master natural scheme of classification. It would be futile, however, to attempt to differentiate within both groups on the basis of the same property; they have too few important things in common. All living matter, in turn, may be classified into two kingdoms, plants and animals, each of which is a distinct unit at a lower level of abstraction. Again it would be futile to attempt to differentiate within both on the basis of a single property. Such contrasting populations are differences in "kind," as Mill used the term (5), and require classification in different systems[4] using different criteria for differentiation at the same level of abstraction. Within the natural phenomenon called "soils," there may be populations that have so few important common properties that a single differentiating characteristic at a given level of abstraction would not frame the important classes in each. The breech between organic and inorganic soils may be of that order of magnitude.

3. *All classes of the same category of a single population should be based on the same characteristics.*

This is a corollary of the "principle of wholeness of taxonomic categories," but it needs some explanation in the light of the preceding concept. The objective of classification is accomplished by arranging the individuals in an orderly manner commensurate with order as conceived in the mind. To differentiate on different properties at the same categorical level complicates the problem of visualizing relationships. It should be resorted to only when two kinds of things have so little in common that the important properties of one do not occur in the other. At some level of abstraction, differences no longer outweigh common properties; at that and all lower levels the "principle of wholeness of taxonomic categories" applies.

4. *A differentiating characteristic in one category must not separate like things in a lower category.*

Every characteristic has a *ceiling of independence* above which it cannot be used to differentiate without separating like things in categories below it. This is illustrated in table 1, where the objective is to show the most important similarities related to color value and reaction. Color value, a property with few accessory characteristics, is used at the highest categorical level. When each of the two classes based on color is subdivided on the basis of

---

[4] The term "system of classification" is restricted in this article to those schemes in which a single differentiating characteristic is used throughout the population in any single category. Two or more such systems for distinctly different populations may be tied together by a common property in a more inclusive scheme.

reaction in category 1, four classes are formed. Two of those classes are acid and two are alkaline. In no category in the system are all acid soils segregated in one group and all alkaline soils in another. Reaction is associated with many accessory characteristics that make similarities in that property more significant than similarities in color value, but individuals of like reaction are separated by differentiation on the basis of color at a higher level. Differentiation at a high categorical level on any basis separates on that same basis throughout all lower categories; consequently, the properties used to differentiate at high levels of generalization must be more important for the objective than those used at lower levels. The importance of a differentiating characteristic must be commensurate with the level of abstraction at which it is used. Failure to observe this principle is a major error in Shaw's (7) classification of soils, in which origin of parent material as a differentiating characteristic in a high category separates like profiles throughout all lower categories.

CLASSIFICATION OF SOILS

The soil as a population

The dimensions of the ultimate individual of the soil population are fixed vertically by the thickness of the soil profile and horizontally by the practical limits of space required for its observation. Vertically, the soil unit must extend from the surface into the parent material; any lesser depth would divide the complete natural body. Horizontally, the limits are not sharply defined; the unit must extend in two directions far enough to allow sampling and accurate determination of the properties that can be observed in the field. It must be large enough to permit observation of relationships of horizons to the rooting of plants. An almost infinite number of these indivduals make up the population—the soils of the world.

It is impractical, however, to attempt to deal with all these small units in any system of classification or in most practical problems involving use of the land. The range of the properties of each is too narrow to be significant; their numbers are too great to allow individual treatment; the area represented is too small to serve as a practical land unit in most operations. These individuals, however, are the sampling units upon which must depend definition of soil of larger areas feasible of treatment for most problems. These more inclusive units, as typified by the soil type or phase, may be considered classes of the lowest category of soil classification. They are more than a categorical grouping of individuals, however, because their variation is fixed partly by the geographic association of the ultimate individuals in areas large enough to be feasible of treatment. The definition of these classes is a problem in sampling to determine the characteristics of the modal profile and the deviations from it. Just as in the definition of a species of plants, no attempt is made to study all individuals and group them into a class of the lowest category. A sample is drawn, and from it are estimated the properties of the class.

Sources of criteria

For the most part, the criteria used to classify are those that can be observed or determined rapidly by simple tests in the field. This is inevitable because the practical limitations imposed by requirements of an adequate sample (1) of such a variable population preclude laboratory determinations as criteria for application on an areal basis. Consequently, until techniques are developed that will allow rapid chemical determinations in the field, one should expect homogeneity of chemical properties of soil units only to the extent that they are accessory to the observable characteristics used to classify in the field.

One is not confined to direct observation, however, in choosing criteria or their class intervals. Controlled experiments establish relationshisp that enable selection of those criteria to which many properties that cannot be observed are accessory. They not only indicate important criteria but also indicate significant class intervals.

For every controlled experiment there are thousands of natural experiments from which criteria of properties that cannot be observed readily may be deduced. Data from these experiments are not in orderly arrangement; they must be collected in many places, classified, and applied to soils by the correlation methods of science. The correlation of natural vegetation types with soil conditions, for example, is a principle criterion for the placement of soil boundaries. The behavior of crops under known management may indicate soil properties that cannot be observed easily.

Enough is known about soil formation to establish broad factors of the process and some of their effects. We expect a difference between two soils if any one of the factors varies. A change in one of the factors is a warning to look at the soil again if no differences have been observed. Genetic factors, though not criteria of classification in themselves, are indexes to soil properties that are criteria.

Criteria of classification

Soil classification is passing through a series of approximations in which the system is being built from the lowest category upward by a process of reducing homogeneity in each successively higher category. If our knowledge were complete, we should be able to choose the differentiating characteristic of each category, but at present not only must we establish the importance of many known relationships but undoubtedly we are still unaware of many relationships that will be discovered. Our choice of differentiating characteristics in the higher categories is limited by our knowledge not only of soil properties but also of relationships among soil properties. As a direct result of requirements imposed by the "principle of a ceiling of independence of differentiating characteristics," the limitations imposed by lack of knowledge are greater, the higher the level of generalization in the system. Let us examine the possible criteria at our disposal.

No one has yet improved appreciably upon the following list of criteria proposed by Marbut (3), for differentiation among soils at the level of the soil type:

1. Number of horizons in the soil profile.
2. Color of the various horizons, with special emphasis on the surface one or two.
3. Texture of the horizons.
4. Structure of the horizons.
5. Relative arrangement of the horizons.
6. Chemical composition of the horizons.

7. Thickness of the horizons.
8. Thickness of the true soil.
9. Character of the soil material.
10. Geology of the soil material.

Each of the 10 factors listed varies narrowly in a soil type within the limitations of precision of their measurement. Long experience with the soil type has proved that it, as defined, is a unit adequately homogeneous for most of the practical problems of the land so far as the soil profile is concerned. No better unit has been suggested at that categorical level. Cases of inadequate homogeneity are the result either of lack of precise application of one or more of the criteria listed or of the importance of features other than characteristics of the profile, such as slope or degree of erosion. Lack of precision is not a fault of the system but of its application; techniques for feasible evaluation of chemical properties, for example, are not adequate. Such characteristics as slope are differentiated in the soil phase or its equivalent in practical problems. One must conclude, therefore, that Marbut's criteria provide a workable basis for classification of soil profiles in the lowest category; when properly applied, they provide units about which the number, precision, and importance of statements are adequate for practical problems in soils.

Now consider the implications of the "principle of accumulating differentia" in relation to the proposition that the properties listed by Marbut are adequate for differentiation at the level of the soil type. All of those properties except the one used to differentiate soil types within a soil series must be accumulated from the differentiating and accessory characteristics of higher categories. The list undoubtedly is not a complete accumulation of all those characteristics, but each property in the list must be either differentiating at a higher level or accessory to a property that is differentiating. It would be sheer accident if any property could be found to differentiate at higher categorical levels that is not either (a) a characteristic listed by Marbut, or (b) a characteristic to which one of the properties listed by Marbut is accessory.

The properties listed by Marbut were not intended as bases of differentiation in higher categories, and careful consideration will show that no single characteristic listed could be used at a level higher than one category above the soil series without exceeding its ceiling of independence. Its use would separate like things in lower categories and defeat the objective of a natural classification. One must conclude, therefore, that these properties are accessory characteristics of those attributes whose ceilings of independence are high enough to justify their use in the higher categories. This is one important reason why the "formula" type of system in which individual soil properties are accumulated in a symbol in some specified order is not a natural classification.

The first step in the selection of differentiating characteristics of higher categories, therefore, is to define the characteristics to which the properties listed by Marbut are accessory. Most of those listed are characteristics of individual horizons. All of them, collectively, plus inferences from them, however, define the whole soil, not only in terms of all of its horizons but also in terms of rela-

tionships among horizons.   It is to this definition of the soil type that one must
turn for criteria whose ceilings of independence permit their use at high categorical levels.   A Podzol, for example, is not just any soil with a bleached horizon;
it is a complete natural body definable in terms of all of its horizons *and the relationships among them*.

At a given level of abstraction, for example, the kind and sequence of horizons
defines a specific "kind of profile."   This is a characteristic of the whole soil and
has a ceiling of independence far above that of soil texture.   The degree of expression of those horizons is also a property of the whole soil; its ceiling of independence is obviously below that of "kind of profile" but still above that of soil
texture.   Kind of parent material is reflected throughout the whole soil; its
ceiling of independence is below that of degree of expression of horizons but
slightly above that of soil texture—in fact, it is roughly commensurate with the
level of abstraction of the soil series as used at present.   It is from properties
such as these that the criteria for differentiation in the various categories above
the soil type must come.   Their definition is controlled by the state of knowledge
about specific soil characteristics and especially about relationships among
horizons.   We may expect many approximations before knowledge will be adequate to frame, in final form, the generalizations necessary for definition of
appropriate criteria at the various categorical levels.

REFERENCES

(1) CLINE, M. G.   1944   Principles of soil sampling.   *Soil Sci*. 58: 275–287.
(2) KELLOGG, C. E.   1937   Soil survey manual.   U. S. Dept. Agr. Misc. Pub. 274.
(3) MARBUT, C. F.   1922   Soil classification.   *Life and Work of C. F. Marbut*, pp. 85–94.
      Artcraft Press, Columbia, Missouri.
(4) MARBUT, C. F.   1935   Atlas of American Agriculture, Part III.   Soils of the United
      States.   U. S. Government Printing Office, Washington, D. C.
(5) MILL, J. S.   1874   A System of Logic, ed. 8, vols. I and II.   New York.
(6) ORVEDAL, A. C., AND EDWARDS, M. J.   1942   General principles of technical grouping
      of soils.   *Soil Sci. Soc. Amer. Proc.* (1941) 6: 386–391.
(7) SHAW, C. F.   1947   The basis of classification and key to the soils of California.   *First
      Internatl. Cong. Soil Sci. Proc. and Papers. Comn.* V: 65–103.

6

Changes in the Criteria Used in Soil Classification Since Marbut[1]

E. P. Whiteside[2]

ABSTRACT

The changes in the criteria used in soil classification since Marbut's time are presented. These changes are of three kinds: additions to the list of properties used, refinements of the properties formerly used, and deletions from the list of properties formerly used. The additional properties now being used are mostly properties of the whole soil body. Most important among these are the properties which are independent variables of the soil system. The introduction of these independently variable properties of the soil body, for use in soil classification, presents the poignant possibility of fusing the genetic and descriptive systems of soil classification without abandoning the premise that soils should be classified on the basis of their properties. The affect of the above changes in soil classification on the definition of the soil individual — the soil type — and its nomenclature are discussed.

MARBUT (5) outlined the properties of soils being used in soil classification in 1922. Cline (1) in 1949 pointed out that, "No one has yet improved appreciably upon the following list of criteria proposed by Marbut." Those properties may be grouped according to whether they are properties of the soil body or of its individual horizons, as follows:

Properties of the soil body:
1. Number of horizons in the soil profile
2. Relative arrangements of the horizons
3. The thickness of the true soil
4. Geology of the soil material

Properties of the soil horizons:
5. Texture
6. Structure
7. Color
8. Chemical composition
9. Thickness
10. Character of the soil material

As to whether the improvements in the criteria used in soil classification are appreciable or not may be a matter for debate. However, all would agree that the terminology for expressing soil characteristics has improved greatly in precision since Marbut's time, and these changes are reflected in the recent Soil Survey Manual (10). The techniques for translating these soil differences into maps giving them geographic significance have been greatly improved.

The author believes that the concepts and criteria used in the classification of the soils have also undergone an important evolution during this same period. These changes are of three kinds: additions to the list of properties used; refinements in the properties formerly recognized; and deletions from this list.

Properties of the Soil Body

The major change in concept has been the further development of the idea that the soil is a dynamic natural body. Some of the implications of this in soil classification are as follows:

1. A body has three dimensions. This adds shape to

[1]Journal Article No. 1454, Michigan Agr. Exp. Sta., East Lansing, Mich. Received for publication Nov. 30, 1953.
[2]Associate Professor of Soil Science.

the list of properties of the soil body, e.g. slope and terrace phases.

2. In considering the properties of a dynamic body, one is soon impressed by its constituent physical phases — solid, liquid, and gaseous. This certainly requires that one seek to define at least two of these phases in order to completely define the soil system. So one needs to deal with the nature of the more dynamic phases, soil moisture and soil air. Whether both will be necessary or only one as independently variable properties is not yet clear. Recently, temperature and moisture of the soil body have been proposed as criteria to be used in the classification of soils (3, 9). Age or degree of development of soils was used by Shaw (7) to sub-divide his families of soils. The biological properties may yet be found to be useful (4). However, ways for expressing these dynamic properties in their most helpful forms have yet to be devised.

3. The introduction of shape, moisture, temperature, age, and the biology of the soil body as characteristics to be used in soil classification makes possible the measurement as soil properties of what were formerly considered to be soil formation factors external to the soil. This presents the very poignant possibility of fusing the genetic and descriptive soil classifications without abandoning the premise that soils should be classified on the basis of their properties.

4. An over-emphasis of slope and erosion by some soil scientists and the over-emphasis of the soil profile by others has led to the idea that soil, slope, and erosion are being shown on detailed soil maps today. However, this is a redundant expression because all three are based on soil properties or interpretations of soil properties, and one need only to say that these maps are detailed soil maps.

Properties of Soil Horizons

A property that has been added to the list of properties of the soil horizons for use in soil classification is mineralogical composition. This has resulted largely from improved techniques for the study of small mineral grains and might be considered as a refinement of a characteristic formerly called vaguely "character of the soil material". By the addition of this property it is possible to delete the earlier one just mentioned from the list; because with the aid of mineralogical composition the "parent material' as used in the Soil Survey Manual (10) can be described as another soil horizon or as parent rock (11). Consistence has also been added to the list of properties of the soil horizons used in soil classification (10).

The introduction of mineralogical composition in soil classification also permits a more concise statement of one of the properties influencing the whole soil body. The vague expression, "Geology of the soil material," can probably be replaced by the "Mineralogical and chemical composition, the texture, and the structure or fabric of the parent rock (11)."

Properties Currently Used in Soil Classification

To recapitulate, the criteria now used in soil classification may be grouped according to whether they are properties of the soil body or of its individual horizons, as follows:

I. Properties of the soil body:
1. Number of horizons in the soil profile.
2. Relative arrangement of the horizons.
3. The thickness of the true soil.
4. Mineralogical and chemical composition of the parent rock.
5. Texture of the parent rock.
6. Structure or fabric of the parent rock.
7. Shape.
8. Temperature.
9. Moisture.
10. Degree of development.
11. Age.
12. Air?
13. Biology?

II. Properties of the soil horizons:
14. Texture.
15. Structure.
16. Consistence.
17. Color.
18. Chemical composition.
19. Mineralogical composition.
20. Thickness.

The above list of soil properties that are currently being used in soil classification[3] includes soil properties that are independent variables of the soil system and dependent variables or s properties as pointed out by Jenny (2). However, at the present stage of development of soil science, neither group of properties alone is adequate for the development of the most satisfactory system of soil classification. The s properties are used for defining and describing the soil individuals or group of individuals but the "independent variables that define the soil system" and the relations of the s properties to them are very important in showing the relationships of the soil individuals to each other. Enough is not yet known about what the independent variables of the soil system are or how they are related to the more easily observable s properties so that they alone can be used in a soil classification system.

The Soil Individual

In soil classification, as in any other geographic science, man must first define the individuals of which he speaks since they occur in nature as a continuum rather than as discrete individuals such as a plant or an animal. Secondly, he seeks to show the genetic relationships of these soil individuals to one another so that their occurrence in nature can be predicted and given geographic significance by the projection of a small number of actual soil observations to a much larger landscape (6), and so that the relationships of the properties of the individuals can be more easily recalled for the purpose of predicting their management needs for specific purposes or their behavior with specific kinds of management.

A soil individual is distinctly different in its range of differentiating properties from every other soil individual (10, pp. 123–126). Because of this difference in properties it frequently, but not necessarily, differs significantly in its use potentialities, its management requirements for given purposes, or its productivity under given management conditions from other soil individuals. Soil individuals that are conspicuously different in one or more properties from other soils are defined as separate individuals because of the differences in properties, even though these have no known significance in utility. Failure to recognize soil differences known to be significant to land use as separate individuals is not permissible in the taxonomic classification of soils. It may be impractical to delineate such individuals on a soil map of a given scale. However, even in this instance it behooves the soil scientist examining the area to observe the relative proportions of these different soil individuals present and their distribution or pattern of occurrence. Here the taxonomic soil individuals serve as adjectives for describing the area.

The allowable limits in the properties of each soil individual are now fairly well known for particular combinations of properties over a large part of the United States. This is a result of the experiences of the past 50 years in soil classification work. Consequently, it seems more correct today to say that soil individuals should be defined in terms of the ranges of their differentiating properties rather than by a modal individual which would vary with the relative areas of the different portions of the range of soil properties.

In the light of these changes in the criteria and concepts used in soil classification, it has been proposed by the Soil Survey Sub-committee of the North Central Regional Soil and Fertilizer Research Committee[4] that the definition and nomenclature of the soil type — the soil individual created by man — be changed to read as follows:

"The soil type is the lowest category in the natural system of soil classification. It is a subdivision of the soil series based on the texture of the surface soil (or degree of decomposition of organic soils), shape or inclination of the surface, stoniness, degree of erosion, depth of the solum, depth to layers of unconformable material, and of the salinity of the soil. It is the taxonomic unit concerning which the greatest number of precise statements and predictions can be made about soil use, management, and productivity.

"The soil type name consists of the series name plus a short descriptive designation of the above mentioned variations of properties within the series."

This nomenclature fits well the soil units as named

---

[3]It should be pointed out, as Marbut mentioned in 1922, that this list is probably still incomplete, e.g. Marshall and Hazeman (Soil Science Soc. Proc. (1942) 7:448–453, 1943) have demonstrated how such properties as the clay content or organic content of soil profiles may be determined quantitatively, and how the changes in these properties of the soil body, compared to the parent rock, may be assessed.

[4]A proposal prepared under date of Sept. 20, 1951 and transmitted to the current terminology committee of the Soil Science Society of America.

in recent detailed soil maps. However, the second sentence of the definition is not very elegant and expresses no basic philosophy for differentiating the soil type from the soil series. A statement of a more general philosophical nature is greatly needed. Would the following statement be better? "It is a sub-division of the soil series based upon significant variations in the properties observable at or near the surface of the soil body." If this was accepted as the definition of the soil type, then the soil series would be defined as soil individuals which vary in the range of the differentiating characteristics of the portion of the body beneath the surface.

The desirability of giving serious consideration to these relationships among the soil series, soil type, and soil phase are mentioned in the new Soil Survey Manual (6, p. 290). It is necessary that workers in soil classification crystallize their views on this fundamental unit for the taxonomic classification of soils, the soil individual. If there is disagreement with the views presented here, what would be a more reasonable solution?

Literature Cited

1. Cline, M. G. Basic principles of soil classification. Soil Sci. 67:81, 1949.
2. Jenny, Hans. Factors of soil formation. McGraw Hill Book Company, New York. 1941.
3. ———. Arrangement of soil series and types according to functions of soil forming factors. Soil Sci. 61:375, 1946.
4. Kubiena, W. L. Micropedology. Collegiate Press Inc., Ames, Iowa, 1938.
5. Marbut, C. F. Life and works of C. F. Marbut. (pp. 85–94) Artcraft Press, Columbia, Mo., 1922.
6. Muckenhirn, R. J., *et al.* Soil classification and the genetic factors of soil formation. Soil Sci. 67:93, 1949.
7. Shaw, C. F. Some California soils and their relationships. Univ. of Cal. Syllabus Series JD, Univ. of Cal. Press, Berkeley and Los Angeles, 1937.
8. Simonson, R. W. Lessons from the first half century of soil survey: I Classification of soils. Soil Sci. 74:249, 1952.
9. Smith, G. D. A proposed taxonomic classification of soil of the United States. (mimeographed). U.S.D.A. Soil Survey Div. 1952. (Personal communication).
10. ———. Soil Survey Manual. U.S.D.A. Handbook No. 18. G.P.O., Washington, D.C., 1951.
11. Whiteside, E. P. Some relationships between the classification of rocks by geologists and the classification of soils by soil scientists. Soil Sci. Soc. Amer. Proc. 17:138, 1953.

7

Reprinted from *J. Soil Sci.* **10**:14–26 (1959)

THE RUSSIAN APPROACH TO SOIL CLASSIFICATION AND ITS RECENT DEVELOPMENT

J. J. BASINSKI[1]

Summary

Due largely to the conditions in which they were working, Russian pedologists were the first to establish pedology as an independent science. From the beginning they regarded soil as an independent body with a definite morphological organization, expressed mainly in the structure of the profile, and resulting from pedogenetic processes determined and directed by environmental factors. This concept of soil led them to adopt a genetic approach to problems of soil classification. Russian soil classifications differed according to the basis accepted, whether bioclimatic, geographical conditions, factors of pedogenesis, pedogenetic processes, or soil evolutionary history. In recent years attempts have been made to construct classification systems based on all these aspects of pedogenesis. The current Soviet trends in soil taxonomy must be regarded mainly as a further development of the traditional approach. Measures are taken to standardize soil nomenclature and improve methods of recognizing (diagnosing) and describing soil types, which are regarded as basic taxonomic units. The evolutionary-genetic approach is considered the only proper approach to soil-classification problems. The importance of organic aspects of soil evolution and formation processes is emphasized. More attention is given to the genetic subdivision of soil types into smaller and better defined soil groups. Interest in the systematics of cultivated soils is also growing.

I. *Introduction*

EVER since Dokuchaev (1879) published his first classification of the soils of Nizhnyi Novgorod Province in 1879, the minds of Russian pedologists have been exercised by the problems of soil taxonomy, probably far more than those of their Western counterparts. Russian literature on this subject is so voluminous that to summarize it for the English reader would involve prolonged and intensive study but for a number of excellent reviews published recently, mainly in one of the journals of the Soviet Academy of Sciences, *Pochvovedenie* (*Pedology*).

The number and extent of quotations from early works found in contemporary Soviet literature on soil taxonomy and used in support of current arguments are striking to the Western reader. It is, however, the best evidence of the evolutionary rather than revolutionary development of Russian thought on the subject. All recent developments are so closely linked with past achievements that it is impossible to consider them apart from the traditional Russian school of pedology and the conditions under which it developed.

II. *Traditional Russian Approach to Soil Science as a Whole*

Russian pedologists have been fortunate in the geographical position of their country—in the middle of the Eurasian continent—and its vast extent. They have been able to study a wide variety of environments and vegetation types as well as soils, and to make extensive comparisons. This has also undoubtedly led to the broad outlook which has persisted

[1] Division of Land Research and Regional Survey, C.S.I.R.O., Canberra, A.C.T.

in their thinking. In addition, the presence of soil undisturbed by man in a wide range of environmental conditions makes it possible for them to be studied in their natural state and form.

The Russian school of pedology in general and soil classification in particular must also have been influenced appreciably by the fact that it developed under the stimulus of the necessity of classifying and mapping land for utilization purposes often without the opportunity of studying it intensively.

The mental climate of Russia in the late nineteenth and early twentieth centuries must also be taken into account. It was the time when modern scientific thinking, seeking not only to establish and record the facts and laws of nature, but also to explain them, received its main impetus. The teaching of Darwin and his disciples often provided the basis of explanation, and hence the evolutionary approach was fashionable.

While western European and American scientists were still considering soil as a geological material (Hilgard) or a complicated mixture of chemical compounds (Liebig), the early Russian pedologists were the first to regard soils as an independent body with a definite morphological organization reflected in the profile. They were the first to show the broad but definite inter-relationship between environment, soils, and vegetation, and the first to recognize soil as the product of a process governed by environmental factors rather than by soil parent material. This led to recognition of the fact that soil is a dynamic and not a static body, a living organism allied in the plane of development to biological bodies.

This Russian concept of soil and its place in nature has, in the words of Marbut (1936), 'established [the study of soils] firmly as an independent science with criteria, point of view, method of approach, process of development applicable to soil alone and inapplicable to any other series of natural bodies'. It also had a predominant influence on the development of Russian taxonomy.

III. *Traditional Russian Approach to Soil Classification*

One of the primary aims of soil-classification systems as proposed or employed by Russian pedologists since the time of Dokuchaev's first soil classification is to group the soils according to their productivity and utilization problems, and to determine, compare, and explain their fertility differences. Dokuchaev (1893), writing of his first classification, stressed 'the close connexion existing between soil types on the one hand and yields, kind of crops and husbandry methods on the other'. In many recent papers on the subject, Russian pedologists continue to stress the importance of soil productivity in taxonomic work. For example, Ivanova (1956) states, 'fertility is an essential property of the soil and consequently soil classification must reflect this property'. She writes further, 'soil world groups, constituted on the basis of radiation energy, define the main ways of utilizing natural resources in national economy. Classes and subclasses define the type of husbandry. Soil types, regional groups, families, subtypes and forms permit more concrete zoning of types of agriculture and husbandry systems, give indispensable amelioration measures, define the order of soil suitability for cultivation, use in agriculture etc.' This stress on utilization aspects in the Russian approach to soil taxonomic work is often forgotten by those outside observers who lose themselves in the theoretical and philosophical arguments of Russian pedologists.

The Russians have always recognized the 'genetic soil type' as the fundamental unit. Their classification schemes are essentially a series of attempts to link soil types into progressively larger groups and to sub-divide them into progressively more defined variants. The definition of soil type is usually based on three criteria first established by Sibirtsev towards the end of the last century, namely:

1. Sameness of genetic properties of the soil.
2. Sameness of the pedogenetic processes determining these pro-perties.
3. Sameness of pedogenetic factors defining and directing the pedo-genetic processes.

The choice and extent of the 'genetic soil properties' taken into account when defining 'soil types' have differed from time to time, although the morphological structure of the profile has always been of primary importance. Moreover, since the days of Sibirtsev, Russian pedologists have considered that only the combination of the attributive and genetic approach can lead to an identification of soil types and their classification which will permit the proper appraisal of soil fertility. This point of view is still generally accepted amongst them (Rozov, 1957).

Because one of the fundamental concepts of pedology as developed mainly in Russia is that soil is a product of geographical environment which determines the process of its formation and which is in itself a resultant of separate environmental factors, the attention of Russian soil classifiers has, since early days, focused on environment, environmental factors, and pedogenetic processes rather than on the soil itself.

It is true that the father of Russian soil taxonomy, Dokuchaev, stressed in connexion with his first classification that 'grouping of the soils should always be based on the sum of existing characters of a given body' (its profile, clay content, chemical characteristics, &c.). As pointed out recently by Rode (1957), however, 'many of Dokuchaev's followers, amongst whom were some eminent scientists, did not always adhere to this principle, and developed their own classification schemes, basing them mainly on conditions of pedogenesis'. Even when the properties of the soils themselves provided the basis of a classification system, the classifications as such were based on genesis rather than on soil attributes, on the assumption that all soil-material characteristics are merely the reflexion of their development processes as directed by environment.

Also, because soils and their pedogenesis are regarded as a single system, it is often very difficult for outside students of Russian pedology to distinguish whether the reference or argument concerns soil as such or the process of its formation, which to a Russian are one and the same thing. This is well illustrated by a recent review of soil-classification systems by Gerasimov (1954) who, dealing for example with Kossovitch, Tumin, Glinka, Gedroitz, Neustruev, and Prasolov's systems, describes them as 'based on intrinsic properties of the soil', while in fact they are based on intrinsic soil-forming processes.

The existence of a definite relationship between geographical environment, soil, and its vegetative cover on which Russian classifications are based, justifies also the inclusion of environmental (geographical) and ecological terms in the classification systems.

The evolutionary character of soil formation makes it possible to explain soil differences as expressions of different stages of their evolution as well as expressions of different conditions governing evolutionary processes. This concept of soil as a product of evolution has played an increasingly important part in the thought of Russian pedologists and in their approach to soil classification.

The experimentation in soil genesis and evolution has not yet been developed as a well-established method of attack on the problems involved. Consequently, reasons for the soil differences found in the field or laboratory must be reached by interpretation of correlations between soil types and factors influencing its development. In this deductive approach to soil taxonomy, Soviet scientists are prepared to go much farther than Western pedologists.

IV. *Differences in the Russian Approach to Soil Classification*

All Russian soil classifications are based on a genetic approach. However, within this general approach five different categories of classification systems may be recognized, according to whether the systems are based on conditions of pedogenesis, the factors governing pedogenesis, the character of pedogenetic processes or evolutionary stages in these processes, or a single process. This grouping of then existing classification systems has been adopted by Gerasimov (1954) in his recent review. His terminology, however, when translated is apt to confuse the Western reader, and has not been adopted here in its entirety.

(a) *Geographic-environmental classifications*

In these classification systems the soil types clearly related to bioclimatic conditions of regions where they are found are classed as 'normal' or 'zonal'. Those not so related are classed as 'transitional' and 'abnormal' or 'intrazonal' and 'azonal'.

Thus, Dokuchaev (1896) in his final classification grouped glacial, light grey podzolic, brown forest, chernozemic, dark chestnut, light greyish brown, red, dark brown swampy and whitish secondary solonetz soil types into the class of 'normal' or 'zonal' soils. He also recognized a class of 'transitional' or 'intrazonal' soils, and a class of 'abnormal' or 'cosmopolitan' soils, including aeolian, alluvial, and swampy soil types.

Sibirtsev (1895, 1898, 1900) classified soils into classes and types. The class of 'zonal' soils included the following types: lateritic, aeolian loess, desert-steppe soils, chernozemic, grey forest, grassland, podzolic, and tundra soils. The 'intrazonal' class included solonetzic, swamp, and humic-carbonate soil types. The class of 'azonal' soils included skeletal, coarse, and riverain types.

In the classification proposed by Afanasev (1922, 1927) who in the

1920's developed further the ideas of Dokuchaev and Sibirtsev, each of five climatic belts (cold, cool, temperate, subtropical warm, and tropical hot) is divided into two complexes, maritime and continental. Each soil type has an equivalent in the corresponding complex. Further subdivision is based on soil changes as influenced by the type of vegetation (forest, grassland-forest, and grassland). Finally, differences due to different phases of evolution influenced by degradation and salinization processes are also considered.

In constructing classification systems in this category, the authors—as stated by Sibirtsev (1895)—attempted 'to explain the soil cover of the earth's surface in its genetic entity and geographical variability'.

(b) Factorial classifications

These classifications are based on separate factors of pedogenesis or their combinations. Glinka (1902, 1915), who was the first to adopt this approach, divided soils into two main groups, 'ectodynamomorphic' and 'endodynamorphic'. The former is the result of external (mainly climatic) factors, while the character of the latter is determined by the parent material. Further classification is based on the moisture régime.

Vilenskii's (1925) classification is typical of this category although, like Dokuchaev, he divides soils into 'zonal' and 'intrazonal'. Zonal soils in his system are divided horizontally into five belts, based on Köppen's temperature zones (polar, cold, temperate, subtropic, and tropic) and five corresponding soil divisions (hydrogenic, phytohydrogenic, phytogenic, thermophytogenic, and thermogenic). Each of these belts is subdivided vertically into columns according to Weigner's humidity regions (arid, semi-arid, weakly arid, semi-humid, and humid). Intrazonal soils are divided horizontally into five divisions (halogenic, phytohalogenic, hydrohalogenic, thermohalogenic, and thermohydrogenic). The first two divisions correspond to the temperate zone and the last two to the subtropical and tropical zones. The first two are also divided vertically according to humidity regions. In addition, the soils of the mountain region are grouped in the orogenic division.

Zakharov's (1927) system is essentially similar, zonal soil complexes being divided into climatogenic, orogenic, hydrohalogenic, fluviogenic, and lithogenic divisions, and classified further according to climatic and ecological conditions.

Vysotskii (1906) also divides soils into zonal, intrazonal, and undeveloped, subdividing them further according to climate and parent material.

The most recent factorial classification is that of Volobuev (1955) who groups the soils into climatic communities, subdividing each community into two orders: automorphic and hydromorphic. Swamp soils and solonchaks are regarded as intrazonal soil types. Further classification is based according to vegetation communities associated with the soils.

(c) Process classifications

These classification systems are based on differences and similarities in the process of soil formation. As has already been pointed out, owing to the lack of distinction between the processes and their result, some of the Russian soil taxonomists who adopted this approach claimed that their classifications were based on intrinsic soil properties, although in fact they were based on the pedogenetic processes.

Kossovitch (1910), for example, wrote, 'genetic soil classification should be based on intrinsic properties and characteristics of the soils themselves'. In his classification, however, he divided soils according to types of pedogenesis. In the group of 'genetically independent' (eluvial) soils he recognized the following types of pedogenesis related to bio-climatic conditions: desert type, semi-desert (solonetzic) type, steppe (chernozemic) type, humid-cool (podzolic) type, humid-hot (lateritic) type, and polar (tundra) type. The class of genetically dependent (illuvial) soils embraced the corresponding types of pedogenesis as affected by impeded drainage. In the case of each type of pedogenesis, transformation and translocation of mineral and organic substances were considered, special attention being given to changes in organic matter.

Tumin's (1907) classification is basically the same. His classes were divided into families according to quantitative expression of the pedo-genetic process (degree of development), and some families were sub-divided into forms, e.g. podzols were divided into podzols proper, podzolic and weakly podzolic, while chernozems were divided into chernozems proper and southern chernozems. There was also division into groups according to parent material.

Glinka (1922, 1924), in his later work, recognized five basic types of pedogenesis (lateritic, podzolic, chernozemic, solonetzic, and swampy) and stressed that since pedogenetic processes lead to transformation of organic matter and decomposition of mineral matter and translocation of the products of these two processes, the soil properties are born in pedogenesis itself.

In Neustruev's (1926) classification there are two broad divisions of soil-forming processes, 'automorphic' active under normal moisture conditions, and 'hydromorphic' active under conditions of excessive moisture. The processes are further subdivided according to their effect on the state of decomposition of the mineral constituents, transformation of organic matter, and translocation of the products of pedogenesis.

Prasolov (1934) classifies the processes into eluvial, salinizing, and desalinizing, and hydromorphic. According to his system, eluviation results in three principal soil groups: soils of the humid regions, soils of the semi-humid and semi-arid regions, and soils of the arid regions. These groups are further divided into genetic soil types defined on the basis of interrelation between biological elements (mainly vegetation) and parent material on the one hand and pedogenetic processes on the other.

(d) Evolutionary classification

Although his classification in its final form was based on types of pedogenesis Kossovitch (1906), in his earlier work, regarded soils as phases in evolutionary development governed by two processes based on acid and alkaline weathering. His ideas undoubtedly influenced other Russian soil taxonomists.

In Polynov's uncompleted classification, the influence of Kossovitch's concept can easily be recognized. Polynov (1933) considered that soil evolution proceeds along two lines, one governed by eluviation and one by salinization or desalinization. In the eluvial evolution, Polynov recognized two basic processes, acid and alkaline weathering, and considered that soil types merely reflect evolution's many stages due to progressive de-alkalinization. Thus alkaline weathering produces alkaline, pre-chernozemic, and chernozemic phases, while acid weathering results in prepodzolic, podzolic, and swampy phases. With increased humidity, alkaline weathering may give place to acid weathering and thus the two series may be regarded as continuous. Halogenic evolution consists of phases corresponding to transition from the solonchak group through the carbonic group to the swamp group.

Until recently, Viliams's (1939) concept of a single pedogenetic process—a part of his much wider theory of environment and life evolution—completely dominated Soviet pedology. This, according to Rozov (1957), is at least partly responsible for 'many unclear and undecided theoretical problems in [contemporary Soviet soil] science'. The wide concept of Viliams undoubtedly stimulated thought, but, being advanced dogmatically, it was also placing thought in 'chains'.

Viliams regarded soil evolution as a phenomenon connected with geological history of life on earth, as a part of a single process of life development. Since living organisms affect soil development directly and indirectly through their effect on environment and pedogenetic processes, life was also considered as the main moving force in pedogenesis. Consequently, the processes of synthesis and decomposition of organic matter are regarded as particularly important in soil formation.

In addition, Viliams's theory postulates that soil evolution proceeds along a single path determined by the development of relief and shifts of climatic conditions and vegetation zones. For example, he thought that 'all the great territory in the U.S.S.R. in turn passed through all the phases of pedogenetic evolution, beginning with tundra'. Thus, according to his theory, all existing soils represent merely different phases in a single evolutionary development, being simply a function of time.

Recently this approach has been strongly criticized by Soviet pedologists. However, during the time when Viliams's views reigned supreme, a number of attempts were made to classify soils entirely in accordance with his principles. Tsyganov's (1955) classification may serve as an example. It is based on a single process of evolution connected with the geological growth and drying out of continents. The swampy, podzolic, grassland steppe and desert soils are regarded as evolutionary zonal phases with intrazonal phases based on the development of relief.

V. *Recent Development in the Russian Approach*

Recent trends in Russian pedology in general and soil taxonomy in particular cannot be described as revolutionary. They constitute merely a further development of the traditional school of thought of Dokuchaev and his early successors, and of Viliams and his followers. The genetic-evolutionary approach still universally accepted is, however, based on a much firmer foundation, made possible by the ever-increasing volume of pedological material. Especially since the 1939–45 war, Soviet soil-survey and mapping activities have been considerably increased, to embrace the lesser known polar and tropical regions of the Soviet Union. Close co-operation with eastern European and Chinese scientists has extended the zone of activities of Soviet pedologists even further and has brought new, often valuable minds to the Russian school of pedology. Material for soil inventories has been greatly extended and, since the war, the compilation of such inventories has taken a place of great importance in Russian work. Field and laboratory studies of different soil types also have been intensified and made more detailed.

(a) *Standardizing nomenclature and methods of diagnosis of soil taxonomic units*

As may easily be understood from the preceding sections, which describe the development of Russian pedology and soil taxonomy, soil nomenclature and methods of recognizing and defining taxonomic units were, until comparatively recently, allowed to develop without any check or co-ordination. This inevitably led to a great profusion of ill-defined technical terms and diagnostic methods, resulting in constant misunderstandings and general confusion. Since the war, work has begun, mainly in the Soil Institute of the Academy of Sciences of the U.S.S.R., to bring order to this chaos in nomenclature and methodology. This culminated in the appointment by the Academy of Sciences in 1956 of a special permanent commission including forty leading Soviet pedologists. Opening the first session of the Commission in February 1957, the Chairman, Academician Tyurin, stated that 'the main task of the Commission appears to be unification of soil nomenclature, systematics and methods of diagnosis. These topics cannot, however, be discussed apart from classification problems. The Commission, although it is unable to undertake the task of developing classification schemes, which is beyond its powers, can nevertheless direct this work by considering them and making the necessary recommendations.'

The minutes of this first meeting (Rozov et al., 1957) stress the considerable discrepancy in the meanings of many pedological terms in the minds of different workers. The Commission began its work by defining soil types and subordinate units:

Type

Major soil group, developing in a single type of bioclimatic and hydrological conditions, characterized by a clear manifestation of the basic processes of soil formation, possibly in combination with other pedogenetic processes. Characteristic points of soil type are defined by:
  (1) same type of accumulation of organic matter, its rate and character of distribution;
  (2) same type of processes in decomposition of mineral substances and synthesis of new mineral and organic-mineral products;
  (3) same type of translocation of soil materials;

(4) same type of structure of soil profile;

(5) same direction of measures for increasing and maintaining soil productivity.

Sub-type

Groups of soils within the range of soil type differing qualitatively in the manifestation of one of the superimposed soil-forming processes and in the intensity with which they reflect the main pedogenetic process. Measures for increasing and maintaining soil fertility are more alike for a sub-type than for a type.

In differentiating a sub-type, the subzonal as well as regional change in environmental conditions is taken into account. (Zonal conditions in the Russian approach are those determined by latitude or altitude—i.e. mainly temperature; regional conditions are determined by the type of climate, continental or maritime.)

Family (Rod)

Group of soils within a sub-type, with qualitative peculiarities dependent on local conditions, for example parent material (including chemistry of ground water), past history of soil development (relics of previous stages), &c.

Form (Vid)

Soils within family divisions differing in the degree of development of pedogenetic process (degree of podzolization, quantity of humus and strength of humic horizon, degree of salinization, &c.).

The Commission also recommended further study and discussion of methods of recognition of soil types and their description as the basis of future standardization. Since its first meeting Rode (1957) has published a very interesting paper on this subject in which he stresses the necessity of basing genetic soil classification on a proper recognition ('diagnosis') of different soil types. This diagnosis, he advocates, should be based on a wide use of the methods of chemical and physical analysis, as well as on field observations. He also recommends, as far as possible, a quantitative approach to the definition of soil taxonomic units. According to him, these units should be based on similarity of a wide variety of soil attributes. These should be presented in the form of curves illustrating changes in different properties within the soil profile.

He considers that each soil type or lower taxonomic unit should have its own 'passport', a document containing the following parts:

1. Morphological column illustrating schematically the morphological structure of the soil profile by the use of agreed symbols.
2. A very short morphological description of separate horizons corresponding to their illustrations in the morphological column.
3. Graphs illustrating distribution in the profile of various soil characteristics, e.g. clay fraction, humus, carbonates, total salts, exchangeable bases, pH, &c.
4. Graph of mechanical composition.
5. Graph of moisture and other physical properties.
6. Graph showing root distribution.
7. Graph showing distribution of main groups of organic matter.
8. Graphs showing schematically the characteristic points of water and heat régime, &c.

Rode considers that these soil 'passports' should play a most important part in the preparation of soil-classification systems. Finally, he advocates the establishment of soil collections which should have a similar function to that of herbaria in plant taxonomy.

GROUPS	Sub-classes / Orders / Classes	Biogenic Soils — Atmospheric wetting	Biogenic Soils — Periodic ground wetting	Biogenic Soils — Permanent ground wetting	Biohalogenic Soils — Atmospheric wetting	Biohalogenic Soils — Periodic ground wetting	Biohalogenic Soils — Permanent ground wetting	Biolithogenic Soils — Atmospheric wetting	Biolithogenic Soils — Periodic ground wetting	Biolithogenic Soils — Permanent ground wetting
SUB-BOREAL AND BOREAL	Class *A* Arctic tundra soils	*1A* Arctic *2A* Tundra *3A* Sub-arctic turf		*4A* Tundra swampy	—	—	*5A* Tundra solonchak-like		—	
	Class *B* Seasonally frozen taiga soils	*1B* Taiga iron enriched *2B* Taiga pale yellow	*3B* Taiga gleyed pale yellow	*4B* Seasonally frozen swampy	*5B* Seasonally frozen solods	*6B* Seasonally frozen gleyed solods				
	Class *C* Taiga forest soils	*1C* Podzolic *2C* Grey forest	*3C* Podzolic swampy *4C* Grey forest gleyed	*5C* Swampy		—	—	*6C* Turf carbonaceous	*7C* Gleyed turf carbonaceous	*8C* Lowland swampy
	Class *D* Moist forest soils	*1D* Acid forest (non-podzolic) *2D* Brown forest *3D* Prairie chernozemie	*4D* Gleyed acid forest *5D* Gleyed brown forest *6D* Gleyed prairie chernozem-like	*7D* Swampy		—		*8D* Humic carbonaceous	*9D* Gleyed humic carbonaceous	
	Class *E* Steppe soils	*1E* Chernozems *2E* Chestnut	*3E* Meadow chernozems *4E* Meadow chestnut	*5E* Meadow swampy	*6E* Steppe solonetz	*7E* Solods *8E* Meadow solonetz	*9E* Solonchaks	—		—
	Class *F* Desert soils	*1F* Brown semi-desert *2F* Gleyed brown desert	*3F* Brown meadow desert		*4F* Desert solonetz	*5F* Desert meadow solonetz *6F* "Takyr"	*7F* Desert solonchaks			
SUB-TROPICAL	Class *G* Sub-tropical moist forest soils	*1G* Yellow earths *2G* Red earths	*3G* Gleyed yellow earths *4G* Gleyed red earths	*5G* Sub-tropical swampy	—	—	—			
	Class *H* Sub-tropical dry forest and savannah soils	*1H* Buff *2H* Grey buff	*3H* Meadow buff *4H* Meadow grey buff	*5H* Sub-tropical meadow swampy	*6H* Sub-tropical solonetzic	*7H* Sub-tropical meadow solonetzic				*8H* "Smolnitzy" (Tar-like soils)
	Class *I* Sub-tropical desert soils	*1I* Grey earths *2I* Sub-tropical desert	*3I* Meadow grey				*4I* Sub-tropical solonchaks	—		—
TROPICAL	Class *J* Tropical moist forest and savannah soils	*1J* Lateritic *2J* Red of tall grass savannah	*3J* Gleyed lateritic *4J* Gleyed red	*5J* Tropical swampy	—	—	—			
	Class *K* Tropical dry forest and savannah soils	*1K* Red buff of dry forests *2K* Red brown of dry savannahs	*3K* Meadow red buff	*5K* Tropical meadow swampy		*6K* Tropical meadow solonetzic		*7K* Tropical black	*8K* Tropical gleyed black	
	Class *L* Tropical desert soils	*1L* Red brown of denudated savannahs *2L* Tropical desert	*4K* Meadow red brown				*3L* Tropical solonchaks	—	—	—

FIG. 1. Tabular representation of Rozov's (1956) scheme for the genetic grouping of soil types showing their evolutionary-genetic relationships.

(b) Approach to classification

The genetic approach is still accepted by Soviet pedologists as the only right approach to soil classification. As we have already seen, Viliams's theory of a single pedogenetic process has lately been subjected to considerable criticism. However, many Soviet pedologists still regard

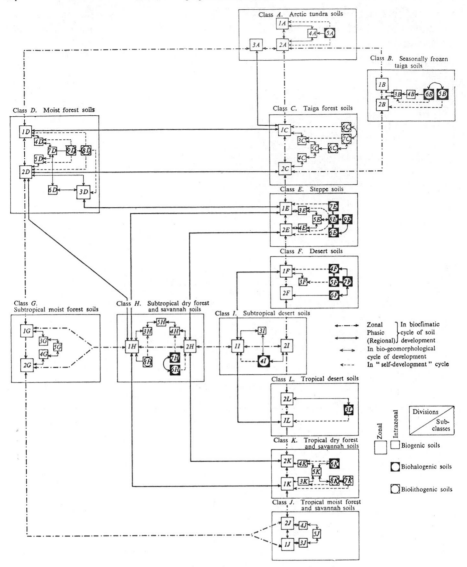

FIG. 2. Diagrammatic form of Rozov's scheme.

it as the best basis for genetic classification systems (Rozov *et al.*, 1957). Moreover, in all current attempts to devise such systems, the evolution concept plays a very important part.

The predominant significance of biological and biochemical processes in soil formation is also universally accepted, though weathering of minerals and translocation of mineral compounds currently receives more attention than in the years when Viliams dominated Russian pedology. At the same time the importance of the direct influence of vegetation on pedogenesis is unquestioned.

Within the framework of the genetic evolutionary approach, the question of whether genetic soil properties and processes or conditions of pedogenesis and pedogenetic factors should be taken as the basis of classification is still alive. New ideas are emerging, however, which envisage the possibility of a combination of all the traditional methods to form a classification system.

The system developed by Rozov (1956) and Ivanova (1956), and shown in Figs. 1 and 2, provides a good example of this tendency. From the geographical-environmental point of view, soils are divided into zonal and intrazonal, and further subdivided into types according to zonal and regional characteristics of their pedogenesis resulting primarily from bioclimatic conditions. From a factorial point of view, the soil types are vertically linked according to the main complexes of pedogenetic factors. For process classification the seventy-seven soil types are differentiated on the basis of differences in their pedogenesis and grouped on the basis of differences in salinization and the character and degree of influence of biological and weathering processes. From the evolutionary viewpoint, this scheme presents the main evolutionary-genetic connexions between soil types and permits them to be regarded as historical-genetic phases.

The same combined genetic approach is advocated by Rode (1957) in the paper mentioned above.

The subdivision of soil types which include a rather wide range of genetically and morphologically similar soils has also received increased attention in the work of Soviet pedologists. The recently standardized definition, quoted above, of sub-type, family, and form illustrates the direction which type subdivision is taking. Further subdivision of soils into variants according to the degree of change of natural profile by cultivation and erosion, and into varieties according to their mechanical composition has been proposed by Ivanova (1956). In this connexion, the growing interest in systematics of cultivated soils should also be mentioned.

REFERENCES

AFANASEV, Ya. N. 1922. Zonal soil systems. Zap. Gorets. sel. Khoz. Inst.
—— 1927. Classification problems in Russian pedology. Proc. 1st Int. Cong. Soil Sci.
DOKUCHAEV, V. V. 1879. Short historical description and critical analysis of the more important existing soil classifications. Trav. Soc. Nat., St. Petersb. **10**, 64–67 (Collected Works, vol. ii, Acad. Sci. U.S.S.R., 1950).
—— 1882. Materials for appraisal of lands of Nizhegorod Province. Collected Works, vol. iv, Acad. Sci. U.S.S.R., 1951.
—— 1893. About the question of overvaluation of lands of European and Asiatic Russia. Collected Works, vol. vi, Acad. Sci., U.S.S.R., 1951.
—— 1896. The catalogue of soil collection of Prof. V. V. Dokuchaev and his students. Collected Works, vol. vii, Acad. Sci., U.S.S.R., 1953.
GERASIMOV, I. P. 1954. Scientific bases of soil systematics and classification. Pedology, **8**, 52–64.
GLINKA, K. D. 1902. Post-tertiary formations and soils of Pskov, Novgorod and Smolensk Provinces (Classification of Soils). Annu. Geol. Miner. Russ. **4–5**.
—— 1915. Pedology. St. Petersburg, 1st ed.
—— 1922. Soil, its Properties and the Natural Laws of Distribution. Edition of N. Dereven, Moscow.

26 J. J. BASINSKI

GLINKA, K. D. 1924. Différents types d'après lesquels se forment les sols et la classification de ces derniers. Comité Int. Pédologie, IV. Commis. No. 20.

IVANOVA, E. N. 1956. An experiment in general soil classification. Pedology, No. 6, 82–102.

KOSSOVITCH, P. S. 1906. The problem of soil genesis and the bases of genetical soil classification. Russ. J. Exp. Agric. 7 (2).

—— 1910. Pedogenetic process as the basis of soil classification. Russ. J. Exp. Agric. 11 (5).

MARBUT, C. F. 1936. In 'Introduction to Pedology' by J. S. Joffe. Rutgers Univ. Press.

NEUSTRUEV, S. S. 1926. An experiment in classification of soil development processes in connection with soil genesis. Izv. Geogr. Inst. 6.

POLYNOV, B. B. 1933. Bases of constructing soil classification—preliminary thesis. Trud. Sov. Sekts. M.A.P. Kom. V. 2 (1).

PRASOLOV, L. I. 1934. The problem of soil classification and nomenclature. Trav. Inst. Dokoutchaiev, 8.

RODE, A. A. 1957. Problems of organization of work on nomenclature, systematics and classification of soils. Pedology, No. 9, 89–95.

ROZOV, N. N. 1956. Problems and principles of constructing genetic soil classification. Pedology, No. 6, 76–81.

—— 1957. Provisional results of the discussion on the problems of soil classification in the journal 'Pedology' in 1955–1956. Pedology, No. 4, 71–79.

——, KARAVAYEVA, N. A., and RODE, A. A. 1957. First plenary session of the Commission on nomenclature, systematics and classification of soils attached to the Academy of Sciences of U.S.S.R. Pedology, No. 8, 60–65.

SIBIRTSEV, N. M. 1895. The basis of genetical soil classification. Mem. Inst. Agron. Novo-Alex. 9 (7).

—— 1898. Short reviews of the main soil types of Russia. Mem. Inst. Agron. Novo-Alex. 11 (3).

—— 1900. Pedology. Collected Works, vol. i, 1951.

TSYGANOV, M. S. 1955. Fundamental principles of genetic classification and nomenclature of soils. Pedology, No. 12, 52–63.

TUMIN, G. M. 1907. Principle of classification schemes of Dokuchaev, Rospolozhenskii, Sibirtsev, Glinka, Selzkin and Kossovitch. Annu. Geol. Miner. Russ. 8 (10).

VILENSKII, D. G. 1925. The classification of soils on the bases of analogous series in soil formation. Proc. Int. Soc. Soil Sci. 1 (14), 224–41.

VILIAMS, V. R. 1939. Pedology. Selkhozgiz, Moscow.

VOLOBUEV, V. R. 1955. Some problems of genetic soil types. Pedology, No. 11, 59–69.

VYSOTSKII, G. N. 1906. Oro-climatic bases of soil classification. Pedology, Nos. 1–4, 1–18.

ZAKHAROV, S. A. 1927. Course of pedology.

Editor's Comments
on Papers 8 Through 11

UNITS OF CLASSIFICATION

One might well imagine, judging from the taxonomic complexity of current systems, that the basic units of classification were clearly delineated in early efforts. Units used to subdivide the soil continuum have been much debated, however. Papers 8 through 11, which bring together interesting and sometimes conflicting points of view as to the nature of basic soil entities, review various concepts of what basic soil units might be, what they actually represent, and how they should be applied. Examples of units that bridge the gap between the soil mantle and soil taxonomic classes and map units are of concern in what follows.

Developers of early soil classification systems did not perceive the need for a generally accepted basic unit of soil. The thing being classified was, of course, soil and that seemed obvious enough. Or was it? Dokuchaev's recognition of soil in terms of organized natural bodies, formed by complex interactions among a number of factors, was a useful conception that brought increased understanding of how soils came to be the way they are. Recent developments in soil classification have renewed interest in soil bodies, coherent segments of the soil mantle. According to Knox (Paper 9), however, the recent search for individuals or discrete units in soil classification is unduly emphasized because, as he contends on quite logically acceptable grounds, continuous universes may be divided into classes without using individuals. This observation poses the dilemma of whether units of classification should be natural or artificial ones, or perhaps more importantly whether any at all are required. Attempts to set up natural soil bodies on the basis of soil properties have been confounded by regional complexities in the landscape. Recognition of some basic volume of soil, however, seems to have certain advantages and leads to alternative use of artificial units. Thus, different observers need to know, as explained by Van Wambeke (Paper 10), exactly what they are studying (classifying), whether it constitutes parts of one and the same soil, parts of different soils, or some other part of a soil continuum. This need becomes pregnant when attempting to compare (correlate) soil units in one system or between systems for, indeed, equivalence can only be determined on the basis of a common denominator.

Unable to consistently agree on the nature of natural soil bodies due to differing points of view, soil taxonomists have created artificial units. Attempting to define some small volume of soil as a basic entity, Cline (Paper 5), by way of one example, utilized the concept of a *soil individual,* which represented "the smallest natural body that can be defined as a thing complete in itself." Proposals specifying precise limits for such small bodies of soil include concepts of tessera (Jenny, 1958), pedon (Soil Survey Staff, 1960; Simonson and Gardner, 1960), and polypedon (Simonson, 1962), the latter being introduced as a replacement for soil individual, which became confused with pedon.

Tessera, a term proposed by Jenny (1958) for small three-dimensional elements of landscape, includes both soil and vegetation. Vertical dimensions were determined by the height of the vegetation plus depth of the soil whereas lateral cross-sections, which fluctuated according to sampling purposes, were on the order of one square meter. The tessera is a basic soil entity that is similar but not identical to the pedon.

A pedon is a real, three-dimensional body of soil (as opposed to the soil profile, a two-dimensional slice through the pedosphere that lacks volume) that shows all soil horizons present and their relationships. Its lower limit is the vague interface between "soil" and "nonsoil." Lateral dimensions permit study of the range of horizon variability that occurs within a small area ranging between one and ten square meters. Where horizons are variable, intermittent, or cyclic the area of the pedon may be expanded to encompass the range of variability. Johnson, in his review of the two terms (Paper 8), concludes that the pedon concept satisfies most requirements of an ideal basic soil unit, one that is: (1) observable, measurable, and complete; (2) independent of all taxonomic systems; (3) delimited by clear natural boundaries; (4) of convenient size for study; and (5) precisely defined. The pedon is, however, so small that is cannot show configuration of the soil surface nor the range of characteristics normally allowed in a soil series. A larger body, the polypedon, proposed by Roy W. Simonson, appears to more generally meet these needs. Polypedons, parcels of contiguous pedons that fall within the defined range of a single soil series, link basic soil entities (pedons) and the soil taxonomic system. They are the objects that are actually classified. Relationships between pedons (and polypedons) and classification (and mapping) are outlined in Paper 11 where Simonson also reviews the development of basic soil entities prior to 1968.

In sum, some sort of basic soil unit seems to be required for the practical purpose of classification. Various types of artificial soil bodies have been proposed as basic entities, with varying degrees of specificity, in efforts to cope with the range of properties in the soil continuum. The concept of the pedon, for example, allows for classification of discrete soil entities based on characteristics of diagnostic horizons. This unit is basic to soil classification in the United States and Canada, and for recognition of the FAO/UNESCO soil map units. Until some means for subdividing the global array of soils into natural taxonomic units is found, segments of the soil mantle contained in an arbitrary volume may serve this function.

REFERENCES

Jenny, H., 1958, Role of the Plant Factor in the Pedogenic Functions, *Ecology* **39**:5–16.

Simonson, R. W., 1962, Soil Classification in the United States, *Science* **137**:1027–1034.

Simonson, R. W., and D. R. Gardner, 1960, Concept and Function of the Pedon, *7th Intern Congr. Soil Sci. Trans.* **4**:127–131.

Soil Survey Staff, 1960, *Soil Classification: A Comprehensive System— 7th Approximation.* U.S. Department of Agriculture, Soil Conservation Service, Washington, D.C., 265 p.

8

The Pedon and the Polypedon[1]

William M. Johnson[2]

ABSTRACT

One of the great difficulties in soil classification is that soils rarely exist as discrete individuals with clear boundaries. Instead, they grade to other soils across broad transition belts and their boundaries are determined by definition. Two new concepts, the pedon and the soil individual, have been proposed to help clarify relations between the soil continuum and soil taxonomic classes.

Pedons are real, natural soil volumes just large enough to show all the soil horizons present and their relationships. Boundaries of pedons do not depend on reference to any taxonomic scheme. A soil individual (polypedon) is also a real soil body; it is a parcel of contiguous pedons all of which have characteristics lying within the defined limits of a single soil series.

Pedons may be considered as building blocks that make up both soil taxonomic classes and soil mapping units. Most pedons are too small to exhibit all the characteristics of a soil individual; for example, usually they do not show shape of the soil nor nature of its boundaries with other soils.

Soil individuals (polypedons) are the subject of soil taxonomy; they are the real objects that are classified. They are comparable to individual pine trees, individual fish, and individual men.

[1]Presented before Div. 5, Soil Science Society of America, at St. Louis, Mo., Nov. 27, 1961. Received Aug. 17, 1962. Approved Nov. 29, 1962.

[2]Principal Soil Correlator, Soil Conservation Service, USDA, Berkeley, Calif.

[3]Pomerening, James A. A test of the differentiation of soil series within the Willamette catena. Ph.D. thesis, Oregon State University, Corvallis. 1961.

It is generally agreed that one of the primary aims of classification is to arrange objects in such an order that we can remember them and so that we can see relationships among them. In soil science we have taken the objects rather for granted. If anyone asks, "What are you trying to classify?" the answer is "Why, soils, of course." But what do we mean by soils? What is *a soil*? What are the real objects (if any) that we are trying to classify?

A simple example of the difficulty is shown in the three photographs, figures 1, 2 and 3. If one specifies that the soil is really as uniform as it appears to be in figure 1, probably most soil scientists would say that it is indeed one single soil. Figure 2, with its cyclic pattern of thickness of the dark surface horizon poses a more difficult question. How many soils are represented, one, two, or several? Figure 3 is perhaps most puzzling. Here all the soil horizons except the thin surface layer are interrupted at intervals by solid rock, and all the subsurface horizons vary greatly in thickness. Again we must try to find the answer to the question: one soil, or two, or several?

To classify soils, I think, one must agree that some sort of *units* exist within the soil continuum, or at least, that we can mark off some arbitrary units for classification. It would be convenient if natural soil units with clear boundaries existed, and perhaps they do, if only we knew how to recognize them[3]—but even if they don't, the idea of primary units seems absolutely necessary. We have, furthermore, a need for some device to relate soil mapping units to classes of our taxonomic system, and a basic soil entity would serve such a purpose. Finally, a concept of *units of soil* is helpful in carrying out investigations, including sampling for research.

Earlier Concepts of Soil Units

Simonson and Gardner pointed out (4) that at one time map units consisting of a single soil type were considered to be "homogeneous in all observable features" and

Figure 1—A Brown Grumusol in Yavapai County, Arizona.

Figure 2—Ft. Garry clay near Manitoba, Canada. Scale in feet.

Figure 3—A Reddish-Brown soil on basalt in Washington County, Utah.

that the entities so delineated were the "units" that were grouped to form the lowest taxonomic classes. Today we know that soil map entities, even those delineated at large scales, are not completely homogeneous. Almost invariably each area of soil shown as a single delineation actually includes some small bodies of different kinds of soil—the so-called mapping inclusions. To consider map entities, then, as the units of the lowest taxonomic classes requires that the taxonomist set up an enormous number of taxonomic classes, or that he pretend that the mapping inclusions don't exist.

Robinson (3) used the soil profile as the basic soil entity. More recently, W. N. Townsend (1) wrote: "Soil as represented by the soil profile is a natural object; and while no two individual profiles are precisely alike, groupings of profiles are possible to enable a more facile interpretation of their properties and a better appreciation of causal relationships." If we follow this line of thinking, we are immediately beset by two practical difficulties. First, a soil profile is two-dimensional, and therefore cannot be sampled; in fact, structural units (peds) cannot be measured and completely described. Second, a soil profile cannot exhibit the shape (surface configuration) that may be characteristic of the soil.

In the Soviet soil classification system, "The basic unit of classification is the genetic type of soil distinguished on the basis of conjugate studies of the properties of the soils, the processes which create them and the factors which determine them." (2) This statement is made in spite of the fact that the "soil type" in the Soviet system is not the lowest category. A taxonomic class, whether it be series, great soil group, genetic soil type, or order, is not a real object but simply an idea. It is, in other words, an abstraction reached by synthesizing a "model" or mental concept from a number of field and laboratory data. To designate such abstractions as basic soil entities does not help to solve the classification problem at all, because it does not provide any way of relating actual soil bodies to classes in the taxonomic system. In fact, it completely evades the questions "What is the basic soil unit or individual? How large, or how small, can it be?"

A Basic Soil Unit

Inasmuch as previous attempts to use soil map entities, soil profiles and soil taxonomic units as the basic units of classification have revealed weaknesses in each, it may be helpful to try to outline the general features of a better unit. Such a unit should probably satisfy the following requirements:

1. It should be a real object that is observable and measurable in three dimensions and it should include the whole vertical thickness of the soil.
2. It should be independent of all taxonomic systems.
3. It should have clear, natural boundaries.
4. It should be of a size convenient for study, measurement, and sampling.
5. It should be susceptible of reasonably precise definition, so that pedologists everywhere may use it consistently.

We have already noted that the soil mantle is a continuum, and it is well known that different kinds of soil rarely have clear boundaries, or, at any rate, they rarely have boundaries that are clear to present-day pedologists.

There is some evidence that some natural boundaries do exist, in the sense of significant differences in selected properties or spectra of properties. For the most part, though, it seems that we are insufficiently prepared to define a basic soil unit in terms of its natural boundaries (except, of course, in those instances where the boundary is one with "not-soil"). Apparently, then, some degree of arbitrariness will characterize the definition that we can write now of the boundaries of our basic soil unit.

Since we are, for the most part, unsuccessful in specifying the natural boundaries of the unit, perhaps we can find some criteria that will enable us to specify the minimum and maximum size limits, in terms of diameter, area of the surface, or volume. In some discussions several years ago it was suggested that the basic soil unit might be that volume of soil occupied by the root system of one plant. Such a definition of extent has some advantages, certainly, particularly in soil morphological and genetic studies. It does have some difficulties, though. What kind of plant shall be chosen? In a mixed vegetation population, one may have to choose among carrots, with a very limited root spread, grasses, shrubs, and trees, with increasingly extensive root areas. And, if the native vegetation has been cleared, and cultivated crops are being grown, then which crop plant shall we select, the sugar beet or potato, corn, or alfalfa? Pleasing as the prospect may be in theory, it just doesn't seem practical to define our basic soil unit in terms of volume occupied by plant roots.

Others have suggested that the soil unit should be just large enough to permit observation of the largest ped, or, alternatively, a minimum of five adjacent peds. Such a definition, of course, would be meaningless in structureless soils.

If one considers the objective of recognizing basic soil units, though, some criteria do become apparent. We want our unit to be thick enough to include the whole profile, down to the boundary with "not-soil." And we want it to be large enough laterally to permit observation, study, measurement, and sampling of all the horizons; not just the horizons as they appear in a two-dimensional profile 2 or 3 feet wide, but as they vary in three dimensions over considerable distances. As long as the horizons are relatively uniform in character, thickness, and boundaries no difficult problem exists (figure 1). If the horizons are quite variable or are interrupted, though, one is faced with the problem of deciding whether one soil, or more than one soil, is represented along any given exposure (figure 3). It will not do to say arbitrarily that two soils are present if a critical property or horizon is present in some spots and absent in others. Some arbitrary limit of size must be set. Otherwise, a single loose, large stone or boulder that interrupts an horizon forces us to recognize two different kinds of soil; such a problem would arise constantly when one works in an area mantled by glacial till. Similarly, a vertical hole made by a burrowing animal, or by decay of a tree root, becomes "not-soil"; then, if the hole becomes filled with earth, it would have to be recognized as a different kind of soil, even though its lateral dimensions did not exceed 1 inch.

As pointed out in the *7th Approximation* (5), to take the opposing view that areal limits of the soil unit cannot be set leads to conclusions that are not only ridiculous, but, more important, prevent rational classification of soils. For example, if no areal limits are established, one is perfectly justified in considering clay coatings on prisms as one kind of soil, the ped cores as another. To follow this line of thinking to its ultimate conclusion, each ultimate soil particle (sand, silt, clay) could be considered as a different kind of soil, since no two grains are exactly alike. The most reasonable alternative to these untenable concepts, at least for the time being, is to establish arbitrary size limits for the basic soil unit.

Returning for a moment to the photograph of a rather uniform soil mantle (figure 1), the size of the basic soil unit here is not critical, beyond a certain minimum volume. That is, one arbitrary size would do about as well as another for purposes of observation and classification. For sampling purposes, though, it is important that the unit be large enough to permit taking the necessary volume of soil from even the thinnest horizon. If one assumes that a 50-pound sample is large enough, and that the thinnest horizon is about 1 inch thick, the soil unit must have an area of at least 6½ square feet.

Looking again at the photograph of nonuniform soil horizons (figure 3), we are faced here with the dual problems of minimum and maximum areal limits, The requirement of at least 6½ square feet of area still holds, at least for sampling purposes. What about the maximum size? Several possibilities suggest themselves: *(a)* an area just large enough to include a half-cycle from a soil profile with minimum allowable thickness of horizon(s) to a profile of the same series with maximum allowable thicknesses of horizons; *(b)* an area which includes a full-cycle continuum between two soil profiles with minimum allowable thicknesses of horizon(s); *(c)* an area more arbitrarily determined. We are faced with a dilemma. Soil volumes up to some arbitrary dimensions do not truly represent the soil, as statistical analyses of critical properties show (pronouncedly skewed distribution, say of horizon thickness; or a monstrosity so far as frequency distribution is concerned). On the one hand, to allow natural variability to define the limits of the soil unit would result in some tiny units and some enormous ones; to determine the limits might well require excavating a trench several miles long in some soils. On the other hand, establishment of arbitrary limits creates some rather artificial units, units that will be displeasing to someone. No one has yet provided a completely satisfactory solution to this problem; probably none exists at this moment.

As an approximation to the concept of an ideal basic soil unit, the *pedon*[4] has been proposed (6). Let us now examine this concept to see how well it satisfied the requirements mentioned previously.

The Pedon

Paraphrasing the *7th Approximation*, the pedon may be defined as follows: a small volume of soil that extends downward to the lower limit of common rooting of the dominant native perennial plants, or the lower limit of the genetic horizons, whichever is the deeper; roughly hexagonal in cross-section, with a surface area of 1 to 10 sq. m.; with a minimum lateral dimension of about 1 m. and a maximum lateral dimension of about 3½ m., depending on variability in the horizons. If the horizons are intermittent or cyclic and recur at intervals of 2 to 7 m., the pedon has a diameter that is one-half the length of the cycle. If the cycle is of wave length < 2 m., or if the horizons are continuous and of uniform thickness, the diameter of the pedon is 1 m.

This definition seems to be "reasonably precise"—thus fulfilling one of the requirements. . . . granted that the specification of lower limit is not so exact as one might wish. It is intended to correspond with the soil- "not-soil" interface. Determination of this interface is easy enough in many instances, but in some soils it has been and probably always will be difficult. The lateral limits are specified exactly enough. The questions that will arise in practice are principally these two:

1. At what degree of variability in horizons does one shift from the 1-m. limit to the ½-cycle limit?

---

[4]Term proposed by Guy D. Smith. The word is pronounced with accent on the first syllable, and it rhymes with "head-on."

2. In the instance of continuous, uniform horizons, how does one determine the locations of the boundaries of one pedon?

There are no readily available quantitative answers to the first question. Of course, intermittent, cyclic horizons constitute variability requiring the use of the ½-cycle limit. In fact, any kind of observable variation that is cyclic and of certain wave lengths requires the use of the ½-cycle limit. In practice, I expect, such variations are likely to be fairly obvious; minute differences in thickness and other features will be ignored because of the practical difficulty of determining cyclicity. It is important to remember that the taxonomist is not privileged to say that this or that variation, though cyclic, is insignificant; neither may he evaluate variations in terms of taxonomic class criteria. If the variation is observable, and cyclical, within limits, it constitutes a requirement for use of the ½-cycle limit.

To locate pedon boundaries within a body of uniform soil is a relatively easy matter. First, the boundaries of pedons with "not-soil" (rock, water, etc.) are clear, Second, the boundaries of pedons with unlike pedons, while not always so obvious, can be observed in exposures. The third case, of boundaries among very similar pedons, involves this situation: there is no way of determining where to place the boundaries of the first pedon identified. If each pedon had natural boundaries, i.e., if it were a discrete natural body like a tree, there would be no problem. A consequence of arbitrary size limits, though, is that one has only the meter stick for locating boundaries within a group of contiguous like pedons. To some this will seem to be a weakness in the definition. It means that two independent observers would, in all likelihood, draw the boundaries differently. This does not detract from the usefulness of the pedon concept, though. Since these pedons are all so similar, it just doesn't matter where the boundaries are placed.

It is necessary, though, to call attention to two limitations in the pedon concept that affect its relationship to the taxonomic system and to soil mapping. The pedon is so small that in the majority of cases it cannot show the shape of the soil, that is, the configuration of the soil surface. Neither can a single pedon exhibit the range in characteristics likely to be allowed in a soil series. Some larger soil body must be taken to satisfy these needs. The *polypedon*[5] (plural, polypedons) has been proposed as such a soil body.

The Polypedon

The polypedon provides the essential link between basic soil entities (pedons) and the soil taxonomic system. It is, moreover, a convenient device for relating taxonomic units and soil map entities.

The polypedon is defined as one or more contiguous pedons, all falling within the defined range of a single soil series. It is a real, physical soil body, limited by "not-soil" or by pedons of unlike character in respect to criteria used to define series. Its minimum size is the same as the minimum size of one pedon, 1 sq. m.; it has no prescribed maximum area. Its boundaries with other polypedons are determined more or less exactly by definition. Within a given polypedon, the individual pedons may vary slightly or a great deal. For example, if the soil series is allowed to range in slope gradient from 0 to 25%, a given poly-

pedon conceivably could exhibit this same range in slope gradient; if the series definition permits a range in depth to bedrock from 20 to 48 inches, a single polypedon might exhibit the same range.

Polypedons are the subject of soil taxonomy in the comprehensive system; that is, they are the real objects that are placed in classes of the lowest category of the system. They are comparable to individual pine trees, individual fish, and individual men.

Polypedons are intimately related to the entities delineated on soil maps, but in most instances they are not identical. Inasmuch as polypedons are real soil bodies, it is possible to delineate them on maps at some scale (the scale would have to be very large in some places). Several kinds of differences between polypedons and soil map entities are evident:

1. Most soil map units are "impure"—that is, they include some pedons and polypedons of unlike soils (the so-called "mapping inclusions"). Even if the external boundaries of the map unit and the polypedon coincide, the polypedon ordinarily has "holes" in it—the pedons and polypedons of unlike soils.

2. Large polypedons often are subdivided (into types and phases) on soil maps in order to create map units that are more homogeneous with respect to characteristics and qualities important to use and management of the soils.

3. Also for pragmatic reasons, two or more major polypedons may be combined on soil maps and designated as a soil complex or association.

Summary

Concepts of two kinds of soil entities have been proposed to help clarify relations between the soil continuum, soil taxonomic classes, and soil map units.

Pedons are real, three-dimensional bodies of soil just large enough to show all the soil horizons present and their relationships. The pedon satisfies most, but not all, of the requirements of an "ideal" basic soil unit.

Polypedons are real soil bodies; they are parcels of contiguous pedons all of which have characteristics lying within the defined limits of a single soil series. They are the real objects that we want to classify into series and higher categories.

If all polypedons consisted of soils having uniform and continuous horizons, there would be little advantage (except perhaps for detailed studies and for sampling) in retaining the concepts of both pedon and polypedon. The polypedon would suffice as the basic entity. The pedon concept serves as an essential device for determining the size of "a soil" wherever soil horizons are variable or intermittent or both. Without this concept, as we have seen, rational classification of such soils is quite impossible.

LITERATURE CITED

1. Comber, N. M. Introduction to the scientific study of the soil. Rev. W. N. Townsend. Edward Arnold & Co., London, Ed. 4. 1960.
2. Ivanova, E. N. and Rozov, N. N. Classification of soils and the soil map of the USSR. Trans. Intern. Congr. Soil Sci., 7th Congr. Madison. 4:77-86. 1960.
3. Robinson, G. W. Mother earth. Thomas Murby & Co., London. 1937
4. Simonson, Roy W. and Gardner, D. R. Concept and functions of the pedon. Trans. Intern. Congr. Soil Sci. 7th Congr. Madison. 4:127-131. 1960.
5. Soil Survey Staff, USDA. Soil classification, a comprehensive system, 7th Approximation. U. S. Govt. Printing Office, Washington. 1960.

[5]Roy W. Simonson proposed the term. In the *7th Approximation* the polypedon is described under the heading, "The soil individual."

9

Soil Individuals and Soil Classification[1]

Ellis G. Knox[2]

ABSTRACT

The standing of 8 kinds of soil bodies as soil individuals was evaluated in the light of the following principles, derived from an analysis of basic ideas: 1) Classes are abstract fields, not groups of individuals. 2) Artificial individuals of continuous universes are different from natural individuals of particulate universes. 3) Classes in continuous universes are independent of individuals. 4) Artificial individuals are member-bodies of minimum size. Soil particles, soil horizons, soil landscape units (including polypedons), and delineated soil bodies are natural individuals, but only within nonsoil universes. The universes of soil resulting from any of the common concepts of soil as an isotropic material or an anisotropic body are continuous, and contain hand specimens, soil profiles taken as bodies, "individuals", and pedons as artificial individuals. These may be regarded as soil individuals. There is need for a concept of soil corresponding to some kind of soil landscape unit.

T HE NATURE of individuals is of primary interest in the construction and use of any classification system because individuals, classes, and the universe treated by the system are inherently interrelated. Because individuals are members of classes, and because classes are subdivisions of the universe that is under consideration, individuals, classes, and the universe must be mutually compatible. Accordingly, any confusion in the understanding of individuals extends to the whole classification and is a weak link in the use and further development of the system.

Differences about the nature of soil individuals are apparent (4, 5, 8, 15, 19, 20, 21, 26). These differences suggest that a better basis for understanding the nature of soil individuals is needed. It is the purpose of this paper to clarify the idea of soil individual by means of an analysis of the nature and relationships of individuals, classes, and universes. In addition, 8 kinds of soil bodies are examined as possible soil individuals.

ANALYSIS

The analysis is presented in the following six definitions and in the discussion of their implications. The first three definitions are merely restatements of accepted ideas. The last three introduce new ideas. Taken together they present an improved conceptual model for understanding soil individuals.

Individual

An individual is "the smallest natural body that can be defined as a thing complete in itself . . ." (4). Individuals that are of interest in classification are members of classes.

[1] Technical Paper No. 1814, Oregon Agr. Exp. Sta., Oregon State University, Corvallis. Presented before Div. S-5, Soil Science Society of America, Nov. 17–21, 1964, Denver, Colo. Received May 8, 1964. Accepted Sept. 16, 1964.
[2] Associate Professor of Soils.

Class

A class is the range of applicability of a class concept (12, 24). It is an abstract field created by a class concept. Class concepts, which express the basis for membership in a class, consist of one or more "differentiating characteristics" (4). A differentiating characteristic may be a simple qualitative or quantitative characteristic or it may be complex. Similarity to a norm, type specimen, or "modal individual" (4) and inclusion within a frequency distribution cluster (23) are examples of complex differentiating characteristics. In any case, a class contains everything that meets the requirements expressed in the class concept. It is the class concept that forms the class and determines its membership.

This definition is consistent with developments in logic and mathematics beginning more than 100 years ago. However, it is opposed to the idea, still common in science, that a class is a group of individuals (4, 14, 27). It implies that a class is quite different from the aggregate of its members and that classes with zero membership are empty classes but classes nevertheless. The distinction between a class and its membership is seldom noticed because for most purposes it may be neglected without serious consequences. Nevertheless, there is a real and important difference (12, 24). For example, the class, Astoria series, was created in 1938 and may be destroyed by soil taxonomists. The membership of the class, Astoria soil, has been in existence for hundreds or thousands of years and is more susceptible to men with bulldozers. It follows that the membership of a class does not necessarily consist of individuals. The apparent need for individuals is an artifact resulting from a limiting definition of class.

Universe

A universe is a superclass that contains all the objects and includes all the other classes under consideration (24). A particulate universe, that is, one that contains discrete objects that can be counted, is very much like a population (4). On the other hand, the membership of a continuous universe cannot be numbered readily. For quantitative treatment, arbitrary units of measurement must be applied. Likewise, individuals must be created arbitrarily if a continuous universe is to be treated like a population. A concept of universe is important in classification, because, without it, there is a strong possibility for confusion resulting from unrecognized shifts between different but related universes.

Member-Body

A member-body is defined here to be a body (in a physical universe) that qualifies for membership in a class. The existence of classes does not depend on the existence of member-bodies, although every class that is not empty presents the opportunity for selection of at least one mem-

112

ber-body. Within particulate universes, individuals and member-bodies are identical. They are fixed and independent of the observer's activities. On the other hand, within continuous universes, individuals of interest in classification are member-bodies, but not all member-bodies are individuals. Member-bodies within a continuous universe may be selected arbitrarily. For this reason, and because there is no requirement that they be mutually exclusive, they are infinite in number.

The size of member-bodies within continuous universes may be selected arbitrarily, but only within certain limits. A member-body must be big enough to exhibit all of the components, characteristics, and relationships required for membership in its class. Similarly, a member-body can be no larger than satisfaction of the class concept allows. The maximum limit is set by the size of the largest exemplification of the class in question. Only as much of the universe as is known to satisfy the class concept can be regarded properly as a member-body. However, a larger body can be considered hypothetically. If the uniformity of a body cannot be determined quickly and easily, or if the physical boundary is obscure between adjacent member-bodies of mutually exclusive classes, the location of member-bodies may require successive trials.

Natural Individual

A natural individual is an individual that is discrete and independent of the observer. That is to say, it is a member-body within a particulate universe. Most of the individuals that come to our attention are of this kind. Almost all of the writing about individuals with respect to classes and classification systems has dealt with natural individuals.

Artificial Individual

An artificial individual is a human construct within a continuous universe. It is an individual that, in the absence of natural individuals, is created arbitrarily and for convenience. An artificial individual of interest in classification must be a member-body with respect to some class within the universe under consideration. Otherwise it would not be "complete in itself" (4). Moreover, to be "the smallest natural body" (4), an artificial individual must be a member-body of minimum size.

The preceding analysis may be reduced to the following four principles.

1. Classes are not groups of individuals; they are abstract fields produced by class concepts.
2. In particulate universes, natural individuals constitute the membership of classes. In continuous universes, there are no individuals except as artificial individuals are selected arbitrarily subsequent to the formation of classes.
3. Classes in continuous universes may be formed and used without regard to individuals.
4. Artificial individuals are member-bodies of minimum size.

SOIL BODIES

Soil can be considered from many different points of view. Many universes and many kinds of member-bodies are involved. Eight prominent kinds of soil bodies[3] are

[3] Recent, restrictive definitions of soil body (25, 15) are not followed here. A soil body is simply any body of soil.

considered below with respect to concepts of soil, soil classification systems, and their qualifications as soil individuals.

Primary Particles

The primary particles of soil, such as crystals, crystal fragments, aggregates of crystals, and discrete grains of amorphous material (glass, opal, etc.), may be considered to be individuals (6). These particles have the characteristics of size, shape, mineralogy, and so on. They are natural individuals within the universe of soil particles. The soil separates (22), sand, silt, and clay, and coarse fragments constitute one of many possible classification systems within this universe. Because primary particles exhibit almost none of the characteristics attributed to soil, it is clear that particle size classification is not soil classification and that the universe of soil particles is quite different from any universe of soil.

Hand Specimens

A hand specimen of soil is taken to be a body small enough so that many characteristics can be considered in bulk, and large enough (but no larger than necessary) for the determination of these characteristics. It is of a size to be held conveniently in the hands. A "sampling unit" (6) or a soil sample is a hand specimen taken for laboratory work. A hand specimen may include repeating macroscopic patterns of spatial distribution (such as color mottles and coatings on sand grains), but progressive changes with distance through the specimen are absent or small. Characteristics that may be determined from a hand specimen include 1) The common field morphological characteristics such as color, texture, structure (exclusive of very large peds), consistence, porosity, nature of ped surfaces, content of the smaller coarse fragments, etc. (22), 2) results of the common microbiological, physical, chemical, and mineralogical determinations performed in soil laboratories, 3) micromorphology, 4) fertility relationships as determined by pot experiments with small plants, and 5) mechanical characteristics of interest in engineering. The volume required ranges from about 10 to about 8,000 cm^3 (roughly from a 1- to 8-inch cube). Commonly only the last two items require more than about 1,000 cm^3 (roughly a 4-inch cube). The larger dimensions may also be required by a large, internal, repeating spatial pattern. (Patterns larger than the maximum dimensions cannot be comprehended in a hand specimen. Thus, material with large coarse fragments or large peds cannot be represented by a single hand specimen.) Some of these determinations involve the original, internal, macroscopic, spatial relationships. Others do not. Accordingly, mechanical disturbance destroys the specimen with respect to some determinations, but is inconsequential with respect to others. For some special purposes (criminal investigations, for instance), a hand specimen may be very small, even microscopic.

A hand specimen in itself reveals no profile relationships. Therefore, it can be a member-body only within a universe resulting from a definition of soil as a simple material—as an isotropic body. Soil, as a natural medium for plant growth or as a body developed in place, is not isotropic. A hand specimen, then, cannot be considered to be an individual with respect to any of the pedological or ecological (including agricultural) classifications of soil.

On the other hand, soil, as an engineering material is isotropic. Hand specimens exhibit the characteristics used to differentiate the classes of engineering classifications (16). They are the smallest bodies adequate for the critical determinations, and they may be selected arbitrarily. Therefore, hand specimens are artificial individuals with respect to engineering soil classification systems. Similarly, they are artificial individuals with respect to any universe of soil as an isotropic material.

Hand specimens, of course, are useful as samples of larger bodies. Moreover, there is a universe of those hand specimens that have been removed from the soil and stored as soil samples. The universe of soil samples is quite different from any universe of soil and is trivial in comparison. Unfortunately, it cannot be taken lightly because it is the universe within which too many soil scientists operate.

Soil Horizons

A soil horizon, as a body, is very thin relative to its lateral dimensions. Therefore, its upper and lower boundaries are very extensive and well-known (although from only a small sample) relative to its lateral boundary. Significant horizons have upper and lower boundaries that are determined objectively. They may be placed where one or more characteristics go through a maximum in rate of change with depth, or they may be fixed by other boundary criteria.

Soil horizons have all the characteristics of hand specimens, plus thickness, position relative to other horizons, and the characteristics of the upper and lower boundaries. Because the lateral boundaries are so little known, their characteristics and such characteristics as total volume, total mass, and geographic shape are generally ignored.

Soil horizons tend to be sufficiently uniform to be member-bodies with respect to the classes within the universe of soil taken as an isotropic material. Hand specimens are the smallest of such member-bodies; soil horizons are the largest. Obviously, then, soil horizons cannot be considered to be individuals within this universe.

On the other hand, any given set of criteria for upper and lower boundaries corresponds to a universe of soil horizons, and every soil horizon is a natural individual within some such universe. However, no universe of soil horizons is equivalent to any universe of soil. This discussion about soil horizons may seem strange because the various universes of soil horizons are almost never considered. Indeed, most workers seldom think of horizons as bodies. Instead, we think of them in small transects as components of the soil profile, or as components of some larger unit. Interest in soil horizons as individuals is limited because of their size and shape, and because, like hand specimens, they reveal little about profile relationships.

Soil Profiles

A soil profile is a device for dealing with soil as an anisotropic (7) body. It extends vertically and completely through the soil and comprehends the changes in characteristics with depth. This vertical pattern of characteristics can be considered as a combination of the characteristics of component horizons and geologic layers exclusive of lateral variations. Alternatively, it may be considered as a combination of the depth functions of the various characteristics of hand specimens. Small-scale lateral variation is included in some cases to comprehend the topography of horizon boundaries, stoniness, large peds, discontinuous horizons, etc. Nevertheless, lateral changes within the profile are always secondary to vertical changes. Since genetic interrelationship of soil horizons was first postulated, it has been clear that nothing less than a soil profile can be taken as an individual of pedologic classification. Similarly, soil-plant relationships require consideration of the soil profile in any classification of soil pertinent to plant growth.

A soil profile is either the vertical pattern of characteristics (an abstract sequence) or the concrete, three-dimensional body that exhibits the pattern (15). (In addition, a profile can be considered to be nothing more than a vertical exposure of soil. This may be an appropriate concept, but it is one without implications for the present discussion.) As a body, a soil profile has lateral dimensions equal at least to the dimensions of hand specimens. As an abstract sequence, of course, it has no horizontal extension. These concepts of soil profile are both useful, and either one may be used by itself without confusion. Confusion does result from attempts to combine the two concepts.

Soil profiles taken as abstract sequences cannot be individuals within any universe of soil taken as a body. On the other hand, abstract soil profiles are individuals within the universe of soil profiles (9). Soil classification divorced from soil mapping may tend to become classification of abstract profiles rather than the classification of soil as a body. Indeed, some well known pedological classification systems may be interpreted as classification of abstract profiles. Nevertheless, it is clear that such classification has no meaning with respect to a corporeal universe.

Soil profiles taken as bodies are essentially the same as the smallest of the bodies considered in the next section.

"Individuals" and Pedons

Cline (4) described the "ultimate individual of the soil population" as a three-dimensional body with the same thickness as the soil profile and with horizontal dimensions large enough (but, by implication, no larger than necessary) for observation and sampling. Presumably, the necessary lateral extension varies with the characteristic to be determined. Successively larger "individuals" are needed for observation of 1) the vertical sequence of the characteristics of hand specimens, 2) small-scale lateral variations (stoniness, tonguing horizons, etc.) which are too large to be comprehended by hand specimens, and 3) larger scale features (slope, root relationships, etc.). The smallest "individuals" are identical with soil profiles taken as bodies. Larger "individuals" have all of the characteristics of profiles plus variation among profiles.

"Individuals" (4) may be chosen arbitrarily and they are infinite in number. They are not natural individuals. However, their characteristics include essentially all of the differentiating characteristics used in some common soil classification systems (2, 3, 11, 13, 17, 21, 25). They are member-bodies and, because of their minimum size, artificial individuals with respect to these systems. In so far as soil is considered to be that which is classified by these systems, "individuals" (4) are artificial individuals of soil. Variation in differentiating characteristics among classification systems and among the categories of one system lead to variations in the minimum size of member-bodies. For example, a smaller "individual" can be used in series (22) classification than is necessary for slope phase (22) classification.

A pedon is "the smallest volume that can be called 'a soil' " (21). Pedons are "individuals" (4) within an arbitrary range in size. The minimum size limit provides for the routine observation of the features and patterns of small-scale lateral variation. By providing for the comprehension of this variation, it eliminates the possibility of having very close-spaced boundaries between different kinds of soil. The maximum limit insures that a pedon is no larger than necesary to comprehend any small-scale pattern of lateral variation. (In fact, it could be argued that a full cycle of variation rather than a half cycle should be included.) By keeping a pedon internal with respect to geographic patterns of variation, the maximum limit allows for detailed geographic differentiation and high intensity mapping. Pedons are the smallest member-bodies and the artificial individuals with respect to classification systems that used small-scale lateral variations as differentiating characteristics or which are otherwise restricted to bodies no smaller than pedons. They are smaller than the bodies required for ordinary measurements of slope and the determination of other larger scale features. Therefore, they are not member-bodies with respect to classes that use these features as differentiating characteristics. The minimum size of a pedon may seem large to those who work with soil that does not exhibit much small-scale lateral variation. However, it should be recognized that the absence of this feature is an important characteristic in itself and cannot be observed in a smaller body.

The mutual exclusiveness of "individuals" and of pedons has been assumed by other writers (4, 8) without comment. This assumption is probably derived from the fact that close observation of a pedon destroys it and from the concept of a class as a group of individuals—a concept that is not adequate for the purposes of this discussion. The definitions of "individual" and of pedon do not require that these bodies be mutually exclusive. The arbitrary selection of "individuals" and pedons may be extended to the selection of overlapping bodies without known disadvantage. Accordingly, "individuals" and pedons are infinite in number.

Soil Landscape Units

A *soil landscape unit* is an objective (not arbitrary), geographic body of soil. It is a spatial aggregate of "individuals" (4) and pedons with thickness equal, of course, to the depth of the constituent soil profiles. Lateral boundaries are determined by the geographic pattern of change in soil characteristics according to objective boundary criteria. That is, within the scope of this discussion, soil landscape units may not have arbitrary boundaries. To be sure, the boundary criteria may be chosen arbitrarily, but with reference to any given criterion, lateral boundaries are fixed by the soil. Similarly, with respect to any given criterion, soil landscape units are mutually exclusive. Soil landscape units have all of the characteristics of pedons and "individuals", geographic shape, boundary relationships, and so on. Area is a much more important characteristic than it is for pedons and "individuals".

Without regard to the particular subject matter of this paper, the most nearly universal boundary criterion is discontinuity. It is used in preference to all other criteria seemingly in every case in which it is discernible and present at a convenient scale. In most cases we do not even consider the alternatives. "Individual pine trees, individual fish, and individual men" (8) are recognized by this criterion. The overriding appeal of discontinuity as a boundary criterion was assumed in the first part of this paper in discussion about particulate universes and natural individuals. It should be recognized that discontinuity is a relative feature. It is simply the extreme degree of concentration of change into maxima in rate of change with distance (relative to bodies) or with time (relative to events).

Sharp lateral discontinuity is present in soil landscapes (1, 18), but it is rare. Very few soil landscape units can be determined according to this criterion. However, less extreme maxima in the lateral rate of change in soil characteristics are common. Where these form closed figures they can be used as boundaries for soil landscape units. Where such maxima are correlated with surface features they provide a basis for accurate and efficient soil mapping. Soil landscape units with boundaries determined by the criterion of maximum lateral rate of change are not necessarily homogeneous. For example, one unit may include the complete range of a drainage catena. Accordingly, such units are not in general member-bodies with respect to any common classification system. However, they are natural individuals within the universe of those soil landscape units that have boundaries of maximum lateral rate of change. This universe is seldom used and within the knowledge of the writer no classification system has been erected within it.

Alternatively, class concepts may be used as boundary criteria. Muir's "soil area" (15) is a soil landscape unit formed in this way. Soil landscape units determined according to soil series concepts have been called "soil individuals" (21) and polypedons (8). The first name offers so much chance for confusion that only the second will be used here. By definition, polypedons are member-bodies—the largest possible member-bodies—with respect to soil series. They are also member-bodies with respect to classes of higher categories. They may be classified into soil series and classes of higher categories, although, contrary to other opinions (5, 8, 21) such placement is a mere exercise after the fact rather than a significant placement in one of several alternative classes. Their placement in some one, particular, known soil series is prerequisite to our consideration of them. By the time that we have a polypedon to classify, we necessarily already know the classification. Accordingly, polypedons seem to be of little significance or utility with respect to the placement of soil into classes.

On the other hand, polypedons are of extreme importance in the creation of soil series. In large part, soil series are judged by the polypedons that they produce. Desirable characteristics of polypedons include 1) identity with or close correspondence to soil landscape units determined according to the boundary criterion of maximum lateral rate of change, 2) large size, 3) relatively low perimeter to area ratio, 4) correlation with easily observed surface features, and 5) homogeneity with respect to genetic factors and use interpretations. We do not insist that polypedons be mappable at the map scales commonly used in soil survey work, but series that produce few or no mappable polypedons must be very strongly justified in some other way. Series concepts that produce unsatisfactory polypedons are altered or discarded. The importance of the polypedon in the trial and error process of creating a significant series classification cannot be overemphasized. However, as indicated above, the polypedon is of essentially no significance in the classification of soil—the placement of soil into classes.

Polypedons are natural individuals within the universe of polypedons. This universe is completely dependent on soil series concepts. Because polypedons have no existence apart from series classification, their significance as individuals seems less than the significance of "individual pine trees, individual fish, and individual men." No formal classification system has been constructed within the universe of polypedons, although there are probably many poorly defined, personal classifications based on such characteristics as size (area), shape, and mappability.

The absence of formal classification systems that take soil landscape units as individuals should not be interpreted to mean that such systems are impossible. It may be possible to develop a concept of soil that corresponds to some universe of soil landscape units. A classification system built within such a universe would have many advantages, chiefly those suggested by Cline in his discussion of the genetic bases of classes (5). Extremely complex problems can be anticipated, because soil landscape units have many more characteristics than the bodies which are classified by our present systems.

Delineated Soil Bodies

A delineated soil body is a geographic body of soil that corresponds to a delineation on a soil map. It has lateral boundaries as represented on the map. In landscapes with many maxima in lateral rate of change, delineated soil bodies tend to approximate soil landscape units determined by the boundary criterion of maximum lateral rate of change. In other landscapes, delineated soil bodies tend to approximate polypedons or parts of polypedons. Almost all delineated soil bodies include parts of more than one polypedon. Accordingly, delineated soil bodies have the characteristics of their constituent polypedons and partial pedons, plus the characteristics of relative proportions of different kinds of polypedons and interrelationships of polypedons.

Even though a delineated soil body has no existence apart from its parent soil map, it may be considered to be a natural individual within the universe of delineated soil bodies created by that map. With respect to one map, delineated soil bodies are mutually exclusive and finite in number. A map unit (22) may be considered to be a class that contains delineated soil bodies as members. Emphasis on the distinction between classification units and map units obscures the fact that both are classes, but classes within entirely different universes.

The land-capability classification system (10) classifies soil map units. Delineated soil bodies, then, are natural individuals with respect to this classification system. Accordingly, the universe of the land-capability classification is quite different from the universe of the common soil classification systems.

The nature of delineated soil bodies may be determined in part by the system of soil classification and the selected level of map complexity, but only within limits set by the pattern of soils in the landscape. The influence of a soil classification system can be great, to the point that in the United States it seems appropriate to provide names for map units from the classification system through the process of soil correlation (19). The overriding influence of the landscape is indicated by the necessity for such devices as complexes, associations, undifferentiated units, and land types (22). Homogeneity of map units can be increased by increasing the complexity of the boundaries between delineated soil bodies. Commonly this means reducing the size and increasing the number of delineated soil bodies. However, the particular relationship between map complexity and map unit homogeneity in a given survey area is fixed by the soil pattern itself.

Soil Types

Marbut (13) and Whiteside (26) considered soil types to be the individuals dealt with in soil classification. A soil type, in this context, is not a class but is an aggregate of bodies each somewhat analogous to a polypedon. (The distinction between map units and classification units was not drawn sharply in Marbut's time.) Marbut wrote that creation of soil units (soil types) was a step prior to classification. Nevertheless, a soil type, if not a class, is clearly the exemplification of a class, as Marbut himself recognized by including a type category in his classification system. Creation of soil types as bodies cannot be separated from creation of corresponding classes.

The constituent bodies of soil types, like polypedons, are "soil areas" (15) and are equal to polypedons in their standing as individuals. That is, they are natural individuals within a little-used universe. Soil types, taken as aggregates of such bodies, are not individuals of any kind.

CONCLUSIONS

The universes of soil particles, soil horizons, soil landscape units, and delineated soil bodies contain natural individuals. None of these universes correspond to any common concept of soil. Therefore, none of their constituent individuals are soil individuals.

The universe of soil taken as an isotropic material contains hand specimens as artificial individuals. The universes of soil taken as an anisotropic body contain profiles taken as bodies, "individuals" (4), and pedons as artificial individuals. Thus, depending on the concept of soil, hand specimens, profiles taken as bodies, "individuals", and pedons may be regarded as soil individuals.

The emphasis on individuals in soil classification results largely from the idea that a class is an aggregate of individuals. The idea that a class is the range of applicability of a class concept eliminates this unnecessary emphasis. The absence of any real need for individuals is indicated by the lack of a clear, generally accepted concept of individual on the part of users and developers of soil classification systems.

Soil as an isotropic material or as an anisotropic body is almost entirely continuous rather than particulate. Therefore, like water and quartz, it cannot be numbered. Strictly speaking, there is no such thing as *a soil*. The mental division of soil into *soils* results from the emphasis on individuals and is unnecessary. The term *soils* can always be replaced by a more accurate term. For example, *bodies of Miami soil* is more accurate than *Miami soils*. In other settings, *classes of soil* is more accurate than *soils*. In any case, the use of *soils* should be regarded as a short cut in language. The plural form has no meaning in a strict sense.

Future developments may lead to a soil classification system based on some kind of soil landscape units, but at present the difficulties seem overwhelming.

LITERATURE CITED

1. Anderson, J. U. 1955. Characterization of some halomorphic soils and their normal associates in the Yakima Valley. Soil Sci. Soc. Amer. Proc. 19:328–333.

2. Aubert, G. and Duchafour, P. 1956. Projet de classification des sols. Trans. Intern. Cong. Soil Sci. 6th, Paris. E:597–604.
3. Baldwin, Mark, Kellogg, C. E., and Thorp, James. 1938. Soil classification. *In* USDA Yearbook of Agriculture, Soils and Men. pp. 979–1001.
4. Cline, M. G. 1949. Basic principles of soil classification. Soil Sci. 67:81–91.
5. ————. 1963. Logic of the new system of soil classification. Soil Sci. 96:17–22.
6. ————. 1944. Principles of soil sampling. Soil Sci. 58:275–288.
7. Jenny, Hans. 1941. Factors of Soil Formation. McGraw–Hill, New York. pp. 3, 4.
8. Johnson, W. M. 1963. The pedon and polypedon. Soil Sci. Soc. Amer. Proc. 27:212–215.
9. Jones, T. A. 1959. Soil classification—a destructive criticism. J. Soil Sci. 10:196–200.
10. Klingebiel, A. A. and Montgomery, P. H. 1961. Land-capability classification. USDA Handbook 210. Soil Conservation Service.
11. Kubiena, W. L. 1953. The Soils of Europe. Thomas Murby and Company, London.
12. Langer, Susanne K. 1953. An Introduction to Symbolic Logic. Dover Publications, Inc., New York. Ed. 2. pp. 116, 117.
13. Marbut, C. F. 1935. Soils of the United States. Part III of the Atlas of American Agriculture. U. S. Govt. Ptg. Off. Washington.
14. Mill, John Stuart. 1874. A System of Logic. Harper & Brothers, New York. Ed. 8. Book I, Chap. V, Sec. 3 and Chap. VII, Sec. 1, and Book IV, Chap. VII.
15. Muir, J. W. 1962. The general principles of classification with reference to soils. J. Soil Sci. 13:22–30.
16. Portland Cement Association. 1956. PCA Soil Primer. Chicago.
17. Report of the meeting of the soil survey committee of Canada, 1960. Canada Department of Agriculture, Research Branch, Ottawa.
18. Sandoval, F. M. Jr., Fosberg, M. A., and Lewis, G. C. 1959. A Characterization of the Sebree–Chilcott soil series association (slick spots) in Idaho. Soil Sci. Soc. Amer. Proc. 23: 317–321.
19. Simonson, R. W. 1963. Soil correlation and the new classification system. Soil Sci. 96:23–30.
20. ———— and Gardiner, D. R. 1960. Concept and function of the pedon. Trans. Intern. Cong. Soil Sci. 7th, Madison. 4:127–131.
21. Soil Survey Staff. 1960. USDA Soil Classification, A Comprehensive System, 7th Approximation. U. S. Govt. Ptg. Off. Washington.
22. ————. 1951. Soil survey manual. USDA Handbook 18.
23. Sokal, R. R. and Sneath, P. H. A. 1963. Principles of Numerical Taxonomy. W. H. Freeman and Company, San Francisco. Chap. 7.
24. Stebbing, L. Susan. 1961. A Modern Elementary Logic. Barnes & Noble, New York. Ed. 5. University Paperbacks. pp. 78, 79, 93.
25. Taylor, N. H. and Pohlen, I. J. 1962. Soil survey method. A New Zealand handbook for the field study of soils. New Zealand Soil Bur. Bull. 25.
26. Whiteside, E. P. 1954. Changes in the criteria used in soil classification since Marbut. Soil Sci. Soc. Amer. Proc. 18: 193–195.
27. Wilson, E. B. Jr. 1952. An Introduction to Scientific Research. McGraw–Hill, New York. p. 151.

10

Reprinted from *Soils Fertilizers* **29**:507–510 (1966)

SOIL BODIES AND SOIL CLASSIFICATIONS

by

A. VAN WAMBEKE

(State University of Ghent, Belgium)

Classifications can be looked upon as devices that enable men to remember the properties of a large number of objects or the parts of complex universes. In another perspective, the systems, once established, are tools for the identification of particulars. As a whole, taxonomy leads from the confrontation of acquired knowledge with new facts, to the discovery of laws and principles.

The process of making a classification involves framing classes either by subdividing a universe or by grouping objects. The classes may be arranged into categories set up at different levels of abstraction. Defining and naming are operations connected with classifications, which may serve purposes of teaching, extension, interdisciplinary research, correlation, and other forms of transfer of knowledge.

Classes are mental constructions. As far as soils are concerned, their relation with reality has been discussed by Robinson (1950).

Membership of classes is determined by criteria. In some systems the criteria are properties of the objects classified; in others, they are properties of related or associated bodies. Some use location and a few use concepts as a basis for grouping.

In the remainder of this paper, unless otherwise stated, any coherent segment of the soil mantle will be called a soil body. The term individual, apart from its use in quotations from other authors, will only be used in the sense of the smallest body that can be defined as a thing complete in itself. Classes will always be regarded as abstract fields, and members of classes may be either bodies or classes of lower categorical level.

The Soil Universe

A universe may be defined as a superclass which contains all the objects and classes under consideration (Knox, 1965).

It has been debated (Robinson, 1950; Knox, 1965) whether the universe of soils comprehends discrete physical bodies, large enough to be complete soils, or whether it should be considered as a continuum, without particulate objects (Simonson and Gardner, 1960).

The need for recognizing independent soil bodies was stressed by Kellogg (1949). He argued that men do not deal with "soil" but with "soils". He defined "a soil" as "an individual in a continuum". A "soil individual" (Kellogg, 1949) was considered a member or a coherent aggregate of members of one class and its boundaries depended on the definition of the taxa. It was not a mapping unit (Kellogg, 1963). Neither was it the smallest natural body that could be called a soil, and was therefore not identical to the concept of an individual that Cline (1949) introduced in soil classification.

Cline (1949) considered an individual as "the *smallest* natural body that can be defined as a thing complete in itself". By this definition one individual may only contain one complete object of the universe under consideration. It cannot be divided without destroying it (indivisibility). It was a concept constructed to meet the needs of taxonomists in accordance with the definition of a class as a group of objects which can be treated statistically.

Recent developments in soil classification (USDA, 1960) increased the interest in soil bodies; obviously, for purposes of correlation, different observers require to know whether they are studying parts of one and the same soil, or parts of different soils or just a part of the soil continuum.

Knox (1965) expressed the contrasting view, that the need for individuals in soil classification has received unnecessary emphasis. He contends that continuous universes may be divided into classes without using individuals, and that "strictly speaking, there is no such thing as *a soil*".

It seems logically acceptable that, in continuous universes, classifications without individuals may achieve all the purposes of the transfer of knowledge among people who use the *same* system. But no accurate comparisons between segments of the soil universe distinguished by different systems can be made consistently, unless there is one unit which cannot be subdivided by any classification without losing its quality of being a complete soil. The physical thing which corresponds to such a unit seems necessary, albeit for correlation purposes alone. Among classifications, individuals, as reference-bodies, are as necessary as are classes in taxonomy.

In this discussion, three kinds of soil bodies will be dealt with:

(1) Natural Soil Bodies, in the sense of Knox (1965), being "an individual that is discrete and independent of the observer";

(2) Artificial Soil Bodies, the dimensions of which depend on the definitions of the classes to which they belong;

(3) Arbitrary Individuals, a segment of the soil mantle, having fixed conventional dimensions which depend neither on soil properties nor on class limits. They should be big enough to comprehend a complete soil.

Natural Soil Bodies

Natural soil bodies are soil bodies the size and shape of which are set by nature. Their boundaries are independent of the viewpoint of the classifier and are not determined by class limits.

Most people engaged in mapping are inclined to think of soils as discrete bodies separated by boundaries, where maximum rate of lateral change in soil characteristics occurs (Knox, 1965). Where these concentrations of change coincide with topographic features, they provide a basis for efficient mapping. Recognition of such natural bodies often leads to the concept of three-dimensional pieces of landscape (Kellogg, 1949), which the soil surveyor identifies with a soil.

Muir (1962) called a segment of the soil mantle, the limits of which are determined by nature, and not by man, a "Soil Body". He contended that "it, itself, should be the object to be classified when constructing systems of soil classification"; he admitted, however, that they were very complex and that it seemed "unlikely that a system of classification of Soil Bodies could be elaborated" at present.

Attempts to define and classify natural soil bodies on the basis of soil properties have usually failed. Only when small regions are considered, is it possible to adjust class limits to natural boundaries, and define accordingly the limiting values of class criteria (Pomerening and Knox, 1962). However, when these critical values are transferred to other regions, they often cease to match the boundaries of natural soil bodies. In the soil universe, discontinuities occur ("natura facit saltus)", but only in small places and not at equal critical values.

If the existence of natural soil bodies could be consistently demonstrated, there would be no need for other kinds of soil bodies, except as sampling units.

Artificial Soil Bodies

Recent trends, however, indicate that soil taxonomists and mappers approach the problem through arbitrary or artificial soil bodies, especially when large areas are considered, and when soils are to be grouped on the basis of soil properties.

The size of artificial soil bodies depends on the definition of the taxa to which they belong. Artificial soil bodies are not independent of taxonomic systems and do not necessarily possess the indivisibility of individuals.

Artificial soil bodies have been proposed as member-bodies of classes. Three-dimensional units include the Soil Individual (USDA, 1960), or Polypedon (Simonson, 1962; Johnson, 1963) and the Soil Area (Muir, 1962).

Polypedons are member-bodies of classes of the lowest category in the USDA comprehensive classification system. They are bodies which contain only pedons (discussed in the following section) of one series. The Soil Area (Muir, 1962) is a three-dimensional part of a soil body and contains limiting profiles belonging to one or more primary classes.

An artificial soil body can only be an individual if its boundaries are determined by class limits at the *lowest* categorical level. The polypedon meets this requirement. If the pedon were not accepted as a complete soil (because it lacks slope), the polypedon, or parts of it, large enough to indicate slope, could be thought of as individuals. In that case shape and slope are taken as differentiating characteristics and as essential parts of the complete soil. Such soil bodies lack indivisibility, their size depends on class limits, and they can only with difficulty serve as links between different systems.

Artificial soil bodies are, however, not necessarily restricted to member-bodies of the lowest classes. In fact, and without concern for the problem of individuals, any coherent segment of the soil mantle which matches certain specifications is an artificial soil body; its size depends on the criteria, which can be soil properties (polypedon, soil area), location as related to topographic or climatic features (some mapping units, soil geographic zones), or any other property taken as a criterion. Some are four-dimensional and include all soils which, at some time, present, past or future, have, had or will have certain characteristics.

Artificial soil bodies are mutually exclusive when the ranges in properties used for class definitions do not overlap. If they do, a complete soil may be a part of two different artificial soil bodies.

Artificial soil bodies may serve many purposes. When they are based on criteria of high rank, they are often used to reduce the number of soils or the variability in the soil continuum, by eliminating parts of the universe under consideration. In comprehensive systems they may divide a continuum into independent kingdoms, in each of which a given set of criteria may prove more significant than in the entire population. Most first divisions of soil maps create artificial soil bodies, i.e., soils of uplands; soils of valleys, etc.

Artificial soil bodies have shape, including slope, which small arbitrary bodies do not. They give geographic meaning to soil properties, because they can be mapped on the basis of these properties. Their extension can be correlated with the geographic distribution of soil-forming factors, and they are useful in developing systems in search of genetic classes (Cline, 1963a, b).

Arbitrary Individuals

Arbitrary individuals have been created, and are widely used in soil investigations, mainly as a result of the taxonomists' failure to recognize or to agree upon a natural or an artificial soil body.

An arbitrary individual is a segment of the soil mantle that has fixed, conventional dimensions.

As an individual, it may only contain, but should contain, one complete soil; it cannot be divided without destroying it; being complete, it should have all the attributes which permit its classification, and no arbitrary individual should remain unclassified.

Arbitrary individuals are not mutually exclusive, although one individual may belong to only one class of the same category in a system which only recognizes mutually exclusive classes. However, a part of an individual may at the same time be a part of another. Small shifts in location of the limits which circumscribe the arbitrary volume correspond to a transition to another individual, which may belong to another class. As pointed out by Knox (1965) in the case of pedons (Simonson and Gardner, 1960; USDA, 1960), the number of arbitrary individuals in a given segment of soil is infinite, and it is consequently impossible to investigate all of them. Therefore, all finite sets of arbitrary individuals, even when they are closely packed and cover the complete area under investigation, cannot be considered as other than a sample of an infinite population. In addition to being physical bodies, arbitrary individuals function as sampling units of larger bodies. In soil science, they may sample mapping units which may be named in terms of the dominant taxa of the individuals at all levels of abstraction, including the lowest in all systems of the same universe. Arbitrary individuals may sample artificial soil bodies in order to evaluate the variability of properties which are not diagnostic for the class.

It should be pointed out, however, that the indivisibility of arbitrary individuals is only achieved by setting arbitrary dimensions: this should be kept in mind when assigning an individual to its proper class: to be diagnostic, each criterion should be dominantly present over competing characteristics in the area covered by the individual. By this method no individual will remain unclassified.

Arbitrary individuals are reference-bodies between different classification systems, if their boundaries are independent of soil properties and class limits.

In soil classification, arbitrary volumes have been proposed to approximate the concept of an individual, being the smallest body that can be defined as a thing complete in itself (Cline, 1949). The pedon is one of them (Simonson and Gardner, 1960; USDA, 1960). Most of the criticism of the pedon is related to one of the two principal attributes of the soil individual: completeness. The pedon is said to be incomplete because it is not large enough to possess slope and the depth of the pedon is still as vaguely defined as the soil itself.

As it is at present defined (USDA, 1960) the size of the pedon is not independent of soil properties: for cyclic soils its dimensions depend on the length of the cycle (Arnold, 1964); it can consequently, as an individual, serve only one system, and the introduction of variable-size pedons restricts its value as a carrier and reference-body of indivisible information between different classifications.

Conclusions

Present trends in soil classification indicate that soil mappers and taxonomists do not consistently accept the existence of natural soil bodies, the boundaries of which are independent of the viewpoints of men. On the other hand, artificial soil bodies are too dependent on class limits to be transferred without alteration from one system to another.

Considering soil as a continuum, without discrete bodies, one may question the need to recognize an individual, as it seems logical to assume that a classification without individuals may satisfactorily achieve all the purposes of the transfer of knowledge among people who use it. All segments of the continuum may be described in terms of the established classes expressed as fractions of areal units.

However, when two or more classifications are used, or compared, an indivisible and complete body of soil is the only unit which can always be classified without distortion at all levels by all systems. All classifications, to be compared and correlated, should have a common denominator. This would be an important function of the soil individual as conceived by Cline (1949). In the absence of a workable definition of such an individual in terms of its constituent parts, a

120

segment of the soil mantle contained in an arbitrary volume may serve this function. A fixed-size pedon would be such a body; it is essentially a physical thing, an indivisible but complete soil, which may be used as a universal sampling unit of larger bodies. It should be pointed out, however, that it is an artifact, created to serve certain objectives. For most purposes a volume circumscribed by 1 square metre, down to two metres depth or shallower unweathered solid rock, would include an adequate reference soil body.

A classification which classifies arbitrary individuals obviously does not classify natural, particulate things, comparable with plants or animals. Nobody, however, has ever seen all the soils he classifies, or examined the complete exposure of a mapping unit. All surveys originate from samples, and all classifications, at some stage of passing from real things to mental constructions, deal with sampling units, which lead to other conceptual or real bodies. A classification of arbitrary individuals does not necessarily disregard natural groupings or subdivisions.

Acknowledgments

Acknowledgment is extended to Professor R. W. Arnold of Cornell University for helpful discussions on the subjects reviewed in this paper, and constructive criticism of the manuscript.

REFERENCES

ARNOLD, R. W. 1964. Cyclic variations and the pedon. *Soil Sci. Soc. Am. Proc.* **28**, 801-804.

CLINE, M. G. 1949. Basic principles of soil classification. *Soil Sci.* **67**, 81-91.

CLINE, M. G. 1963a. Logic of the new system of soil classification. *Soil Sci.* **96**, 17-22.

CLINE, M. G. 1963b. The new soil classification system *Cornell Univ. Agron. Mimeo*, No 62-6.

JOHNSON, W. M. 1963. The pedon and the polypedon. *Soil Sci. Soc. Am. Proc.* **27**, 212-215.

KELLOGG, C. E. 1949. Soil classification, introduction. *Soil Sci.* **67**, 77-80.

KELLOGG, C. E. 1963. Why a new system of soil classification? *Soil Sci.* **96**, 1-5.

KNOX, E. G. 1965. Soil individuals and soil classification. *Soil Sci. Soc. Am. Proc.* **29**, 79-84.

LEEPER, G. W. 1956. The classification of soils. *J. Soil Sci.* **7**, 59-64.

MUIR, J. W. 1962. The general principles of classification with reference to soils. *J. Soil Sci.* **13**, 22-30.

POMERENING, J. A., KNOX, E. G. 1962. A test for natural soil groups within the Willamette catena population. *Soil Sci. Soc. Am. Proc.* **26**, 282-287.

ROBINSON, G. 1950. Soil classification. *J. Soil Sci.* **1**, 150-155.

SIMONSON, R. W., GARDNER, D. R. 1960. Concept and functions of the pedon. *Trans. VIIth Internat. Congr. Soil Sci.* **IV**, 127-131.

SIMONSON, R. W. 1962. Soil classification in the United States. *Science* **137**, 1027-1034.

U.S.D.A., SOIL SURVEY STAFF. 1960. Soil Classification. A Comprehensive System. Washington D.C., U.S.A.

11

CONCEPT OF SOIL

R. W. Simonson

[*Editor's Note:*   In the original, material precedes this excerpt.]

II. Basic Soil Entities

The nature of basic soil entities is discussed in this section. Their functions in the mapping and classification of soils are also considered. Relationships of the basic soil entities to prevailing conceptions of soil are explored because of the important reciprocal effects of the conceptions or working models of soil and the understanding of basic entities in the minds of scientists. Before possible basic entities are considered, however, a current conception of soil is outlined. That conception in turn underlies the concepts of basic soil entities.

Several years ago in writing about the changing model of soil, Cline (1961) reviewed several conceptions of soil that had been or were held in the United States. He gave primary attention to the conception outlined by Marbut (1935) and to that in the monograph on the 7th Approximation (Soil Survey Staff, 1960). These two conceptions are separated by twenty-five years. The earlier conception stressed the place of soil as the outer layer of the earth's crust whereas the latter stressed soil as a collection of natural bodies. Both conceptions are within a concept of soil as something at the land surface, but the two conceptions lead to obvious differences in the way of thinking about soils as objects for study.

The importance of the concept of the objects under study to processes of thought is stated by Cline (1961) as follows:

> Within the framework of its accumulated knowledge, every science develops a mental image of the thing with which it is concerned. This model of a science is the

organized aggregate of accumulated facts, and laws and theories based on those facts; it is a mental picture of that which is known viewed in organized perspective through verified quantitative relationships, which we call laws, with varying degrees of distortion by virtue of theories that attempt to explain the observed relationships. The picture is not the same to all who work in the science, for it is composed of knowledge and the extensions of theory from knowledge into the unknown, and different men know, or think they know, different things.

Several conceptions of soil, each reflecting the state of knowledge in its day, have been held in the past. These conceptions were reviewed earlier in the article. It is evident from the history of the several conceptions that more than one is in use at any given time. The existence and application of several conceptions or working models is not peculiar to soil science; one need look back only a short way for the wave and corpuscular theories of the propagation of light. Changes in the working model or in the conception of the objects under study follow as knowledge itself changes, for whatever reason.

Since soil first became a subject for deliberate study, the single most profound change in the conception or working model was the introduction of the idea that soil was an independent natural body with genetic horizons (Cline, 1961). Steps in the early development and spread of this conception in Russia and the United States have been discussed in previous sections. By comparison with the recognition of soils as organized natural bodies further changes are much less far-reaching though they also have importance.

During the last three decades the principal modification in the conception of soils as organized natural bodies has been the growing recognition that the basic entities are three-dimensional. They consist of volumes or polyhedrons of some kind. It is true that soils form a continuum over the land surface with few sharp breaks, but soils differ from place to place (Simonson, 1957). Because of the differences from place to place the continuum can be considered a mosaic or patchwork consisting of many polyhedrons (van Wambeke, 1966).

Soil mantles the land surface of the earth generally. This mantle may be looked upon as a collection of organized natural bodies that contain living matter and either have horizons or are subject to horizon differentiation. The morphology of these bodies reflects many paths of horizon differentiation. Horizonation also has a wide range, taking all segments of the mosaic as a group. Some kinds of soils have few and faint horizons, whereas others have prominent horizons.

All soils share a number of characteristics. All are three-phase systems composed of the same major constituents. All are open systems to which substances may be added and from which substances may be lost. All have profiles, some with more distinct horizons than others. Local kinds of soils merge with one another as a general rule; gradations are the normal mode of change.

Spatially, basic soil entities must be three-dimensional, i.e., have length, breadth, and depth. Each stands in relation to the earth much as a small piece of rind stands in relation to the whole of an orange. Conceptually, Crowther (1953) considered soils multidimensional, meaning that many characteristics must be specified to define soils. To restate the argument, soils are as complex as they are commonplace.

A. Nature and Dimensions of Basic Soil Entities

The major difficulty in defining soil individuals or basic entities follows from the existence of soil as a continuum. Discrete individuals comparable to those of plants and animals do not exist. The problems in defining some small volume of soil as a basic entity is thus similar to those faced in defining some basic unit for classification of bodies of rock and rock formations.

As part of a discussion of basic principles of soil classification, Cline (1949) offered the following definitions: "The smallest natural body that can be defined as a thing complete in itself is an *individual*. All the individuals of a natural phenomenon, collectively, are a *population*."

The definition does not provide limits for a soil individual or basic entity. What the smallest natural body of soil can most appropriately be must still be spelled out in some way. Various ways have been tried (Simonson and Gardner, 1960).

The term "pedon" has been proposed as a collective noun for small basic soil entities (Soil Survey Staff, 1960; Simonson and Gardner, 1960). As a generic term, pedon would thus parallel the word "tree" as a collective noun covering oaks, pines, elms, and other kinds.

The pedon is in a sense an abstraction or soil unit that is a creation of the mind, to use the terminology of Marbut (1921). According to the terminology of Knox (1965), the pedon is an artificial rather than natural individual. Moreover, Knox (1965) questions the desirability of trying to define either natural or artificial soil individuals. He argues that natural individuals do not exist in a universe that is not particulate. He also believes that classification and mapping of soils can proceed without defining some volume of soil as an individual.

Recognition of some physical entity, some volume of soil as basic in mapping and classification does have certain advantages, at least for most people. It is imperative that mapped soil bodies be related to classes in a general system to permit the transfer of knowledge about the nature and

124

behavior of soils from one place to another with reasonable assurance that the transfer will be valid. Relating the mapped bodies to classes at some level in a system will be easier for most people if they can think in terms of some physical entity rather than exclusively in terms of an abstraction. This ia a prime argument for attempting to define some small volume of soil as a basic entity.

Each pedon consists of a small volume of soil that is part of the continuum mantling the land surface. Each is a tiny segment of the rind of the earth, an irregularly shaped solid or polyhedron. Each pedon begins at the surface and extends downward to include the full set of horizons or to some arbitrary depth corresponding approximately to the vertical dimensions of a set of horizons. The upper boundary is clear enough, but the lower boundary remains vague, as a rule. The perimeter is gradational from one pedon to its neighbors. Distinguishing pedons is thus like setting apart different kinds of climate. Sharp changes in climate are uncommon though differences exist over the face of the earth. Moreover, some of the differences are substantial. Similar statements apply to the soil mantle.

A concept broader than, but parallel in some ways, to that of the pedon was proposed a few years earlier by Jenny (1958). He suggested the term "tessera," which was to be a small three-dimensional element consisting of soil plus vegetation. Lateral dimensions might be of the order of 1 square meter or as small at 8×8 inches, with the dimensions to be determined by the purpose of each study. The vertical dimension or thickness of the tessera was to be the combined height of the vegetation and depth of the soil. The proposed tessera includes not only the soil, but also the vegetation growing on that soil. Concepts of the tessera and pedon are thus related but not identical since the latter does not include vegetation.

Certain requirements must be met by any definition of the pedon. It must be large enough to be observable, to be sampled, and to exhibit a full set of horizons. At the same time it should be as small as possible while meeting the above requirements. A volume of soil or a polyhedron could be very small and still be observable. With care a person could examine a vertical core no more than a few centimeters in diameter. A volume of soil large enough to be sampled with an auger can be examined, but full observation of the relationships between horizons is not possible. Consequently, some larger volume is required.

Setting the minimum dimensions for the pedon can be attempted by analogy to the defining of unit cells in crystallography. The unit cell can be defined in crystals of minerals such as kaolinite and mica, though the cell is an abstraction and does not exist alone. The unit cell can be recognized by the succession and spatial arrangement of ions in the lattice. Each unit cell is like the next. The analogy between the unit cell in a crystal and the pedon in a soil body can readily be carried too far. Crystals have more regularity in structure than does the soil mantle in its morphology.

A pedon must be a large enough volume of soil to include a full set of horizons and permit observation of the boundaries between them. For example, a single pedon would have to consist of a volume of soil that includes the whole set of features associated with a coarse prism in a fragipan. A single pedon would consist of a volume of soil expending from the middle of the coarse prism to the middles of the neighboring prisms. Thus, the study of horizons and their interrelationships in place largely govern possible minimum and maximum dimensions of the pedon.

Small differences in the nature of horizons must be permitted within a pedon. For example, a krotovina passing through a horizon would not be the basis for identifying a second pedon.

Cyclical variations are also permitted within pedons because they are essential parts of relationships between horizons in some kinds of soils. These cyclical variations are also keys to the genesis of some soils. For example, the downward tonguing of both A and B horizons is a normal feature of Spodosols (Podzols) formed in sandy materials with good drainage.

Study of requirements that must be met demonstrate that the best definition now possible for the pedon is in terms of lateral dimensions. The pedon must be large enough to be sampled. It should be large enough to show the relationships between horizons. It should also be as small as practicable in order to minimize variability within the unit.

> The best definition of the pedon that can be offered now is not complete but represents a step toward full definition. A pedon consists of a small volume of soil which includes the full solum and the upper part of the unconsolidated parent material (or a volume of comparable size if horizons are faint), is usually less than 2 meters in depth, and has a lateral cross section that is roughly circular or hexagonal in shape and between 1 and 10 square meters in size. The smallest of these lateral dimensions is proposed for use in most soils. The larger dimensions, up to the maximum of 10 square meters, are proposed for use where needed to cover the full amplitude of one cycle in the arrangement of horizons. Where the nature and arrangement of horizons are cyclical, the full cycle must occur within a lateral cross section that is roughly circular and is 10 square meters in size if the volume of soil under observation is to be considered one pedon. Otherwise, the volume of soil under observation is to be considered as two or more pedons (Simonson and Gardner, 1960).

A single pedon can usually be defined by preparing a description of a soil profile plus statements of the ranges in characteristics of horizons. All ranges are limited in most instances. If cyclical variations occur within a pedon, the definition requires a statement of the nature and amplitude of those variations. To return to the example of the A_2 and B_2 horizons of Spodosols (Podzols), the definition of a pedon would have to indicate the distribution of the horizons, for example, the frequency of tongues and the dimensions of those tongues.

Possible difficulties in applying the suggested lateral dimensions for pedons in soils with cyclical variations have been suggested by Arnold (1964). He analyzed the use of the half-cycle width as the basis, with the sine curve as his model. He concluded that about 80 percent of a cycle would have to be included in each pedon if it were to cover 80 percent of the vertical horizon variability for horizons that were cyclic at linear intervals ranging from 1.3 to 4.3 meters.

Another part of the effort to define basic soil entities is the proposal of polypedon as a term for a larger volume of soil than that of the pedon (Simonson, 1962). The term polypedon was proposed to replace "soil individual" as that was used in the monograph outlining the 7th Approximation (Soil Survey Staff, 1960).

A polypedon consists of a group of contiguous pedons that are within the limits of one soil series (Simonson, 1962; Johnson, 1963). In other words, the set of pedons must fit within the range of one series and occur in a contiguous group to form a polypedon. For the most part, one polypedon is the most extensive in each delineated body of soil shown on a detailed map. Such bodies normally include parts of other polypedons. These parts of the second, third, and fourth polypedons constitute mapping inclusions.

The relationships between a polypedon and its constituent pedons in a single body of soil are illustrated in Fig. 1. In the diagram, part of the soil body is split into pedons. All but a few of these pedons are of the same kind and represent a single polypedon. A part of one pedon, however, shown at the left edge of the diagram, is dark-colored to the bottom. That pedon and several others lack the A2 horizon characteristic of pedons constituting the dominant polypedon. The one darkened pedon along the edge plus others along the small drainageway extending into the diagram from the left side represent a mapping inclusion. They are part of a second polypedon differing from the dominant one. As soils occur naturally, parts of several polypedons are commonly inclusions within bodies of soil that can be delineated on detailed maps. To keep the illustration simple, however, a single mapping inclusion has been shown in the diagram.

127

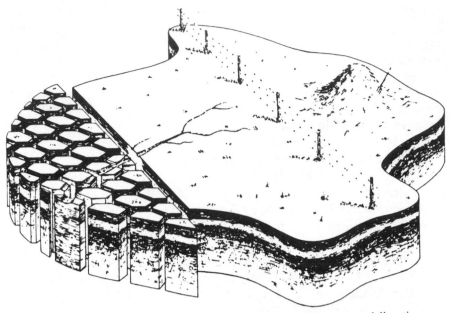

FIG. 1. Schematic diagram of a body of soil that would be shown as one delineation on a detailed soil map and named as one phase of a soil series. Part of the body is split into pedons, a few of which represent a mapping inclusion or part of a polypedon different from the dominant one within the body as a whole. (Sketch by Walter M. Simonson.)

B. RELATING THE MAPPING AND CLASSIFICATION OF SOILS

Some classification of soils is applied deliberately or otherwise by everyone who uses them. The classification may be simple or sketchy, taking few features into account. For example, it was believed by homesteaders filing their claims in north central North Dakota just prior to 1900 that land supporting clumps of silverbrush (*Eleagnus argentea*) was better for growing wheat than land without patches of the shrub. The single criterion used in that classification of soils was a kind of vegetation, the presence of which on a given spot could have been due to any one of several factors. More sophisticated classifications have since been developed, and these have greater value in appraising the usefulness of the soils.

The combination of maps showing distribution of kinds of soils and a classification of them is most effective. The bodies of soil shown as delineations on maps are related to a classification system, usually to classes of low rank, through the process of correlation (Simonson, 1963). Existing knowledge about the nature, origin, and behavior of soils can then be brought to bear more effectively on specific tracts of land.

Existing knowledge may be applied to further the understanding of soil genesis. More commonly, existing knowledge is applied to achieve better use of soil resources in the satisfaction of human wants. The ultimate objectives in the process of relating mapped soil bodies to a classification system are the same as those of soil science generally, i.e., to predict what will follow from the use of soils in different ways and to predict the long time effects of such use on the soils.

The making of predictions and the transfer of results of experience and research from soils of one locality to those of another are greatly facilitated by adequate classification and reliable maps. The two must be tied together to be most useful. The basic soil entities discussed in the immediately preceding section are intended to simplify the task of relating mapped bodies of soils to classes in a general system.

Each class of low categoric rank in a classification system, e.g., the soil series in the United States, can be looked upon as a group of similar pedons or polypedons. All pedons belonging to one class, i.e., one soil series, are required to be alike within narrow limits in the sequence, arrangement, and character of definitive horizons except the surface layer. If horizons are few or faint, the uniformity is required in a fixed portion of the soil between depths of 10 and 40 inches (25 and 100 cm.). To state these requirements in another way, the ranges in thickness, color, texture, consistence, and the like, of individual horizons must be small or they must be cyclical within a limited lateral cross section.

The soil series category is roughly comparable to that of the species in the classification of plants, animals, and minerals. The word "roughly" is used because analogies can only be illustrative for sets of natural objects such as a collection of pedons and a collection of orthoclase grains. Even so, it may be helpful in grasping the concept of a series to think of each one as a group of closely similar pedons paralleling the group of oak trees known as *Quercus alba* or the group of mineral grains known as orthoclase.

The pedons or polypedons with the specified definitive features are members of one class, a single series, regardless of their geographic occurrence. The basis for grouping the pedons into one class is the degree of similarity among them rather than how or where they occur. Their occurrence is independent of their classification into a given series. Pedons of one series need not occur in the same delineated soil body or even in the same county.

In contrast to classes such as the soil series, a mapping entity or single body of soil shown as a delineation on a map must consist of pedons or polypedons and parts of polypedons that occur together. Each delineated

129

body is a bundle of contiguous pedons. The contiguous occurrence is an essential feature of mapping entities. A boundary must be drawn around each entity in order to record on a map the position of that entity on the face of the earth. Every mapping unit consists of a number of these individual entities, each of which is a segment of the soil continuum.

Pedons and polypedons cannot be moved about to facilitate their classification or mapping. It is not possible, for example, to take all pedons of the Sassafras series and arrange them in a clump somewhere so that a single boundary could be placed around the whole lot. In practice, mapping entities must be defined so as to accommodate the existing distribution of pedons and polypedons in the soil continuum. In detailed surveys, mapping entities are defined insofar as practicable to consist of pedons or polypedons representing a single series.

Rarely does a delineated soil body consist of one polypedon. This is illustrated by diagram in Fig. 1. Virtually all delineated soil bodies or mapping entities include pedons representing several series. Most mapping units in detailed surveys consist of delineated soil bodies in which the number of pedons of a single series, i.e., the extent of one kind of polypedon, is greater than that of any other kind.

Conventions have been developed to permit the naming of mapping units — each set of soil bodies delineated on field sheets and identified by the same symbol — to show their relationships to taxonomic classes, primarily the soil series. Thus, a mapping unit recognized in a detailed survey might be named as a phase of the Palouse series. According to present conventions for naming mapping units, this means that polypedons of the Palouse series form major proportions of all the delineated bodies and that included kinds of soils are either closely similar or form small proportions of the whole. Some pedons or parts of polypedons classifiable in other series can be expected in all delineated bodies named as phases of the Palouse series.

If the set of soil bodies comprising a mapping unit consists of appreciably differing pedons or polypedons and these occur in large enough proportions, the names of two or more series are used in the naming of the unit. Multiple component units may be named as complexes, undifferentiated groups, or soil associations, depending on the pattern of occurrence of the component kinds of soil. Conventions for nomenclature are outlined in the *Soil Survey Manual* (Soil Survey Staff, 1951) and are not repeated here.

Summarizing briefly, a taxonomic class of low categoric rank consists of a set of pedons or polypedons considered together because of their similarities. The grouping is independent of the geographic occurrence of the basic entities themselves. A mapping entity, on the other hand, consists of a bundle of contiguous pedons or the whole of one plus parts of several polypedons in place. Classification is a mental operation exclusively and the pedons can then be considered apart from their mode of occurrence. Mapping, however, must accommodate itself to the distribution of pedons in the soil mantle.

The pedon promises to be useful in two ways. It will be of value in relating bodies of soil represented as delineations on maps to the classes of soil in a general system. Pedons will also serve as the entities studied in field observations of soil profiles and in sampling for laboratory investigations.

The polypedon will be useful as a larger volume of soil becomes important. In the production of corn or in the building of a road, the results must be related to a larger volume of soil than that of a pedon. Performance of soils when used for the growing of plants, in the construction of highways, or as foundations for buildings can thus be related to polypedons much better than to pedons.

III. Epilogue

As emphasized earlier, conceptions of soil disappear slowly and persist for a long time, even though they may be largely superseded. All conceptions that have been held during historical times persist to some extent now. Only one of the several discussed in this article has largely faded away, but it has not disappeared completely. Each of the other conceptions that have been discussed is held as a primary one by a share of the present population of soil scientists.

The idea that fire, water, earth, and air were the basic components of all things, introduced by Empedocles about 400 B.C., is no longer seriously held. The idea did persist, however, for many centuries. Echoes of the original idea can be found in one of the present day nursery rhymes for children.

The conceptions of soils as a medium for plant growth, as the surficial mantle of weathered rock, and as organized natural bodies are all in current use. In part the different conceptions are held by different people because their interests lead them to focus on different characteristics and relationships of soils. Certain characteristics and certain relationships to other objects assume major importance in the study or use of soils for different purposes. The features of consequence in the growing of cotton are not the same as those important to the design of a residential subdivision. Men interested in growing cotton will thus look at soils from a different point of view than men designing a residential subdivision. In part the holding of different conceptions of soils also reflect original differences in the education or training of men. Things once learned are are not readily dropped or displaced, as observed by John Dewey (1958).

Reasons for the long history and persistence of the conception of soil as a medium for plant growth are not far to seek. The primary interest of mankind in soils rests on their capacity to support plants that provide food and fiber. This interest is common to all humanity. Production of food and fiber retains first place even in the highly industralized nations of the world, though important shifts in usefulness of soil resources have occurred during the last half century (Simonson, 1966).

It is also easy to see why the conception of soils as a surficial mantle of rocks and weathered rock persists. The great bulk of most soils by weight consists of rock fragments and mineral grains. Furthermore, the accumulation of parent materials in which soils form is due to rock weathering. The weathering of rock fragments and mineral grains does not stop once evident horizons have been formed but may continue long after horizonation is distinct. Thus, the weathering of rock is involved in the formation of all soils, though weathering alone does not result in the formation of soil.

The conception of soils as organized natural bodies, last to be born of the three prevalent ones, has drawn some of its substance from the two older conceptions. Thinking of soils as organized natural bodies implies that their history is of importance to their present nature. How the parts of these bodies are arranged and related, the sequence and relationships between horizons, and their degree of expression reflect the entire history of each pedon. Part of that history is the weathering of rock and the resulting character of those materials. The duration and combinations of processes in horizon differentiation also strongly affect the present nature of pedons. Thus, some characteristics of soils are related to source rock and its weathering and some are not. Some characteristics are due to the effects of living organisms in the addition of organic matter, in the transfer of substances within the system, and in the transformation of substances. Thus, the arguments offered by Dokuchaiev (1948) little less than a century ago that soils are formed through complex interactions among a number of factors remain valid. The conception of soils as organized natural bodies has the greater promise for expanding current understanding of soil genesis—of how soils were formed—and thus acquired their present character.

The recognition of soils as organized natural bodies promises to be more helpful than any other conception in dealing with problems that will

132

have to be faced as society becomes progressively more dependent on science and technology. The development of powerful tractors and other large machines plus the rapid accumulation of capital is encouraging profound modification of soils in some localities in the highly industrialized nations. The trend in recent years can be expected to continue. More drastic modifications of more soils can be anticipated in the future. The cost of these modifications, as for example, the ripping of duripans (hardpans) in some soils in the central valley of California, is high. The costs of normal operations in using soils for farming, forestry, highway construction, and the like are rising. Adjustment of the operations to kinds of soils and their distribution patterns will assume greater significance in the future because of the need to keep costs down and to minimize failures.

Soil resources in the industrialized nations are being used more and more in the construction of highways, as foundations for homes, for the disposal of sewage, and for parks and playgrounds. These uses are expanding as population expands. The uses of soils not directly related to the production of crops, pasture, and trees will require attention to soil characteristics of little consequence in the past. Predictions of what will follow from the use of soils must be expanded from those of crop yields under specified management and of forest growth in given time spans to the probable results when pedons and polypedons are reworked in various ways.

The conception of soils as organized natural bodies with certain characteristics arranged systematically should provide better bases for predictions than can be obtained otherwise. The working model of soil must serve to organize existing fact and theory so that the whole can be mobilized and applied, both in the production of plants and in the use of soil resources for other purposes.

REFERENCES

Afanasiev, J. N. 1927. *Russ. Ped. Invest. No. 5.*
Arnold, R. W. 1964. *Soil Sci. Soc. Am. Proc.* 28, 801–804.
Barber, B. 1961. *Science* 134, 596–602.
Bartholomew, W. V., and Kirkham, D. 1960. *Trans. 7th Intern. Cong. Soil Sci.* 2, 471–477.
Black, C. A. 1968. "Soil-Plant Relationships," 2nd ed. Wiley, New York.
Bonsteel, J. A. 1911. *U.S. Dept.Agr. Bur. Soils Cir.* 32.
Braidwood, R. J. 1961. *Chicago Nat. Hist. Museum Popular Ser. Anthropol. No.* 37.
Braidwood, R. J., and Howe, B. 1962. *In* "Courses Toward Urban Life" (R. J. Braidwood and G. R. Willey, eds.), pp. 132–146. Viking Fund Publ. Anthropol. No. 32.
Brehaut, E. 1933. "Cato the Censor on Farming." Columbia Univ. Press, New York.
Brown, P. E. 1936. *Iowa Agr. Expt. Sta. Spec. Rept.* 3.

Bureau of Soils 1902. "Instructions to Field Parties and Descriptions of Soil Types – Field Season, 1902." U.S. Dept. Agr., Washington, D.C.

Bureau of Soils 1903. "Instructions to Field Parties and Descriptions of Soil Types." U.S. Dept. Agr., Washington, D.C.

Bureau of Soils 1904. "Instructions to Field Parties and Descriptions of Soil Types – Field. Season, 1904." U.S. Dept. Agr., Washington, D.C.

Bureau of Soils 1906. "Soil Survey Field Book, Field Season, 1906." U.S. Dept. Agr., Washington, D.C.

Butler, B. E. 1958. *J. Australian Inst. Agr. Sci.* **24,** 14–20.

Cain, A. J. 1958. *Proc. Linnean Soc. London* **169,** 144–163.

Cain, A. J. 1962. *Symp. Soc. Gen. Microbiol.* **12,** 1–13.

Clark, J. G. D. 1952. "Prehistoric Europe: The Economic Basis." Methuen, London.

Cline, M. G. 1949. *Soil Sci.* **67,** 81–92.

Cline, M. G. 1961. *Soil Sci. Soc. Am. Proc.* **25,** 442–446.

Coe, M. D., and Flannery, K. V. 1964. *Science* **143,** 650–654.

Coffey, G. N. 1912a. *Proc. Am. Soc. Agron.* **3,** 115–129.

Coffey, G. N. 1912b. *U.S. Dept. Agr. Bur. Soils Bull. No. 85.*

Columella, Ca. 60 A.D. "Res Rustica (On Agriculture)" (Transl. from Latin by H. B. Ash). Harvard Univ. Press, Cambridge, Massashusetts, 1941.

Conant, J. B. 1951. "On Understanding Science." New Am. Library of World Lit., New York.

Crowther, E. M. 1953. *J. Soil Sci.* **4,** 107–122.

Davy, H. 1813. "Elements of Agricultural Chemistry." Longmans, Hurst, Rees, Orme, & Brown, London.

Dewey, J. 1958. "Experience and Nature," p. 219. Dover, New York.

Dokuchaiev, V. V. 1893. "The Russian Steppes and Study of the Soil in Russia, its Past and Present" (J. M. Crawford, ed. of Engl. Trans.). Dept. Agr. Ministry Crown Domains, St. Petersburg, Russia.

Dokuchaiev, V. V. 1948. "Selected Works of V. V. Dokuchaiev, Vol. I – Russian Chernozem." (Transl. from Russian by N. Kaner). Israel Program Sci. Transl., Jerusalem, 1967.

Eaton, A. and Beck, T. R. 1820. "A Geological Survey of the County of Albany." Agr. Soc. of Albany County, New York.

Ehwald, E. 1962. *Albrecht-Thaer-Arch.* **6,** 95–110.

Ehwald, E. 1964. *Albrecht-Thaer-Arch.* **8,** 5–36.

Evans, E. E. 1956. *In* "Man's Role in Changing the Face of the Earth" (W. A. Thomas, Jr., ed.), pp. 217–239. Univ. of Chicago Press, Chicago, Illinois.

Fallou, F. A. 1862. "Pedologie oder Allgemeine und besondere Bodenkunde." Schoenfeld, Dresden.

Fippin, E. O. 1911. *Proc. Am. Soc. Agron.* **3,** 76–89.

Fireman, P. 1901a. *Expt. Sta. Record* **12,** 704–712.

Fireman, P. 1901b. *Expt. Sta. Record* **12,** 807–818.

Gilmour, J. S. L., and Walters, S. M. 1963. *Vistas Botany* **4,** 1–22.

Glinka, K. D. 1914. "Die Typen der Bodenbildung." Geb. Borntraeger, Berlin.

Glinka, K. D. 1927. "The Great Soil Groups of the World and their Development." (Transl. by C. F. Marbut). Edwards, Ann Arbor, Michigan.

Glinka, K. D. 1931. "Treatise on Soil Science," 4th ed. (Transl. from Russian by A. Gourevitch). Israel Program Sci. Transl., Jerusalem, 1966.

Hilgard, E. W. 1860. "Report on the Geology and Agriculture of the State of Mississippi." E. Barksdale, State Printer, Jackson, Mississippi.

Hilgard, E. W. 1892. *U.S. Dept. Agr. Weather Bur. Bull.* 3, 1–59.

Hilgard, E. W. 1904. *U.S. Dept. Agr. Off. Expt. Sta. Bull.* 143, 117–121.

Hitchcock, E. W. 1838. "Report on a Re-examination of the Economical Geology of Massachusetts." Dutton and Wentworth, State Printer, Boston, Massachusetts.

Hough, B. K. 1957. "Basic Soils Engineering." Ronald Press, New York.

Hudson, H. 1844. *The Prairie Farmer* 4, 1–172.

Hunt, C. B. 1967. "Physiography of the United States." Freeman, San Francisco, California.

Jarilow, A. 1913. *Intern. Mitt. Bodenk.* 3, 240–256.

Jenny, H. 1958. *Ecology* 39, 5–16.

Johnson, W. M. 1963. *Soil Sci. Soc. Am. Proc.* 27, 212–215.

Kellogg, C. E., and Ableiter, J. K. 1935. *U.S. Dept. Agr. Tech. Bull.* 469.

Knox, E. G. 1965. *Soil Sci. Soc. Am. Proc.* 29, 79–84.

Leggett, R. F. 1953. *Nature* 171, 574.

Liebig, J. 1843. "Chemistry in its Application to Agriculture and Physiology." Campbell, Philadelphia, Pennsylvania, 111 pp.

Lucretius, Ca. 60 B.C. "De Rerum Natura (Of the Nature of Things)" (Engl. transl. by W. E. Leonard). Dutton, New York, 1921.

Lyford, W. H., and Quakenbush, G. A. 1956. *Soil Sci. Soc. Am. Proc.* 20, 397–399.

MacNeish, R. S. 1964. *Science* 143, 531–537.

Mander, A. E. 1947. "Logic for the Millions." Phil. Libr., New York.

Mangelsdorf, P. C. 1958. *Science* 128, 1313–1320.

Marbut, C. F. 1921. *Soc. Promotion Agr. Sci.* 41, 116–142.

Marbut, C. F. 1922. *Am. Soil Surv. Workers Rept.* 3, 24–32.

Marbut, C. F. 1924. *In* "État de l'Étude et de Cartographie des Sols dans divers Pays de l'Europe, Amérique du Nord, Afrique, et Asie" (G. Murgoci, ed.), pp. 215–225. Inst. Geol. Roumanie, Bucharest.

Marbut, C. F. 1925. *Ann. Assoc. Am. Geographers* 15, 1–29.

Marbut, C. F. 1927. *Proc. 1st Intern. Congr. Soil Sci.* 4, 1–31.

Marbut, C. F. 1928a. *In* "Soils of Cuba" (H. H. Bennett and R. V. Allison), pp. 341–354. Tropical Plant Res. Found., Washington, D.C.

Marbut, C. F. 1928b. *In* "The Bureau of Chemistry and Soils: Its History, Activities, and Organization" (G. A. Weber), pp. 91–98. The Brookings Institution, Washington, D.C.

Marbut, C. F. 1935. *In* "Atlas of American Agriculture" (O. E. Baker, ed.), Part III. U.S. Dept. Agr., Washington, D.C.

Marbut, C. F., Bennett, H. H., Lapham, J. E., and Lapham, M. H. 1913. *U.S. Dept. Agr. Bur. Soils Bull. No. 96.*

Mitchell, J., Moss, H. C., and Clayton, J. S. 1950. *Saskatchewan Soil Surv. Rept. No. 13.*

Müller, P. E. 1887. "Studien über die Naturlichen Humusformen." Springer, Berlin.

Muir, A. 1961. *Adv. Agron.* 13, 1–56.

Neuss, O. 1914. *Intern. Mitt. Bodenk.* 4, 453–495.

Olson, L., and Eddy, H. 1943. *Geograph. Rev.* 33, 100–109.

Odhiambo, T. R. 1967. *Science* 158, 876–881.

Ping-Hua Lee, M. 1921. *Columbia Univ. Studies History, Economics, Public Law,* 99, pp. 33–40.

Polanyi, M. 1963. *Science* 141, 1010–1013.

Raychaudhuri, S. P. 1953. *Indian Council Agr. Res. Rev. Ser. No. 4.*

Ruffin, E. 1832. "An Essay on Calcareous Manures." Campbell, Petersburg, Virginia.

Russell, E. W. 1961. "Soil Conditions and Plant Growth," 9th ed. Wiley, New York.

Semple, E. C. 1921. *Ann. Assoc. Am. Geographers* 11, 47–74.

Shaler, N. S. 1877. "A General Account of the Commonwealth of Kentucky." *Kentucky Geol. Repts. Progr.* [N.S.] 2, 382–385.

Shaler, N. S. 1891. *U.S. Geol. Survey 12th Ann. Rept.* Pt. 1, pp. 219–345.

Shaw, C. F., and Baldwin, M. 1938. "Bibliography of Soil Series." Am. Soil Surv. Assoc., Washington, D.C.

Sibertzev, N. M. 1951. "Selected Works, Vol. I — Soil Science." (Transl. from Russian by by N. Kaner). Israel Program Sci. Transl., Jerusalem, 1966.

Simonson, R. W. 1952a. *Soil Sci.* **74**, 249–257.

Simonson, R. W. 1952b. *Soil Sci.* **74**, 323–330.

Simonson, R. W. 1957. *Yearbook (U.S. Dept. Agr.)* pp. 17–31.

Simonson, R. W. 1962. *Science* **137**, 1027–1034.

Simonson, R. W. 1963. *Soil Sci.* **96**, 23–30.

Simonson, R. W. 1964. *Trans. 8th Intern. Congr. Soil Sci.* **5**, 17–22.

Simonson, R. W. 1966. *Agr. (Montreal)* **23**, 11–15, 21.

Simonson, R. W., and Gardner, D. R. 1960. *Trans. 7th Intern. Congr. Soil Sci.* **4**, 127–131.

Soil Survey Staff 1951. *U.S. Dept. Agr. Handbook* **18**.

Soil Survey Staff 1960. "Soil Classification — A Comprehensive System. 7th Approximation." U.S. Dept. Agr., Washington, D.C.

Tulaikoff, N. M. 1908. *J. Agr. Sci.* **3**, 80–85.

Tull, J. 1733. "The Horse Hoing Husbandry." Publ. by Author, London.

Usher, A. P. 1923. *Quart. J. Econ.* **37**, 295–411.

Vanderford, C. F. 1897. *Tennessee Agr. Expt. Sta. Bull.* **10**, pp. 31–139.

van Wambeke, A. 1966. *Soils Fertilizers* **29**, 507–510.

Varro, Ca. 35 B.C. "Rerum Rusticarum (On Agriculture)." (Engl. transl. by W. D. Hooper and H. B. Ash). Harvard Univ. Press, Cambridge, Massachusetts, 1934.

Wallerius, J. G. 1761. Dissertation, Univ. of Upsala, Sweden.

Walters, S. M. 1961. *New Phytologist* **60**, 74–84.

Whitney, M. 1892. *U.S. Dept. Agr. Weather Bur. Bull.* **4.**

Whitney, M. 1900. *U.S. Dept. Agr. Rept.* **64.**

Whitney, M. 1901. *Yearbook (U.S. Dept. Agr.),* pp. 117–132.

Whitney, M. 1904. *U.S. Dept. Agr. Office Expt. Sta. Bull.* **142**, pp. 111–117.

Whitney, M., and Cameron, F. K. 1903. *Bur. Soils Bull.* **22.**

Yarilov, A. A. 1927. *Russ. Ped. Invest.* **11.**

Young, T. C., Jr., and Smith, P. E. L. 1966. *Science* **153**, 386–391.

Part III

GENERAL CLASSIFICATION EFFORTS

Editor's Comments
on Papers 12, 13, and 14

12   BALDWIN, KELLOGG, and THORP
Soil Classification

13   IVANOVA
An Attempt at a General Soil Classification

14   SOIL SURVEY STAFF
The Categories of the System

Although no universal or truly comprehensive soil classification scheme has yet been devised, the three papers reprinted here represent notable attempts in this regard. They depict vastly different approaches, especially in the case of Papers 12 and 14, which depict changes in U.S. concepts of classification over the past several decades. The approach outlined in Paper 13, though innovative and provocative, emphasizes the role of soil genesis and geography in classification falling in line with previous Russian efforts.

The scheme presented in the 1938 U.S. Department of Agriculture Yearbook (Paper 12) replaced the one developed by Marbut (1927, 1935) a few years earlier. The 1938 system retained named categories similar to Marbut's but of the six defined classes, only the two lowest remained unchanged; definitions of the other classes were appreciably modified. Category names of Marbut were retained, with the addition of order and suborder. The terms *pedocal* and *pedalfer* were dropped as names of soil orders but retained as informal terms for sets of great soil groups of the zonal order. The new categories were thus formalized in the following sequence, from highest to lowest: order, suborder, great soil group, family, series, and type. Over the years as the soil-survey program drew from experience at home and abroad, the widely used categories of type, series, and great soil group were increasingly revised. Modifications of the system are summarized by Riecken and Smith (1949) and Thorp and Smith (1949) for the first decade of use and by Simonson and Steele (1960) through

1960. The system, as it stood in final revised form in 1960, is outlined in Table 1 for orders, suborders, and great soil groups.

Table 1. Classification of soils according to the 1938 system and subsequent revisions

Order and suborder	Great soil group*
Zonal soils	
Soils of cold zones	Tundra soils
	Subarctic Brown Forest soils
Light-colored soils of arid regions	Desert soils
	Red Desert soils
	Sierozems
	Brown soils
	Reddish Brown soils
Dark-colored soils of semiarid, subhumid and humid grasslands	Chestnut soils
	Reddish Chestnut soils
	Chernozems
	Brunizems
	Reddish Prairie soils
Soils of the forest-grassland transition	Noncalcic Brown soils
Light-colored podzolized soils of timbered regions	Podzols
	Brown Podzolic soils
	Gray Wooded soils
	Sols Bruns Acides
	Gray-Brown Podzolic soils
Lateritic soils of forested warm-temperate and tropical regions	Red-Yellow Podzolic soils
	Reddish-Brown Lateritic soils
	Yellowish-Brown Lateritic soils
	Groups of Latosols
Intrazonal soils	
Halomorphic (saline and alkali) soils of imperfectly drained places	Solonchak (saline) soils
	Solonetz (alkali) soils
	Soloth soils
Hydromorphic soils of marshes, swamps, seep areas, and flats	Humic-Gley soils
	Alpine Meadow soils
	Bog soils
	Low-Humic Gley soils
	Planosols
	Ground-Water Podzols
	Ground-Water Laterite soils
Calcimorphic soils	Brown Forest soils
	Rendzinas
	Grumusols
	Calcisols
Dark soils on volcanic ash	Ando soils
Azonal soils	
(No suborders)	Lithosols
	Regosols
	Alluvial soils

Source: R. W. Simonson and J. G. Steele 1960, Soil (Great Soil Groups), in *McGraw-Hill Encyclopedia of Science and Technology*, vol. 12, McGraw-Hill, New York, p. 433.

* Each great soil group consists of several or many soil series.

Ivanova's review of Russian soil classifications (Paper 13) figures five different approaches, as grouped by Gerasimov. The various genetic classification systems, and there were still others subsequent to this discussion (for example, Kovda, et al., 1967; Rozov and Ivanova,1967; Kovda et al., 1969; Ivanova and Rozov, 1970), are based on: (1) connections between soil genesis and geography (Geographic Approach); (2) factors of soil formation (Factorial Approach); (3) internal or inherent soil characteristics (Morphogenetic Approach); (4) developmental intensity of soil-forming processes (Dynamic Approach); and (5) soil process groupings (Evolutional Approach). The general system described here, a product of the Soil Insitute of the Academy of Sciences of the USSR, incorporates the following taxonomic units, in order of increasingly narrowed class limits: type, subtype, genus, species, and variety. One such unit, the soil type, is defined on the basis of soils that are similar in terms of processes of transformation and translocation, water and thermal regimes, structure, natural fertility, and vegetation. Classification of soil types is thus based on their genesis, on properties that condition their origin, development, and geographic distribution, and relations between soils.

Soil types of the world are comprehended in global groups, called "soil formation groups," which are in turn subdivided into twelve classes, as follows: *Boreal Soil Formation Group* (tundra-Arctic soils, boreal-permafrost taiga soils, boreal-taiga forest soils, sub-boreal humid forest and meadow soils, sub-boreal steppe soils, and sub-boreal desert soils), *Subtropical Soil Formation Group* (humid forest soils; dry forest, savanna, and steppe soils; desert soils), and *Tropical Soil Formation Group* (humid savanna and forest soils, dry forest savanna soils, and tropical desert soils). Ivanova concludes that soils grouped on this basis helps determine ameliorative procedures, suitability for agricultural purposes, and relevant management and cultivation practices that are in the best interest of the state economy.

The current U.S. system, sometimes referred to as the "new soil taxonomy," was preceded, prior to publication as a monograph in 1975, by a sequence of "approximations" that were developed over a period of twenty-five years. Early trials came about after efforts to revise the 1938 system revealed serious deficiencies that could not be easily corrected. Attempts to group series into families, for example, proved to be unsuccessful because much change in one category altered its level of generalization requiring in turn changes in adjacent categories (Simonson, in press). Thus, initial attempts to improve individual categories led to an entirely new system first described in published form as "Soil Classification: A Comprehensive System—7th Approximation" (Soil Survey Staff, 1960). The first full description of the U.S. classification was published fifteen years later.

The new multiple-category system contained six categories called, in sequence from highest to lowest: orders, suborders, great groups, subgroups, families, and series. Although some of the names of categories are the same as in the previous system (order, suborder, family, series), *Soil Taxonomy* makes a significant and radical departure from previous efforts. All categories are used, which had not been done in any earlier system, making the system comprehensive or complete. Further, the system contains class limits that are explicit and quantitative, uses a neoteric nomenclature where terms are coined largely from Greek and Latin roots, and incorporates a number of innovative techniques that assist in the defintion and differentiation of classes.

One aspect of the system that initiated great hew and cry from specialists in other fields (for example, Hunt, 1972, p. 180) concerns the distinctive nomenclature. There is always some resistance to new proposals and especially to those that infiltrate other fields. The names of orders, suborders, and great groups given in Table 2 indicate the nature of the terminology as well as show the general organization of the higher categories. The situation is not as untenable as it may appear, because the names of classes in the four highest categories (three shown here) are systematic. Names of orders contain three or four syllables and every name ends with the suffix *sol*, the plural form being *sols*. Suborder names contain exactly two syllables, a formative from the name of the parent order plus a prefix syllable. Three or four syllable names are used for great groups by adding a prefix to the name of the parent suborder. All names of subgroups consist of two words where an adjective is placed ahead of the name of the parent great group. For example, follow the development up the sequence for Vertic Torrifluvents—a subgroup; Torrifluvents—a great group; Fluvents—a suborder; and Entisols—an order. Each element in a name connotes some specific property or attribute of the soil unit so designated and each name indicates its categorical level in the system.

The strict hierarchial nature of *Soil Taxonomy* and narrowly defined classes have troubled some soil classifiers who are concerned with the placement of intergrade soils (Butler, 1980, p.74). The multiple categories of the system, many as there are, still can not accommodate a full range of soil attributes. It may, after all, be

Table 2. Names of orders, suborders, and great groups in *Soil Taxonomy*

Order	Suborder	Great group	Order	Suborder	Great group
Alfisols	Aqualfs	Albaqualfs.	Histosols	Fibrists	Borofibrists.
		Duraqualfs.			Cryofibrists.
		Fragiaqualfs.			Luvifibrists.
		Glossaqualfs.			Medifibrists.
		Natraqualfs.			Sphagnofibrists.
		Ochraqualfs.			Tropofibrists.
		Plinthaqualfs.		Folists	Borofolists.
		Tropaqualfs.			Cryofolists.
		Umbraqualfs.			Tropofolists.
	Boralfs	Cryoboralfs.		Hemists	Borohemists.
		Eutroboralfs.			Cryohemists.
		Fragiboralfs.			Luvihemists.
		Glossoboralfs.			Medihemists.
		Natriboralfs.			Sulfihemists.
		Paleboralfs.			Sulfohemists.
	Udalfs	Agrudalfs.			Tropohemists.
		Ferrudalfs.		Saprists	Borosaprists.
		Fragiudalfs.			Cryosaprists.
		Fraglossudalfs.			Medisaprists.
		Glossudalfs.			Troposaprists.
		Hapludalfs.	Inceptisols	Andepts	Cryandepts.
		Natrudalfs.			Durandepts.
		Paleudalfs.			Dystrandepts.
		Rhodudalfs.			Eutrandepts.
		Tropudalfs.			Hydrandepts.
	Ustalfs	Durustalfs.			Placandepts.
		Haplustalfs.			Vitrandepts.
		Natrustalfs.		Aquepts	Andaquepts.
		Paleustalfs.			Cryaquepts.
		Plinthustalfs.			Fragiaquepts.
		Rhodustalfs.			Halaquepts.
	Xeralfs	Durixeralfs.			Haplaquepts.
		Haploxeralfs.			Humaquepts.
		Natrixeralfs.			Placaquepts.
		Palexeralfs.			Plinthaquepts.
		Plinthoxeralfs.			Sulfaquepts.
		Rhodoxeralfs.			Tropaquepts.
Aridisols	Argids	Durargids.		Ochrepts	Cryochrepts.
		Haplargids.			Durochrepts.
		Nadurargids.			Dystrochrepts.
		Natrargids.			Eutrochrepts.
		Paleargids.			Fragiochrepts.
	Orthids	Calciorthids.			Ustochrepts.
		Camborthids.			Xerochrepts.
		Durorthids.		Plaggepts	Plaggepts.
		Gypsiorthids.		Tropepts	Dystropepts.
		Paleorthids.			Eutropepts.
		Salorthids.			Humitropepts.
Entisols	Aquents	Cryaquents.			Sombritropepts.
		Fluvaquents.			Ustropepts.
		Haplaquents.		Umbrepts	Cryumbrepts.
		Hydraquents.			Fragiumbrepts.
		Psammaquents.			Haplumbrepts.
		Sulfaquents.			Xerumbrepts.
		Tropaquents.	Mollisols	Albolls	Argialbolls.
	Arents	Arents.			Natralbolls.
	Fluvents	Cryofluvents.		Aquolls	Argiaquolls.
		Torrifluvents.			Calciaquolls.
		Tropofluvents.			Cryaquolls.
		Udifluvents.			Duraquolls.
		Ustifluvents.			Haplaquolls.
		Xerofluvents.			Natraquolls.
	Orthents	Cryorthents.		Borolls	Argiborolls.
		Torriorthents.			Calciborolls.
		Troporthents.			Cryoborolls.
		Udorthents.			Haploborolls.
		Ustorthents.			Natriborolls.
		Xerorthents.			Paleborolls.
	Psamments	Cryopsamments.			Vermiborolls.
		Quartzipsamments.		Rendolls	Rendolls.
		Torripsamments.		Udolls	Argiudolls.
		Tropopsamments.			Hapludolls.
		Udipsamments.			Paleudolls.
		Ustipsamments.			Vermudolls.
		Xeropsamments.		Ustolls	Argiustolls.

142

Order	Suborder	Great group	Order	Suborder	Great group
		Calciustolls.			Fragihumods.
		Durustolls.			Haplohumods.
		Haplustolls.			Placohumods.
		Natrustolls.			Tropohumods.
		Paleustolls.		Orthods	Cryorthods.
		Vermustolls.			Fragiorthods.
	Xerolls	Argixerolls.			Haplorthods.
		Calcixerolls.			Placorthods.
		Durixerolls.			Troporthods.
		Haploxerolls.	Ultisols	Aquults	Albaquults.
		Natrixerolls.			Fragiaquults.
		Palexerolls.			Ochraquults.
Oxisols	Aquox	Gibbsiaquox.			Paleaquults.
		Ochraquox.			Plinthaquults.
		Plinthaquox.			Tropaquults.
		Umbraquox.			Umbraquults.
	Humox	Acrohumox.		Humults	Haplohumults.
		Gibbsihumox.			Palehumults.
		Haplohumox.			Plinthohumults.
		Sombrihumox.			Sombrihumults.
	Orthox	Acrorthox.			Tropohumults.
		Eutrorthox.		Udults	Fragiudults.
		Gibbsiorthox.			Hapludults.
		Haplorthox.			Paleudults.
		Sombriorthox.			Plinthudults.
		Umbriorthox.			Rhodudults.
	Torrox	Torrox.			Tropudults.
	Ustox	Acrustox.		Ustults	Haplustults.
		Eutrustox.			Paleustults.
		Sombriustox.			Plinthustults.
		Haplustox.			Rhodustults.
Spodosols	Aquods	Cryaquods.		Xerults	Haploxerults.
		Duraquods.			Palexerults.
		Fragiaquods.	Vertisols	Torrerts	Torrerts.
		Haplaquods.		Uderts	Chromuderts.
		Placaquods.			Pelluderts.
		Sideraquods.		Usterts	Chromusterts.
		Tropaquods.			Pellusterts.
	Ferrods	Ferrods.		Xererts	Chromoxererts.
	Humods	Cryohumods.			Pelloxererts.

Source: Soil Survey Staff, 1975, *Soil Taxonomy: A Basic System of Soil Classification for Making and Interpreting Soil Surveys*, U.S. Government Printing Office, Washington, D.C., pp. 86–87

unrealistic to expect discrete, clear cut units in the natural soil universe where soils exist as a continuum with one soil grading diffusely into the next. FitzPatrick (1980), for example, does not rely on the unity of a soil profile. Some profiles contain elements of different soil-stratigraphic layers (Finkl, 1980) which complicate the classification process. Attempts to reconcile the concept of polypedons with natural soil entities is another area of concern. Tests of the U.S. system in other countries, especially in tropical regions where the soils are less well known, indicate that certain sections of the scheme could stand improvement (Buringh, 1979). In spite of minor shortcomings, *Soil Taxonomy* seems to be the best sytem so far.

REFERENCES

Buringh, P., 1979, *Introduction to the Study of Soils in Tropical and Subtropical Regions*, Center for Agricultural Publishing and Documentation, Wageningen, The Netherlands, 124 p.

Butler, B. E., 1980, *Soil Classification for Soil Survey*, Clarendon Press, Oxford, 129 p.

Finkl, C. W., Jnr., 1980, Stratigraphic Principles and Practices as Related to Soil Mantles, *CATENA* **7**:169–194.

FitzPatrick, E. A., 1980, *Soils: Their Formation, Classification and Distribution*, Longman, London, 353 p.

Hunt, C. B., 1972, *Geology of Soils*, Freeman, San Francisco, 344 p.

Ivanova, E. M., and M. M. Rozov, eds., 1970, *Classification and Determination of Soil Types*, Nos. 1–5, U.S. Dept. of Commerce, Springfield, Va., 271 p.

Kovda, V. A., Y. V. Zobova, and V. V. Rozanov, 1967, Classification of the World's Soils, *Soviet Soil Sci.* **7**:851–863.

Kovda, V. A., B. G. Rozanov, and Y. M. Samoylova, 1969, Soil Map of the World, *Soviet Soil Sci.* **1**:1–9.

Marbut, C. F., 1927, A Scheme of Soil Classification, *1st Intern. Congr. Soil Sci. Trans.* **4**:1–31.

Marbut, C. F., 1935, Soils of the United States, in *Atlas of American Agriculture, Part III*, O. E. Baker, ed., U.S. Department of Agriculture, Washinton, D.C., pp. 12–15.

Riecken, F. F., and G. D. Smith, 1949, Lower Categories of Soil Classification: Family, Series, Type, and Phase, *Soil Sci.* **67**:107–115.

Rozov, N. N., and E. N. Ivanova, 1967, Classification of the Soils of the USSR, *Soviet Soil Sci.* **2**:147–156; **3**:288–300.

Simonson, R. W., in press, Soil Classification History, United States, in *The Encyclopedia of Soil Science, Part II: Morphology, Genesis, Classification, and Geography*, C. W. Finkl, Jnr., ed., Hutchinson Ross, Stroudsburg, Pa.

Simonson, R. W., and J. G. Steele, 1960, Soil (Great Soil Groups), in *McGraw-Hill Encyclopedia of Science and Technology*, vol. 12, pp. 433–441.

Thorp, J., and G. D. Smith, 1949, Higher Categories of Soil Classification: Order, Suborder, and Great Soil Groups, *Soil Sci.* **67**:117–126.

12

Reprinted from pages 979–1001 of *Soils and Men: U.S. Dept. of Agriculture Yearbook, 1938*, G. Hambridge, ed., U.S. Government Printing Office, 1938, 1232 p.

Soil Classification

By MARK BALDWIN, CHARLES E. KELLOGG,
and JAMES THORP [1]

THE soil is a more or less continuous body covering that portion of the land surface of the earth upon which plants grow. That its characteristics vary from place to place probably was recognized by man as soon as agriculture began. The importance of such variations is emphasized in all the early writings dealing with agricultural affairs. This recognition of different kinds of soils and the application of names to them were early steps in soil classification made to satisfy a definite practical need. According to early Chinese records, a classification of soils, made largely on the basis of their color and structure, was developed by the engineer Yu during the reign of the Emperor Yao about 4,000 years ago (*405*).[2]

The soil type as conceived by the modern scientist represents the combined expression of all those forces and factors that, working together, produce the medium in which the plant grows. The fundamental soil types can be described and their capabilities for use can be defined through the interpretation of experimental data and experience. After these types have been defined, knowledge regarding them can be accumulated and classified; and with their distribution shown on maps, this knowledge may be extended to definite areas of land easily and directly.

Since there are a great number of different kinds of soil varying from one another in different degrees of contrast, it is necessary to group them into progressively higher categories in order that the maximum application of our knowledge may be made.

---

[1] Mark Baldwin is Senior Soil Scientist, Charles E. Kellogg is Principal Soil Scientist, and James Thorp is Soil Scientist, Soil Survey Division, Bureau of Chemistry and Soils.
[2] Italic numbers in parentheses refer to Literature Cited, p. 1181.

EARLY SYSTEMS OF CLASSIFICATION

The early recognition of soil differences was based on local observations and served local or limited purposes. Many were based on single features of the soil, such as texture or color. These differentiations, while incomplete, were scientifically valid, since they dealt with true soil differences. The rise of geology as a distinct science with field methods and the recognition of the close relationship between soil and its parent material (in most instances the geological formation beneath it) led to a classification based on the composition of the underlying formations, such as the one defined by Fallou (107). Other systems of classification, [3] based on features lying outside the soil itself or only partly on soil characteristics, were developed. Some were based on geology strictly, others on physiography, plant ecology, [4] or agricultural quality, or combinations of these. Some of the schemes were fairly complete, in the sense of providing categories [5] and groups for the various features under observation. Thus Richthofen's system (308), based for the most part on the geology of the parent material, was sufficiently broad and complete to encompass practically all of the materials of the earth's surface. But it was not a soil classification; the units and the groups were not defined on the basis of soil characteristics, and the nomenclature was geological.

About 1870 a new school of soil science was founded in Russia under the leadership of Dokuchaiev. The scientists of this school recognized that each soil has a definite morphology, or form and structure, which is associated with a particular combination of vegetation, climate, relief, parent material, and age. They stressed the fact that soil is not a geological formation but an independent natural body, and they developed systems of classification in harmony with this new concept. Sibirtsev's Genetical Soil Classification of 1895 (1) illustrates in brief form the early trend of Russian soil science toward a genetic soil classification:

Division A. Soils wholly developed or zonal
1. Laterite soils.
2. Aeolian-loess soils.
3. Desert-steppe soils.
4. Chernozem.
5. Gray-forest soils.
6. Podzolized soddy (turfy) soils.
7. Tundra soils.

Division B. Intrazonal soils
8. Alkaline soils.
9. Moor-and-bog soils.

Division C. Immature soils
10. Coarse soils.
11. Alluvial soils.

Here the concept of three main groups of soil, zonal, intrazonal, and azonal (immature) was first presented. The first includes those soils having well-developed soil characteristics that reflect the influ-

[3] For a good discussion of some of the early schemes of soil classification see Afanasiev (1).
[4] Plant ecology deals with the mutual relations between plants and their environment, i. e., the relationship of plants to soil, land relief, climate, and other organisms.
[5] The term "category" is used in the sense of a class to which objects of knowledge may be reduced and by which they may be arranged in a system of classification. It is approximately equivalent to "class," but this term has already been appropriated in soil science for the various grades of soil texture, such as loam, sandy loam, and clay loam. It should be clearly distinguished from "group." Similar soils may be placed in a group, the Bellefontaine series; several similar series may be placed in a broader group, the Miami family; and several similar families in a still broader group, the Gray-Brown Podzolic great soil group. Any one particular series, family, or great soil group is a group of soils, but series, family, and great soil group, conceived as separate parts in a system of classification, are categories. Thus Bellefontaine is not a category of soil classification but one particular group of soils in the category "series."

ence of the active factors of soil genesis—climate and vegetation; the second, those soils having more or less well-developed soil characteristics that reflect the dominating influence of some local factor of relief, parent material, or age over the normal effect of the climate and vegetation; and the third, those soils without well-developed soil characteristics.

These early Russian investigators were concerned chiefly with the determination of general characteristics and the recognition of soil units that could be given broad geographic expression. They did not define local soil types or groups within the lower categories. As the Russian workers developed their science, the classification came more and more to be based upon soil characteristics and less and less on the environmental factors that produced them.

DEVELOPMENT IN THE UNITED STATES

About 40 years ago soil survey work was instituted in the United States for the purpose of defining and mapping the important soil types in the country. This research was started in the Department of Agriculture and naturally had agronomic purposes. The investigations have continued, in cooperation with the agricultural experiment stations of the States and Territories, up to the present time. Naturally as the science has progressed and as an increasing amount of knowledge has become available, there has been a continued development toward a system based strictly on soil characteristics—less on the environmental and external features and more on the internal soil morphology. From the beginning, in the field classification of soils all features have been taken into consideration that appeared to the scientist to influence the suitability of soils for crops.

The classification is based mainly upon the physical properties and condition of the soil . . . any chemical feature, such as deposits of marl, or highly calcareous soils, or of highly colored soils is considered as well as character of the native vegetation and the condition of crops. . . . only such conditions as are apparent in the field, such as the texture as determined by the feel and appearance, the depth of soil and subsoil, the amount of gravel, the condition as to drainage, and the native vegetation or known relation to crops, are mapped (459).

A system of nomenclature was set up: "Each well-defined area is established as a class and given a local name." The term "class" referred to the texture of the surface soil. The geographical significance of the word "area" is apparent. The unit of classification, called the type elsewhere in the report, had a definite geographical expression. In fact, the choice of a place name as part of the soil-type name, as Roswell sandy loam, is significant and probably implied the concept of restricted distribution for any given type. The scale of the maps (1 inch = 1 mile) was considered ample to allow the delineation of all features of significance. Thus was laid the groundwork for the establishment of soil units, through the accumulation of soil data in the field, and the delineation of the boundaries of the units.

In the same report there is an implied grouping of these soil types, as indicated by the names Roswell loam and Roswell sandy loam, and the development of the concept of the soil series. The basis of the grouping, however, was not clearly conceived. Later the grouping of soil types into soil series had its basis in common geologic origin (not composition) of the parent material. Thus the Miami soil types were grouped into a single soil series (Miami), because they were "derived from glacial drift."

This grouping of soils on the basis of the geological origin of the parent material from which they were developed led to the grouping of soil types that had few true soil characteristics in common in a single soil series. The result was naturally confusing, and the purpose of classification was defeated. This defect gradually and necessarily was corrected as soil science developed in the United States.

As the work progressed, correlation of soil types between widely separated areas was attempted, and this in itself revealed the necessity of more rigid definitions of the soil units if the work was to have wide value either scientifically or practically. There was a gradual shifting from the geological definition of soil series to one strictly pedological, i. e., one based entirely on soil characteristics, and in 1920 Marbut (236), then chief of the United States Soil Survey, definitely listed eight features of the soil profile necessary to the definition of a soil unit. These features were:

(1) Number of horizons in the soil profile.[6]

(2) Color of the various horizons, with special emphasis on the surface one or two.

(3) Texture of the horizons.

(4) Structure of the horizons.

(5) Relative arrangement of the horizons.

(6) Chemical composition of the horizons.

(7) Thickness of the horizons.

(8) Geology of the soil material.

In his paper Marbut further analyzed the data from the soil surveys, which by that time had extended into most parts of the United States, and empirically and logically proceeded to the building up of the broader divisions or categories of soil classification, basing the definition of the groups entirely on soil characteristics. The hundreds of soil types and phases, used as units of mapping, comprised the first or lowest category and included all soils. He was now prepared to group the soils of the United States in broader categories. Accordingly, the multitude of soil types were combined on the basis of their characteristics into two great groups—since named by him the Pedocals and the Pedalfers—to form the highest category. The names were coined by combining the Greek word "pedo" (ground) with an abbreviation of the Latin word "calcis" or "calx" (lime) and with abbreviations of the Latin words "alumen" and "ferrum" (aluminum and iron). Pedocals are distinguished by the accumulation of carbonates of calcium or of calcium and magnesium in all or a part of the soil profile. Pedalfers are distinguished by the absence of carbonate of lime accumulation and usually by an accumulation of iron and aluminum compounds. Marbut, of course, recognized the relationship of these great soil groups to the climatic zones of the country, the first to the subhumid, semiarid, and arid regions, the second to the humid regions. The Pedocals and Pedalfers were subdivided into groups in lower categories on the basis of their character-

---

[6] Many of the soil terms used are explained in the previous articles, and short definitions are given in the Glossary.

istics. The technical details of the system are set forth in Marbut's last great monograph (*236*). His grouping by categories is summarized in table 1.

Table 1.—*Soil categories*

Category VI............	Pedalfers (VI-1)	Pedocals (VI-2)
Category V............	Soils from mechanically comminuted materials. Soils from siallitic decomposition products. Soils from allitic decomposition products.	Soils from mechanically comminuted materials.
Category IV............	Tundra. Podzols. Gray-brown Podzolic soils. Red soils. Yellow soils. Prairie soils Lateritic soils. Laterite soils.	Chernozems. Dark-brown soils. Brown soils. Gray soils. Pedocalic soils of Arctic and tropical regions.
Category III............	Groups of mature but related soil series. Swamp soils. Glei soils. Rendzinas. Alluvial soils. Immature soils on slopes. Salty soils. Alkali soils. Peat soils.	Groups of mature but related soil series. Swamp soils. Glei soils. Rendzinas. Alluvial soils. Immature soils on slopes. Salty soils. Alkali soils. Peat soils.
Category II............	Soil series.	Soil series.
Category I............	Soil units, or types.	Soil units, or types.

In his publication Marbut discussed the "geologic, topographic, physiographic, climatic, and biologic factors" of soil formation, distinguishing between the dynamic (climatic and biologic) factors and the passive factors. He pointed out the geographic significance of these factors and the consequent geographic distribution of the product of their interaction, the soil. The soil proper (the solum) is distinguished from the underlying material (C horizon), and the features which form the differentiating characteristics of soils are mentioned. The categories are defined, and the groups within the categories are described and named.

The desirability of geographic expression in broad zones, correlated with other broad geographic features, was recognized by Marbut in the emphasis he placed upon the so-called mature (zonal) soils, his category IV.

Those kinds of soils which bear the impress of local features of the environment do not find a fully satisfactory place in the scheme, in spite of the profound differences which distinguish them from their associated normal or mature (zonal) soils. They are listed in category III of Marbut's table. In examining such soils it is evident that they are of two general kinds: (1) Those soils which have definitely developed and in many instances strongly developed profile characteristics that reflect a local but dominating feature of the environment or parent material, such as poor drainage or calcareous parent material; (2) those soils which are without definite profile features, owing to youth, characteristics of parent material, or conditions of relief that prevent or inhibit the development of such features.

These two great groups of soils without normally developed profiles are broadly similar to the groups called intrazonal and azonal (or im-

149

mature) by Sibirtzev in 1895. Glinka (*124*) objected to these names, as well as to the name "zonal," partly because of their geographical connotations. This may be a valid objection to the use of these words in soil classification, but the concepts seem sound, and in the absence of a better nomenclature, these words are used as the names for the groups of the highest category in the system of classification outlined in the following pages.

All the great soil groups listed in Marbut's scheme of classification (categories III and IV) are still recognized, but some changes in names and in the arrangement of categories have been found desirable. The characteristics of the great soil groups are briefly described in the article Formation of Soil and are summarized in table 3 in the Appendix (p. 996). Before proceeding further with a discussion of the higher categories of classification, it will be necessary to define more precisely the lower categories as they are now conceived.

SIMPLE UNITS OF CLASSIFICATION

Three categories are commonly recognized in the classification of soils in the field—(1) series, (2) type, and (3) phase. The grouping of these units in higher categories will be dealt with presently.

The most important of these field units is the soil series—defined as a group of soils having horizons similar as to differentiating characteristics and arrangement in the soil profile and developed from a particular type of parent material. Except for texture, especially of the A horizon, the morphological features of the soil profile, as exhibited in the physical characteristics and thicknesses of the soil horizons, do not vary significantly within a series. These characteristics include especially structure, color, and texture (except the texture of the A horizon, or surface soil) but not these alone. The content of carbonates and other salts, the reaction (or degree of acidity or alkalinity), and the content of humus are included with the characteristics which determine series.

Each soil in a series is developed from parent material of similar character. Parent material for soil is produced from rocks through the forces of weathering. Similar parent materials may be produced from different geological deposits and in different ways, and unlike parent materials may be produced from the same rocks because of differences in weathering. It is the character of the parent material itself which is important.

It follows that the external characteristics and environmental conditions of the soils within a series will also be similar. Each series has its characteristic range in climate and relief. Ordinarily the more strongly the soil characteristics are developed, the narrower is this range in external features. Except for young soils or those owing their distinctive characteristics to some unusual feature of the parent material, all the soil types within a series have essentially the same climate. It is to be expected, of course, that any differences in climate or relief sufficient to influence the native vegetation significantly would be reflected in the internal characteristics of the soil.

Variations in texture, especially of the A horizon, occur within a series. In former years soils having considerable range in texture throughout the entire profile were sometimes included within a series.

Significant differences in the texture of the B horizon or of the parent material are now considered to be sufficient grounds for recognizing new series.

The soil series are given names taken from place names near the spot where the soil was first defined, such as Miami, Hagerstown, Mohave, Houston, and Fargo. Many of the first series recognized in the United States were given such broad definitions that it became necessary to split them into several series after the soils had been studied more thoroughly. For example, several soil series are now recognized for soils included with the Miami and Carrington as first defined. Of course, the definitions of series cannot be made more closely than the limits of observation and measurement with available field techniques. Such techniques have improved considerably during the past 40 years, and there is promise of their further development.

The soil type is the principal unit used in detailed soil researches. The definition of soil type is identical with that of soil series, except that the texture of the A horizon does not vary significantly. Thus, there may be one or more types within a series, differentiated from one another on the basis of the texture of the surface soil, the upper 6 to 8 inches. Since the greater part of the roots of crop plants are in this upper soil layer and since this part of the soil is directly involved in tillage and fertilization, especial emphasis has been given its texture.

Attention has already been directed to the determination and nomenclature of soil textural classes. The class name of the A horizon (or average of the surface soil to a depth of 6 to 8 inches in soils with weakly developed profiles), such as sand, sandy loam, loam, silt loam, clay loam, or clay, is added to the series name to give the complete name of the soil type. For example, Miami loam and Miami silt loam are two soil types within the Miami series. With the exception of the texture of the surface soil, these two soil types have the same differentiating characteristics, both internal and external. In Bog soils the word peat or muck, whichever is appropriate, is added to the series name to give the complete name of the soil type.

During the time when special emphasis was placed on the geological character of the parent material, soil series were defined in terms that allowed a wide range in soil characteristics, and several types were included within a series. In a few instances the texture of the soil beneath the surface layer was given major emphasis in determining the class name of the soil type before the present concepts were so precisely defined. As the definitions of soil series came to be made more accurately in terms of soil characteristics, there were fewer types within each series. This is to be expected, for it is inconceivable that soils varying greatly in texture would be similar in their other characteristics. Young or otherwise undeveloped soils, such as alluvial soils, may have a considerable range in texture, although by no means the whole range from sand to clay, and still fall within the limits of a particular series. Well-developed soils are now being classified in series having but one or two or, at most, three types. As research continues, the series with only one type will become still more common. Within the range permitted in a soil type there may be small differences in climate—frostiness, for example—of much greater significance to crop plants than to the native plants. Similarly differences in relief, of little or no importance to the native vegetation, may be significant in the use of the soil when the land is cultivated. Such differences are recognized and mapped as phases of specific types.

A phase of a soil type, then, is defined on the basis of characteristics of the soil, or of the landscape of which the soil is a part, that are of importance in land use but are not differentiating characteristics of the soil profile. The three most important of such characteristics are slope, stoniness, and the degree of accelerated erosion. For example, from the point of view of land use, there are five principal classes of land defined according to slope, as follows:

(1) Nearly level to level land, on which external drainage is poor or slow. From the point of view of slope there is no difficulty in the use of agricultural machinery nor is there likelihood of water erosion.

(2) Gently undulating land, on which external drainage is good but not excessive. All types of ordinary agricultural machinery may be used with ease, and there is little likelihood of serious water erosion.

(3) Gently rolling lands, on which external drainage is good to free but not excessive. Ordinary agricultural machinery may be used, but the heavier types of equipment with difficulty. On soils subject to erosion there is likelihood of water erosion where intertilled crops are planted.

(4) Strongly rolling land, on which agricultural machinery cannot be used. External drainage is free, but sufficient water is available for a good grass cover. Soil erosion is likely to be serious on land planted to cultivated crops.

(5) Steeply sloping and hilly land, on which external drainage is so excessive that good pasture grasses cannot maintain themselves, although trees may be able to do so.

Frequently soil types have no greater range in slope than that allowed within one slope class, but other soil types have a greater range, and in such instances the variations are recognized as phases. The important criteria of these slope classes is not the percentage of slope but their land-use definitions. In itself slope has a limited significance; its importance can be studied and evaluated only in respect to a definite type of soil. For example, some soils with a 5-percent slope erode easily when devoted to clean cultivation, whereas others erode very little under such treatment, even with slopes in excess of 50 percent.[7]

In a similar way phases are defined for differences in stoniness and accelerated erosion (195).

HIGHER CATEGORIES OF CLASSIFICATION

The soil series are grouped in higher categories according to their characteristics. Of particular importance to our purpose are the great soil groups. Several of the great soil groups in the United States include hundreds of soil series, differing from one another in important ways because of differences in parent material, relief, and age, but all showing the same general sort of profile. Groups of soils between series and great soil groups, or families of closely related soil series,

---

[7] The percentage of slope indicates the number of feet drop for 100 feet in a horizontal plane. A 5-percent slope drops 5 feet in 100, a 50-percent slope 50 feet in 100, and so on.

have been recognized, such as the Miami family, including the Miami, Bellefontaine, Hillsdale, Russell, Fox, and similar soils. On the whole, however, there has been no consistent grouping of all series into strictly defined families intermediate between the soil series and the great soil groups, and based on soil characteristics. This problem may be expected to receive an increasing amount of attention as research proceeds.

The great soil groups, in turn, can be placed in several suborders and three orders—(1) zonal, (2) intrazonal, and (3) azonal.

Except where the continuity of the landscape is interrupted by mountains or large bodies of water, zonal soils occur over large areas, or zones, limited by geographical characteristics. Thus the zonal soils include those great groups having well-developed soil characteristics that reflect the influence of the active factors of soil genesis— climate and living organisms (chiefly vegetation). These characteristics are best developed on the gently undulating (but not perfectly level) upland, with good drainage, from parent material not of extreme texture or chemical composition that has been in place long enough for the biological forces to have expressed their full influence.

The intrazonal soils have more or less well-developed soil characteristics that reflect the dominating influence of some local factor of relief or parent material over the normal effect of the climate and vegetation. Any one of these may be associated with two or more zonal groups, but no one with them all.

The azonal soils are without well-developed soil characteristics either because of their youth or because conditions of parent material or relief have prevented the development of definite soil characteristics. Each of them may be found associated with any of the zonal groups.

The arrangement of the principal groups of soils according to these concepts is shown in table 2 in the Appendix (p. 993). The distribution of the more extensive great soil groups in the United States is shown on the map of soil associations at the end of this Yearbook. In only a few instances are there areas of the intrazonal and azonal groups large enough to separate on a small-scale map, but they occur scattered throughout the regions generally occupied by the zonal soils. In those parts of the country where climate and other conditions change greatly within short distances, as in the far Western States, it is not everywhere possible to separate the zonal groups on small-scale maps.

Although the classification must proceed from the small groups upward to the progressively larger groups differentiated by a decreasing number of characteristics, the details of this process are too voluminous to develop here. As there are several thousand individual soil types in the United States, no attempt will be made here to discuss them. The reader will need to consult the separate soil survey reports of particular areas for the description of local soil types.[8]

The scheme of soil classification outlined in table 2 (p. 993) is designed to make it possible to trace any local soil type logically and directly from the lowest to the highest category. It is believed that all soils will fall into one of the three orders, zonal, intrazonal, or azonal, but

---

[8] The Soil Survey Division has published more than 1,500 individual maps and reports since its initiation in 1899. The Illinois Agricultural Experiment Station and some other research institutions have published a few additional soil maps.

it will probably be necessary to add new suborders and great soil groups from time to time as more is learned about the soils of the world.

Geographic Association of Soil Units

In order that the data of soil science may be understood and made available for the solution of practical problems it is necessary that these units of classification be expressed upon maps. The significance of the data shown by such maps is dealt with elsewhere, but their relationship to the problem of classification may be discussed briefly here. The simple units—series, types, and phases—must be shown upon large-scale maps in order that their relationship to one another, to the other local features of the landscape, and to the detailed pattern of human occupancy may be understood. In order that broader relationships may be understood and regional problems attacked, smaller scale maps showing the distribution of soil groups in the higher categories, especially the great soil groups, must be compiled.

Since the soil is the combined product of climate, living organisms, relief, parent material, and age, each different combination of these factors will produce a different soil. If all variations in each factor were measurable, and had measurable influences on the soil, individual soil types would be so numerous that they would occupy points. In a strict sense each soil profile is individual; no two are identical in every detail. Since there can be some range in the environmental factors without producing measurable differences in the soil, each soil type occupies an area rather than a point. The size and shape of individual areas varies greatly in different places.

By constantly enlarging the scale of a map, individual soil types can be shown separately, regardless of the size of the separate areas or the complexity of their pattern. If the scale is fixed at any practical point, however, certain soil types must be grouped together and shown as complexes. As the scale of the map is decreased, the number of such complexes will increase and the number of individual soil types shown will decrease. The definitions of individual complexes are made in terms of the geographic pattern of the soil units making up the complex.

Thus there is an important difference between geographic groups or associations of soils and groups based strictly on soil characteristics in the system of classification. For example, one may find series, types, and phases of Bog soils, Half Bog soils, alluvial soils, and Gray-Brown Podzolic soils in such close association that the individual great soil groups can be shown only on maps of a scale of an inch to the mile or even larger. On any small-scale map it would be impractical to delineate these groups separately. Since they have no common internal characteristics, it is out of the question to place a soil of the Gray-Brown Podzolic great group and one from the great group of Bog soils in the same order or suborder, to take an extreme example. Although they are intimately associated geographically and have the same climate, their profiles and the chemical and physical properties of their horizons are entirely unlike. The fact that an oak tree and a pine tree may be growing side by side is insufficient reason for placing them in the same species or family. The ecologist must recognize and define particular associations of plants if he wishes to make a generalized or schematic map of vegetation. Similarly an alluvial soil and a Bog soil cannot be classified in the same order, but they can form a part of an association or complex, defined as consisting of certain closely associated soil types with a characteristic pattern of distribution.

The soil complex is used frequently in both detailed and reconnaissance soil mapping and on generalized maps. It is a unit for the purposes of mapping, not a category in soil classification. Two complexes may be quite unlike, yet be composed of the same soil units, in different proportions or in different patterns. The differences between the soil units—series, types, and phases—may be due to differences in any factor responsible for their development. Ordinarily any one complex will lie within the region of one zonal great soil group, and differences between the units composing it will be due to variations in parent material, relief, or age. In mountainous regions, however, where environmental conditions are very complex, the zonal groups of soils may be closely associated in intricate patterns. On any small-scale map, showing the distribution of the zonal great soil groups (Chernozem, Podzol, etc.), soils belonging to intrazonal and azonal groups necessarily are included in the area occupied by a particular group of zonal soils. Thus the great soil groups, as shown on these maps, are, in a sense, complexes of the normal zonal soils and their geographic associates in the intrazonal and azonal groups.

Soils may be grouped in other ways for specific studies. Especially it may be important to group together all the soils in a soil region developed from the same parent material but differing in relief and in degree and character of profile development. Figure 2 (p. 890) shows such a series of profiles, developed from similar parent material. These soils may be expected to occur in association, although not necessarily in equal proportion. They may make an intricate pattern or a simple one. For such a group Milne (*265*) has suggested the appropriate term "catena" (Latin for chain). It may or may not be possible to map the catena, depending upon the uniformity of the factors other than relief. The concept of the catena has proved useful in the United States as a means of facilitating the logical grouping of soil units and for remembering their characteristics and relationships.

NOTES ON NOMENCLATURE [9]

Marbut's terms "Pedocal" and "Pedalfer" have a useful connotation, but they seem to form a better basis for a grouping within the great soil groups than for forming a separate category. It is not feasible, for example, to classify Degraded Chernozem under Pedocal, since it has many of the features of a Pedalfer while still retaining, in some instances, a good part of its carbonate of lime accumulation—one of the principal distinguishing characteristics of the Pedocal. Prairie soils, on the other hand, have no lime accumulation but do have dark-colored, humus-rich A horizons, much like those of the Chernozems, and show but little evidence of podzolization.

Several great soil groups have been renamed in order to eliminate such geographical terms as "southern," "northern," and "eastern." Even if valid for the United States, these terms are inappropriate in Mexico, South America, and other parts of the world. Descriptive terms have been substituted. For example southern brown soils are renamed Reddish Brown soils; northern dark-brown

[9] This section is intended primarily for soil scientists and students of soil science.

soils are now given the European name of Chestnut soils; and southern dark-brown soils are now called Reddish Chestnut soils. Noncalcic Brown soils show several characteristics common to podzolic soils, and, although their profiles are relatively weakly developed, they seem to belong within the outer range of the Pedalfers.

Distinctions among soils of the deserts are not sharp. Characteristics seem to depend largely on vegetation and temperature conditions, and there is much evidence to support the view that age and former relief have been extremely important in determining their character. Brown and Reddish Brown Pedocal soils are indicated as transitional between suborders 1 and 2. Possibly they might comprise another suborder, but it is not certain whether this disposal of them would be justified. They correspond respectively to Marbut's Brown and southern Brown soils.

Chernozem, a literal translation of which is black earth, was formerly described to include not only the nearly black Pedocal soils of the northern Great Plains, but also the dark-brown soils of the southern Great Plains, which have a definitely red or pink tinge, especially in the lower horizons. It is now thought that the latter belong more properly to the southern dark-brown group, for which the new name "Reddish Chestnut soil" is proposed. The subhumid parts of the southern Great Plains, where one would expect to find Chernozem, are underlain by soft marly limestones with black or very dark brown Rendzina soils developed upon them.

The Noncalcic Brown soils seem to owe their characteristics to the wet-dry subhumid climate and to the forest-grass vegetation characteristic of their environment. They were first recognized as a broad group in China under the name Brown soils (*355*), but as this term conflicts with the Brown Pedocal soils, the name "Shantung Brown" soils (*407*) was proposed later. The alternative name of Noncalcic Brown soils has the advantage of eliminating geographic restrictions and at the same time of clearly distinguishing these soils from the Brown Pedocals.

The proposed term "Brown Podzolic" soils covers a great group of soils that have some of the morphological features of immature Podzols. Their recognition as a group of category IV is based upon the wide geographic distribution of their characteristic profile and upon the apparent equilibrium of the well-developed soils with their environment.

Much of Marbut's Southern Prairie belongs to the Rendzina group (as was fully recognized by Marbut himself). Some relatively small areas of true Prairie soils do exist in warm-temperate humid areas of the United States. They have a reddish-brown color and are called Reddish Prairie soils, in conformity with the terminology used for the Pedocals of warm climates. Since much of the Reddish Prairie soil, now recognized in the United States, is developed on red materials, it is not yet known to what extent the color is inherited and to what extent it is developmental. Certainly the Reddish Prairie soils generally contain somewhat less organic matter than the Prairie soils of cooler regions.

It is well-known that there are important areas of Prairie soils in tropical regions, but their characteristics are not yet well enough known to make possible their proper classification.

Table 2 (p. 993) shows an overlap in suborders between the podzolized and lateritic soils. This is because many of the soils whose parent materials are of a lateritic nature show strong morphological and chemical evidences of podzolization. Laterization and podzolization are both active in the humid Tropics. Soils of the Tropics in general, and especially those of the humid and wet-dry Tropics, are not satisfactorily classified. This is more because of a lack of systematized study of data than of a lack of data, although there is still need for a vast amount of field and laboratory work on these soils. For example, we know that there are interrelationships involving Red and Yellow soils, some of which are sticky and plastic, whereas others are granular and friable. Colloids of some have low silica-alumina and silica-sesquioxide ratios, while in others the reverse is true. Marbut proposed the name "Tropical Red Loams" for soils containing a high percentage of friable clays. Although these soils contain much clay, their friable nature gives them a physical character more closely akin to loam. This group corresponds closely to the Reddish-Brown Lateritic soils of the present classification. Yellowish-Brown Lateritic soils have similar physical properties but a decidedly different color. To what extent the chemical properties differ has not yet been determined.

Many soils of the humid wet-dry tropical regions are developed on residual material showing a strong degree of reticulate mottling, apparently caused by

156

partial segregation of iron compounds from clays. This type of mottling is characteristic of the material originally called Laterite by Buchanan, but it is very common throughout the region of lateritic soils even where chemical characteristics are different from those originally recognized as Laterites. Some of the Red and Yellow soils of the tropical regions are characterized by high organic content in the surface soil, while in others there is but little organic material. Although rocks weather very rapidly in humid tropical regions, the factor of time still remains very important in the development of soils; it is not yet known to what extent this factor influences the characteristics of soils that must be recognized in classification.

Much has been written, especially by European pedologists, concerning Terra Rossa (literally translated as red earth). The term has been widely applied to red soils developed under the warm-temperate Mediterranean type of climate, marked by wet and dry seasons. Many writers have preferred to limit Terra Rossa to soils developed on limestones, while some would have it include any red soil in a Mediterranean climate. According to Blanck's Handbuch der Bodekunde (38), Terra Rossas include red soils which vary greatly as to silica-alumina and silica-sesquioxide ratios in the colloidal fraction, and as to lime content. Some analyses show more than 10 percent of calcium oxide, whereas others show only a trace. It is well known that red soils developed on limestones vary in character from strongly podzolized red soils, on the one hand, to true Laterites on the other, and from strongly acid soils to those high in free carbonate of lime. Silica-alumina ratios vary from considerably more than two to much less than one. It seems evident from these facts that Terra Rossa cannot be classified satisfactorily until it has been defined more exactly. At present its only distinction lies in its color.

The terms "halomorphic," "hydromorphic," and "calomorphic" are not entirely satisfactory, since soil genetics rather than soil characteristics are implied. These names were used because they are conveniently short and because certain soil characteristics are associated with high salt content, with wet conditions, and with the presence of absorbed calcium. It would be desirable for these terms to be more descriptive in nature. Broad groups under these suborders more nearly conform to the desirable descriptive name.

The term "Planosol" is being proposed to cover those soils with claypans and cemented hardpans not included with the Solonetz, Ground-Water Podzol, and Ground-Water Laterite. Families of Planosols correspond to associated normal zonal soils. For example, the Grundy family represents the Planosol associated with the Prairie, the Clermont that associated with the Gray-Brown Podzolic, and the Crete that associated with the Chernozem.

Brown Forest soils are here recognized as calomorphic because of their high absorbed calcium. They seem to correspond to Ramann's original Braunerde and are distinguished from the associated Gray-Brown Podzolic soils by lack of evidence of podzolization. They are somewhat leached but have not developed eluvial and illuvial horizons to any appreciable extent. Incomplete evidence indicates that Brown Forest soils may extend well into the Tropics.

Ground-Water Laterites are characterized by hardpans or concretional horizons rich in iron and aluminum compounds and sometimes in manganese. The only family of this group shown in the table is the Tifton, and the Tifton and Caguas series are given as examples. The Caguas series, as mapped in Puerto Rico, includes both Ground-Water Laterite with hardpan and soils with concretionary horizons. These inclusions were made because of scale limitations on the map and not because of a misunderstanding of the character of the soils involved.

Although the azonal soils bear a much stronger imprint of their geological origin than the zonal and intrazonal soils, the fact remains that climatic and vegetation zones have had some influence on their character and development. Azonal soils of desert regions, particularly the Lithosols, are usually alkaline in reaction and often are actually calcareous even where the rock materials do not contain free lime. Lithosols within the Podzol region are likely to be acid in reaction, but this is not always the case and certainly is not true where the parent rocks contain free lime.

Alluvial soils as recognized in the field may or may not have some of the local zonal influences impressed in their characteristics. Their character depends very largely on their source, but local conditions of drainage and vegetation very soon have some influence on their nature after their deposition by streams. Alluvial soils support a larger proportion of the world's population than do any other great soil group. Family separations in alluvial soils depend largely on the character and source of the silts, sands, and clays of which they are composed and, as with other soils, each takes its name from a well-known and representative soil series.

It will be noticed at once that the nomenclature of the great soil groups involves the use of many color terms; indeed, these terms are the sole ones used in many instances. It is recognized that color is not the most important characteristic of soils. In fact some shade or tint of brown is the most common color of soils throughout the world. Yet the use of these color terms has come to mean more than mere color. For example, the term Chestnut implies not only the color of the A horizon but also the prismatic structure of the B horizon, the accumulation of lime in the substrata, and the grassy vegetation under which the soils develop. This implication comes not from the name itself but from its common use among pedologists during the last few decades for soils with a particular combination of characteristics. Similar statements could be made for other soils of the well-recognized groups. It is hoped and believed that the new names proposed in this classification will come to have equal significance when the characteristics of these newly recognized groups become as well known.

Table 3 gives general information concerning the properties of the soils of the great groups and briefly mentions certain features of their environment and their more important uses.

APPENDIX [10]

Table 2.—Classification of soils on the basis of their characteristics

Category VI Order	Category V Suborder	Category IV Great soil groups	Category III Family [1]	Category II Series [1]	Category I Type [1]
	Soils of the cold zone	1. Tundra soils			
	1. Light-colored soils of arid regions	2. Desert soils	Mesa	Mesa	Mesa gravelly loam.
				Chipeta	Chipeta silty clay loam.
		3. Red Desert soils	Mohave	Mohave	Mohave loam.
		4. Sierozem	Portneuf	Reeves	Reeves fine sandy loam.
				Portneuf	Portneuf silt loam.
		5. Brown soils	Joplin	Joplin	Joplin loam.
				Weld	Weld loam.
		6. Reddish Brown soils	Springer	Springer	Springer fine sandy loam.
Pedocals				White House	White House coarse sandy loam.
	2. Dark-colored soils of the semiarid, subhumid, and humid grasslands.	7. Chestnut soils	Rosebud	Rosebud	Rosebud fine sandy loam.
				Keith	Keith silt loam.
		8. Reddish Chestnut soils.	Amarillo	Amarillo	Amarillo fine sandy loam.
				Abilene	Abilene clay.
		9. Chernozem soils	Barnes	Barnes	Barnes very fine sandy loam.
		10. Prairie soils	Carrington	Carrington	Carrington loam.
				Tama	Tama silt loam.
Zonal soils		11. Reddish Prairie soils	Zaneis	Zaneis	Zaneis very fine sandy loam.
		12. Degraded Chernozem soils.		Renfrow	Renfrow silt loam.
	3. Soils of the forest-grassland transition.	13. Noncalcic Brown or Shantung Brown soils.	Holland	Holland	Holland sandy loam.
				Vista	Vista sandy loam.
				Fallbrook	Fallbrook fine sandy loam.
				Sierra	Sierra coarse sandy loam.
Pedalfers	4. Light-colored podzolized soils of the timbered regions.	14. Podzol soils	Placentia	Placentia	Placentia fine sandy loam.
				Ramona	Ramona sandy loam.
			Weihaiwei	Weihaiwei	Weihaiwei loam.
				Tinghsien	Tinghsien fine sandy loam.
			Kalkaska	Kalkaska	Kalkaska loamy sand.
				Au Train	Au Train loamy sand.
			Rubicon	Rubicon	Rubicon sand.
				Roselawn	Roselawn sand.
			Hermon	Hermon	Hermon loam.
				Colton	Colton loamy sand.
				Becket	Becket loam.

[10] Unfamiliar terms in these tables are defined in the Glossary, p. 1162.

Table 2.—*Classification of soils on the basis of their characteristics*—Continued

Category VI Order	Category V Suborder	Category IV Great soil groups	Category III Family	Category II Series	Category I Type
Zonal soils ----- Pedalfers-----	4. Light-colored podzolized soils of the timbered regions (cont'd).	15. Brown Podzolic soils--------	Gloucester----	Gloucester----	Gloucester loam.
					Gloucester sandy loam.
				Merrimac-----	Merrimac sandy loam.
					Merrimac loamy sand.
		16. Gray-Brown Podzolic soils---	Miami-------	Miami-------	Miami silt loam.
				Fox--------	Fox silt loam.
				Bellefontaine--	Bellefontaine loam.
			Plainfield-----	Plainfield-----	Plainfield loamy sand.
				Coloma------	Coloma loamy sand.
			Chester-----	Chester------	Chester loam.
			Porters------	Frederick-----	Frederick silt loam.
				Porters------	Porters loam.
		17. Yellow Podzolic soils--------	Norfolk------	Norfolk------	Norfolk sandy loam.
	5. Lateritic soils of forested warm-temperate and tropical regions.	18. Red Podzolic soils (and Terra Rossa).	Orangeburg---	Orangeburg---	Orangeburg sandy loam.
				Greenville----	Greenville sandy loam.
				Magnolia-----	Magnolia sandy loam.
				Cecil--------	Cecil sandy loam.
		19. Yellowish-Brown Lateritic soils	Coto--------	Coto--------	Coto clay.
		20. Reddish-Brown Lateritic soils	Bayamón-----	Bayamón-----	Bayamón clay.
		21. Laterite soils---------------	Nipe (ferruginous).	Nipe--------	Nipe clay.
				Rosario------	Rosario clay.
	1. Halomorphic (saline and alkali soils of imperfectly drained arid regions and littoral deposits.	1. Solonchak or saline soils-----	Sage--------	Sage--------	Sage clay.
			Lahontan-----	Lahontan-----	Lahontan clay loam.
				Fresno-------	Fresno clay loam.
		2. Solonetz soils---------------	Phillips------	Phillips------	Phillips loam.
				Rhoades-----	Rhoades loam.
		3. Soloth soils-----------------	Beadle-------	Beadle-------	Beadle silt loam.
			Arvada-------	Arvada-------	Arvada clay loam.
				Beckton------	Beckton silty clay loam.
		4. Wiesenböden (Meadow soils)--	Clyde--------	Clyde--------	Clyde silty clay loam.
				Webster------	Webster silty clay loam.
		5. Alpine Meadow soils---------	Duncom------	Duncom------	Duncom silt loam.
		6. Bog soils--------------------	Edwards------	Edwards------	Edwards muck.
			Carlisle------	Carlisle------	Carlisle muck.
				Pamlico------	Pamlico muck.
			Greenwood----	Greenwood----	Greenwood peat.
				Spaulding-----	Spaulding peat.

Order	Suborder	Great soil group	Series	Types
Intrazonal soils	2. Hydromorphic soils of marshes, swamps, seep areas, and flats.	7. Half Bog soils	Maumee	Maumee loam. / Bergland loam.
		8. Planosols	Grundy	Grundy silt loam. / Oswego silt loam.
			Clermont	Clermont silt loam. / Vigo silt loam.
			Crete	Crete silt loam. / Idana silty clay loam.
		9. Ground-Water Podzol soils	Saugatuck	Saugatuck loamy sand. / Allendale sandy loam.
			Leon	Leon sand. / St. Johns loamy sand.
		10. Ground-Water Laterite soils	Tifton	Tifton fine sandy loam.
		11. Brown Forest soils (Braunerde)	Brooke	Caguas clay. / Brooke clay loam.
			Burton	Burton loam. / Houston clay.
	3. Calomorphic	12. Rendzina soils	Houston	Soller clay loam. / Bell clay.
			Aguilita	Aguilita clay. / Diablo clay.
Azonal soils		1. Lithosols	Underwood	Underwood stony loam. / McCammon loam.
			Muskingum	Muskingum stony silt loam.
			Dekalb	Dekalb stony loam.
		2. Alluvial soils	Wabash	Wabash clay loam. / Cass loam.
			Laurel	Laurel fine sandy loam. / Sarpy very fine sandy loam.
			Sharkey	Sharkey clay.
			Genesee	Genesee silt loam. / Huntington silt loam.
			Gila	Gila very fine sandy loam. / Pima silty clay loam.
		3. Sands (dry)	Hanford	Hanford loam. / Yolo loam.

¹ Families, series, and types listed are intended only as examples to illustrate the system of classification. When all of the soils of the United States (and of the world as a whole) are studied and classified, many more families, a few thousand series, and thousands of local soil types will have to be recognized.

161

Table 3.—General characteristics of soils and their environs

ZONAL

Zonal soils	Profile	Native vegetation	Climate	Natural drainage	Soil-development processes	Productivity (crop plants)	Present use
Tundra	Dark-brown peaty layers over grayish horizons mottled with rust. Substrata of ever-frozen material.	Lichens, moss, flowering plants, and shrubs.	Frigid humid	Poor	Gleization and mechanical mixing.		Pasture and a few short-season crops. Hunting and trapping are associated enterprises.
Desert	Light-gray or light brownish-gray, low in organic matter, closely underlain by calcareous material.	Scattered shrubby desert plants.	Temperate to cool; arid.	Good to imperfect.	Calcification	Medium to high, if irrigated.	Grazing in large units. Intensively farmed in small units where irrigated. Crops specialized in many places.
Red Desert	Light reddish-brown surface soil, brownish-red or red heavier subsoil closely underlain by calcareous material.	Desert plants, mostly shrubs.	Warm-temperate to hot; arid.	do	do	do	Do.
Sierozem	Pale grayish soil grading into calcareous material at a depth of 1 foot or less.	Desert plants, scattered short grass, and scattered brush.	Temperate to cool; arid.	do	do	do	Do.
Brown	Brown soil grading into a whitish calcareous horizon 1 to 3 feet from surface.	Short-grass and bunch-grass prairie.	Temperate to cool; arid to semiarid.	Good	do	High, if irrigated.	Large farms of small grain (if unirrigated). Ranching in large units.
Reddish Brown	Reddish-brown soil grading into red or dull-red heavier subsoil and then into whitish calcareous horizon, either cemented or soft.	Tall bunch grass and shrub growth.	Temperate to hot; arid to semiarid.	do	do	Moderate to high, if irrigated. Not suited to dry farming. Grazing good.	Grazing in large units. Small specialized farms where irrigated.
Chestnut	Dark-brown friable and platy soil over brown prismatic soil with lime accumulation at a depth of 1½ to 4½ feet.	Mixed tall- and short-grass prairie.	Temperate to cool; semiarid.	do	do	Medium. High where irrigated.	Cereal grains, especially wheat and grain sorghums throughout the world. Excellent grazing in large units.

Soil	Profile	Vegetation	Climate		Soil-forming process	Productivity	Agricultural use
Reddish Chestnut.	Dark reddish-brown cast in surface soil. Heavier and reddish-brown or red sandy clay below. Lime accumulation at a depth of 2 feet or more.	Mixed grasses and shrubs.	Warm-temperate to hot; semiarid.	do	do	do	Cereal grains and cotton. Excellent grazing in large units.
Chernozem.	Black or very dark grayish-brown friable soil to a depth ranging up to 3 or 4 feet grading through lighter color to whitish lime accumulation.	Tall- and mixed-grass prairie.	Temperate to cool; subhumid.	do	do	Medium to high. High to very high where irrigated.	Small grains and corn in moderate-sized or large units.
Prairie.	Very dark-brown or grayish-brown soil grading through brown to lighter colored parent material at a depth of 2 to 5 feet.	Tall-grass prairie.	Temperate to cool-temperate, humid.	do	Calcification with weak podzolization.	High.	Medium to small farm units. General farming, with emphasis on corn, hogs, and cattle.
Reddish Prairie.	Dark-brown or reddish-brown soil grading through reddish-brown heavier subsoil to parent material. Moderately acid.	Tall- and mixed-grass prairie.	Warm-temperate, humid to subhumid. Possibly some tropical conditions.	do	do	Medium to high.	Wheat, oats, corn, cotton, hay, and forage crops.
Degraded Chernozem.	Nearly black A_1, somewhat bleached grayish A_2, incipient heavy B, and vestiges of lime accumulation in deep layers.	Forest encroaching on tall-grass prairie.	Temperate and cool; subhumid to humid.	do	Calcification followed by podzolization.	Medium to high. Low where strongly degraded.	Agriculture intermediate between Chernozem and Podzol. Of little importance in the United States.
Noncalcic Brown (Shantung Brown).	Brown or light-brown friable soil over pale reddish-brown or dull-red B horizon.	Mostly deciduous forest of thin stand with brush and grasses.	Temperate or warm-temperate; wet-dry, subhumid to semiarid.	do	Weak podzolization and some calcification.	Medium. High were irrigated.	Grazing, dry farming with small grains, specialized irrigated crops including fruits.
Podzol.	A few inches of leaf mat and acid humus, a very thin dark gray A_1 horizon, a whitish-gray A_2 a few inches thick, a dark or coffee-brown B_1 horizon, and a yellowish-brown B_2. Strongly acid.	Coniferous, or mixed coniferous and deciduous forest.	Cool-temperate, except in certain places where the climate is temperate; humid.	do	Podzolization.	Usually low. Medium under good practices.	Small subsistence farms, including dairying. Wood lots and pasture important.

163

Table 3.—*General characteristics of soils and their environs*—Continued

ZONAL—Continued

Zonal soils	Profile	Native vegetation	Climate	Natural drainage	Soil-development processes	Productivity (crop plants)	Present use
Brown Podzolic	Leaf mat and acid humus over thin dark-gray A, and thin grayish-brown or yellowish-brown A2 over brown B horizon which is only slightly heavier than surface soil. Solum seldom more than 24 inches thick.	Deciduous or mixed deciduous and coniferous forest.	Cool-temperate; humid. Effective moisture slightly less than in Podzol.	Good	Podzolization	Low to medium. High where heavily fertilized and well managed.	Small subsistence farms, dairying, wood lots, and pasture. Specialized truck crops near large cities are important.
Gray-Brown Podzolic.	Thin leaf litter over mild humus over dark-colored surface soil 2 to 4 inches thick over grayish-brown leached horizon over brown heavy B horizon. Less acid than Podzols.	Mostly deciduous forest with mixture of conifers in places.	Temperate; humid.	do	do	Medium. High where well fertilized and managed.	Small farm units with general farming. Wide variety of crops. Some specialization. (Much industrial activity.)
Yellow Podzolic.	Thin dark-colored organic covering over pale yellowish-gray leached layer 6 inches to 3 feet thick over heavy yellow B horizon over yellow, red, and gray mottled parent material; acid.	Coniferous or mixed coniferous and deciduous forest.	Warm-temperate to tropical; humid.	Imperfect to good.	Podzolization with some laterization.	Poor. Responsive to good management and fertilization.	Small to medium-sized farm units. Subsistence crops. Cotton, tobacco, peanuts, and some fruit and vegetables. Few livestock. Many wood lots and forested areas.
Red Podzolic	Thin organic layer over yellowish-brown or grayish-brown leached surface soil over deep-red B horizon. Parent material frequently reticulately mottled red, yellow, and gray; acid.	Deciduous forest with some conifers. (With cogonales—burned over areas covered with cogon, tall coarse grasses.)	do	Good	Podzolization and laterization.	Medium. Responsive to fertilization and good management.	Small to medium-sized farms, with cotton, peanuts, tobacco, and subsistence crops. Much waste land and forests.
Yellowish-Brown Lateritic.	Brown friable clays and clay loams over yellowish-brown heavy but friable clays. Acid to neutral.	Evergreen and deciduous broad-leaved trees. Tropical selva. (Some cogonales.)	Tropical; wet-dry. High to moderate rainfall.	Good externally. Good or excessive internally.	Laterization and some podzolization.	Low. Medium with irrigation and fertilization.	Small farm units, with subsistence and some specialized crops. Some forests.

164

Soil	Profile	Native vegetation	Climate	Factors responsible for development	Natural drainage	Soil-development processes	Productivity (crop plants)	Present use
Reddish Brown Lateritic.	Reddish-brown or dark reddish-brown friable granular clayey soil over deep-red friable and granular clay. Deep substrata reticulately mottled in places.	Tropical rain forest to edge of savannah. (Some cogonales.)	Tropical; wet-dry. Moderately high rainfall.		Good externally and internally.	Laterization with little or no podzolization.	Low to medium. Medium to high where fertilized and irrigated.	Small farm units with subsistence crops. Plantations of citrus, pineapple, sugarcane, etc. Some forest.
Laterite	Red-brown surface soil. Red deep B horizon. Red or reticulately mottled parent material. Very deeply weathered.	Tropical selva and savannah vegetation. (Some cogonales.)	Tropical; wet-dry. High to moderate rainfall.		Good externally. Good or excessive internally.	Laterization and a little podzolization.	Low. Medium to high with heavy irrigation and fertilization.	Small farm units with wide variety of subsistence crops. Some specialization on plantations. Large areas of waste land and forest. Mined for iron and aluminum in places.

INTRAZONAL

Intrazonal soils	Profile	Native vegetation	Climate	Factors responsible for development	Natural drainage	Soil-development processes	Productivity (crop plants)	Present use
Solonchak	Gray thin salty crust on surface, fine granular mulch just below, and grayish friable salty soil below. Salts may be centrated above or below.	Sparse growth of halophytic grasses, shrubs, and some trees.	Usually subhumid to arid. May be hot or cool.	Poor drainage with evaporation of capillary water. Salty accumulations.	Poor or imperfect.	Salinization	Very low except where washed free of salts.	Some grazing. Much waste land. Used for producing salt and saltpeter in places.
Solonetz	Very thin to a few inches of friable surface soil underlain by dark, hard columnar layer, usually highly alkaline.	Halophytic plants and thin stand of others.	do	Improved drainage of a sodium Solonchak.	Imperfect	Solonization (desalinization and alkalization).	Low (medium where reclaimed).	Same use as associated normal soils.
Soloth	Thin grayish-brown horizon of friable soil over whitish leached horizon underlain by dark-brown heavy horizon.	Mixed prairie or shrub.	Usually subhumid to semiarid. May be hot or cold.	Improved drainage and leaching of Solonetz.	Imperfect to good.	Solodization (dealkalization).	Low to medium.	Do.
Wiesenböden (Meadow).	Dark-brown or black soil grading, at a depth of 1 or 2 feet, into grayish and rust-mottled soil.	Grasses and sedges.	Cool to warm; humid to subhumid.	Poor drainage	Poor	Gleization and some calcification.	Generally high or very high when drained.	Used in connection with associated normal soils. Drained areas similar to Prairie.

165

Table 3.—General characteristics of soils and their environs—Continued

INTRAZONAL—Continued

Intrazonal soils	Profile	Native vegetation	Climate	Factors responsible for development	Natural drainage	Soil-development processes	Productivity (crop plants)	Present use
Alpine Meadow.	Dark-brown soil grading, at a depth of 1 or 2 feet, into grayish and rust soil, streaked and mottled.	Grasses, sedges, and flowering plants.	Cool-temperate to frigid (alpine).	Poor drainage and cold climate.	Poor.	Gleization and some calcification.	Yields limited by cool climate.	Mostly summer pasture.
Bog.	Brown, dark-brown, or black peat or muck over brown peaty material.	Swamp forest or sedges and grasses.	Cool to tropical; generally humid.	Poor drainage. Water-covered much of the time.	Very poor.	Gleization.	Low or medium. High in some places where drained.	Special crops when drained. Undrained areas in forest - swamp or marsh plants.
Half Bog.	Dark-brown or black peaty material over grayish and rust-mottled mineral soil.	do.	do.	do.	do.	do.	Medium to high when drained. Some low.	Used in connection with normal soils for pasture or forest, where undrained, and for special crops where drained.
Planosols.	Strongly leached surface soils over compact or cemented claypan or hardpan. Some have normal A and B horizons above the claypan or hardpan—a secondary profile.	Grass or forest.	Cool to tropical; humid to subhumid.	Flat relief, imperfect drainage, and great age.	Imperfect or poor.	Podzolization, gleization. Also laterization in tropics.	Medium to low.	Crops, pasture, and forest trees, varying with regions in which they occur.
Ground-Water Podzols.	Organic mat over very thin acid humus, over whitish-gray leached layer up to 2 or 3 feet thick, over brown or very dark-brown cemented hardpan or ortstein. Grayish deep substrata.	Forest of various types.	Cool to tropical; humid.	Imperfect drainage and usually sandy material.	do.	Podzolization.	Low to medium in cool areas. Low in warm areas.	Forest. Some land planted to regional crops and pasture grasses.

166

Soil	Profile	Vegetation	Climate	Parent material and age	Drainage	Soil-development processes	Productivity	Present use
Ground-water Laterites.	Gray or gray-brown surface layer over leached yellowish-gray A₂ over thick reticulately mottled cemented hardpan at a depth of 1 foot or more. Hardpan up to several feet thick. Laterite parent material. Concretions throughout.	Tropical forest.	Hot and humid; wet and dry seasons.	Poor drainage and considerable or great age.	do	Podzolization and laterization.	Low to medium.	Subsistence crops, sugarcane, and forest trees.
Brown Forest (Braunerde).	Very dark brown friable surface soil grading through lighter colored soil to parent material. Little illuviation. High absorbed calcium.	Forest, usually broad-leaved.	Cool-temperate to warm-temperate; humid.	High calcium colloids and youth.	Good	Calcification with very little podzolization.	High	Subsistence and regional crops, pasture and forest.
Rendzina	Dark grayish-brown to black granular soil underlain by gray or yellowish, usually soft, calcareous material.	Usually grassy. Some broad-leaved forest.	Cool to hot; humid to semiarid.	High content of available lime carbonate in parent material.	do	Calcification	do	Regional and special crops. Pasture grasses.

AZONAL

Azonal areas	Profile	Vegetation	Climate	Drainage	Soil-development processes	Productivity and present use
Lithosols	Thin, stony surface soils—little or no illuviation. Stony parent materials.	Depends on climate.	All climates. Most characteristic of deserts; least so of humid Tropics.	A wide range, mostly good to excessive.	Those characteristic of the region. Little effect has been made.	Forestry, grazing, barren. Some agriculture on limited areas. Low productivity.
Alluvial soils	Little profile development. Some organic matter accumulated. Stratified.	do	All climates except extremely frigid ones.	A wide range, mostly poor to good.	do	Practically all crops of world represented. Yields vary from very high to very low. Large proportion of world's population supported by production from alluvial soils. Both subsistence farms and large plantations on these soils.
Sands, dry	Essentially no profile. Loose sands.	Scanty grass or scrubby forest. Much of land has no vegetation.	Humid to arid, temperate to hot.	Excessive.	do	Very seldom used except for grazing.

167

LITERATURE CITED

[*Editor's Note:* Only those references cited in the preceding excerpt are reproduced here.]

Afanasiev, J. N., 1927. The classification problem in Russian soil science. *Russ. Acad. Sci., Pedol. Invest.,* 5, 51pp.

Blanch, E., 1930. Die Mediterran-Roterde (Terra Rossa). IN *Handbuch der Bodenlehre,* v. 3, pp. 194–257, illus. Berlin.

Fallou, Friedr, Alb., 1862. *Pedologie oder allgemeine und besondere Bodenkunde.* 488pp., illus. Dresden.

Glinka, K., 1914. *Die Typen der Bodenbildung, Ihre Klassifikation und Geographische Verbreitung.* 365pp., illus. Berlin.

Kellogg, Charles E., 1937. *Soil Survey Manual.* U.S. Dept. Agr. Misc. Pub. 274, 136pp.

Marbut, C. F., 1921. The contribution of soil surveys to science, *Soc. Prom. Agr. Sci. Proc. (1920)* **41**:116–142, illus.

Milne, G., in collaboration with Beckley, V. A., Jones, G. H., Gethin, Martin, W. S., Griffith, G., and Raymond, L. W., 1936. *A Provisional Soil Map of East Africa (Kenya, Uganda, Tanganyika, and Zanzibar) with Explanatory Memoir.* 34pp. illus. London.

Richthofen, Ferdinand Freiherr von., 1886. *Führer für Forschungsreisende. 745pp. London.*

Shaw, Charles, F., 1930. The soils of China. A preliminary survey. *Natl. Geol. Survey of China Soil Bull.* 1, 38pp., illus.

Thorp, James, 1936. *Geography of the Soils of China.* 552pp., illus. Nanking.

——— and Tschau, T. Y., 1936. Notes on Shantung soils: A rconnaissance soil survey of Shantung. *Natl. Geol. Survey of China Bull.* 14. English part, 132pp., illus. (Also in Chinese)

13

This paper was translated by the Israel Program for Scientific Translations for the National Science Foundation and U.S. Department of Agriculture from Pochvovedenie
6:82–102 (1956)

AN ATTEMPT AT A GENERAL SOIL CLASSIFICATION*

(Opyt obshchei klassifikatsii pochv)

E. N. Ivanova

The basic principles of the first scientific soil classification were established by V. V. Dokuchaev and N. M. Sibirtsev. The starting point of their classification was the genetic approach to the soil, considered to be an independent natural body whose development is determined by natural factors: the soil-forming rock, the climate, the activity of the organisms, the age and the topography.

They maintained that the way to solve the classification problem would be by ascertaining the methods of formation (soil genesis) and the regularities conditioning their geographical distribution. They assumed that provided it were possible to arrive at such a solution, the soil classification would have followed naturally and this would have been the only indisputably correct one.

These principles were established and elaborated upon by all the Russian soil scientists in forming their classifications. With the appearance of each new soil classification system, a complete revision of all the accumulated data was carried out.

In the postwar years, the Soviet soil scientists carried out extensive work for surveying the available soil-geographical material, accompanied by the compilation of survey soil maps of individual areas and of the entire territory of the Union. Data from foreign publications were also collected for the compilation of a new soil map of the world. During these years new soil investigations of insufficiently studied areas of the USSR were undertaken and, in addition, Soviet soil scientists carried out soil investigations and surveys in a number of foreign countries. All this contributed to an increase in the data on the genesis and geography of soils and brought about the necessity for new theoretical conclusions.

The present article is devoted to an attempt at compiling a general soil classification, based on the contemporary knowledge of the genesis and geography of soils**

---

* The basic statements of this article were read at the All-Union Conference of Soil Scientists held in Moscow from 28 January to 2 February 1956.

** The observations made while discussing the data in the Soil Institute of the Academy of Sciences of the USSR and at the All-Union Conference of Soil Scientists were taken into consideration in compiling this article. (These observations were made by I. V. Tyurin, I. P. Gerasimov, V. A. Kovda,

169

E. N. Ivanova

Bibliographical Survey

I. P. Gerasimov /16/ summed up all the soil classifications found in Russian soil science, and grouped them in the following way: 1) Geographic – genetic, in which the connection between genetic types is established on the most significant properties of their geographic distribution. 2) Factor – genetic, combining soils according to the factors of soil formation. 3) Genetic classifications, based on the internal properties of the soils. 4) Dynamic – genetic, based on the stage development of the soil-formation processes from the alkaline to the acid stages; 5) Evolutional, combining the soil types in a single soil formation process.

The first scientific soil classification suggested by Dokuchaev /24/ was a geographic-genetic one. This was reported at the meeting of the geologic and mineralogic section of the Petersburg Society of Natural Research on 14 April 1879. The brief account of this meeting indicated that they divided all soils into two parts: regular and irregular soils. The sections are further subdivided into classes, and within the classes, types are isolated. The section of regular soils includes two classes: Class I, continental-vegetational soils: a) gray northern soils; b) chernozem soils; c) chestnut soils; d) reddish solonchakic soils, and Class II, continental swampy soils. The irregular soils are divided into two classes: Class III, of flushed soils, and Class IV, of deposited soils.

In 1886 /26/, this classification being fully verified, it was supplemented and published. According to this classification soils are divided into three sections: regular, transitional and irregular.

(1) The regular soils are divided into three classes:

Class I – continental-vegetational soils consisting of five types: 1) northern light gray, 2) gray (transitional) soils, 3) chernozem, 4) chestnut (transitional) soils, 5) brown solonetsic soils.

Class II – the continental swampy soils including the soils of non-floodland meadows and of the dark ramens*.

Class III – the swamp soils, including peat-bogs, tundras and flood areas.

(2) Transitional soils including two classes: the flushed soils and the terrestial deposited soils.

(** – continued from previous page):

N.I. Bazilevich, A.A. Zavalishin, V.V. Egorov, B.A. Kalachov, Yu.A. Liverovskii, E.V. Lobov, M.M. Kononova, N.A. Nogina, V.A. Nosina, A.N. Rozanova, A.A. Rode, I.N. Skrynnikova, V.M. Fridland and others).

The principal statements were discussed with, and agreed upon, by N.N. Rozanov. The classification scheme has been elaborated with his collaboration.

* [Ramen' – Spruce wood on fresh (slightly humid) loamy soil; the forest is highly productive and the wood material is excellent. Widespread especially in the taiga zone. The name "Ramen" is connected with the spread of agriculture in forest clearings: cultivated fields are, so to say, "framed" in a "frame" (Russian rama) of spruce trees. Sometimes ramen designates other "dark coniferous" woods (e.g., cedar, etc)].

(3) Irregular soils representing a single class: the typical deposited soils (alluvial and aeolian).

While evaluating the significance of this classification the author writes: "The basic principles of the first (not only in our country, but abroad as well) strictly scientific genetic soil classification were defined for the first time by me only in the year 1886. Later, in 1893, this soil classification was substantially supplemented and enlarged by Professor Sibirtsev, and in 1896 it was again revised finally and corrected by me. In addition, and this being in my opinion the most important point, the genetic soil types and the soil zones were compared and referred to – natural-historical belts and zones (Dokuchaev) /29/.

In the classification scheme of 1896 /28/ the soils are subdivided into three classes which were subdivided again into 12 sections.

Class I – regular (vegetative-terrestrial or zonal soils): 1) glacial (tundra), 2) light gray podzolized, 3) brown-nutty (forest), 4) chernozem, 5) dark-chestnut, 6) light-brown, 7) krasnozem [red earth] (laterites), 8) dark-brown terrestrial-swampy, 9) whitish secondary solonetses.

Class II – transitional or intrazonal soils.

Class III – irregular or "cosmopolitan" soils: 10) aeolian, 11) alluvial, 12) swamp soils.

Though such terms as "regular, transitional and irregular" are still used in this classification, their meaning here is completely different.

Dokuchaev's classification is based on genetic principles – namely the origin of the soils while being subjected to soil-forming factors. In addition, it reflects the spatial relations between the various soils. "It shows the soil cover of the earth in its genetical integrity and in its geographical variability" /46/.

Dokuchaev, however, considered that the classification be based also on the inner properties of the soils themselves. In his opinion, "the natural grouping of soils must be indisputably based on the sum total of all essential characteristics of the given body". However, to determine all these features, "to describe them . . . sufficiently would be difficult at present – this remains the task of future investigations". However, Dokuchaev himself establishes the principles of soil classification according to physical-chemical properties: according to the humus content, to the zeolitic clay and to the mechanical composition /26/.

The soil groups isolated by Dokuchaev according to their mechanical composition were described by Sibirtsev as "soil genera" /49/.

In Sibirtsev's classification /46, 47, 48/ the soils are subdivided into classes, or sections and into soil types. The class of zonal or fully developed soils includes 7 types: 1) laterites, 2) aerial-silty, 3) desert-steppe soils, 4) chernozem, 5) gray-forest soils, 6) turfy-podzolized soils, 7) tundra soils. The class of intrazonal (semizonal) soils is subdivided into three types: 8) solonetsic, 9) marshy soils, 10) humus-carbonated.

The class of azonal, not fully developed soils is divided into subclasses: the subclass of non-floodland soils: 11) skeletal, 12) coarse soils, and the subclass of the alluvial soils, 13) floodland soils. All in all, 13 soil types were isolated. The soil types are subdivided into subtypes ("according to the nature and extent of the dynamic soil-forming processes, which impart to the soil the common features of the particular genetic type") and into petrographic groups – according to the mechanical composition /48/.

These classifications were elaborated further in the works of Ya. N. Afanas'ev.

Afanas'ev /1, 2/ expressed the opinion that soil classifications, being logical schemes, must represent the natural soil zone within a system of genetic series and he therefore, in his classification, isolates particular systems of zonal soil complexes in the regions having a continental type of climate and in the regions having a marine type of climate. Each member of the complex and the complete complex as a whole, with an even succession of climates,is a repetition, a simile,of the preceding zone, its climatic analogue. The author has isolated 5 climatic zonal belts: the cold, the moderately cold, the moderately warm, warm subtropical, and the hot tropical. The change of soils induced by vegetation is considered for every belt: forest, turfy-forest and turfy. Evolutional groups (processes of degradation and desalinization) are given for some soils.

The factor-genetic classifications are based on the effects exercised by individual factors or by a combination of them: the rocks, moisture conditions, climate, topography and vegetation (Glinka, Vilenskii, Zakharov, Vysotskii, Gerasimov – Zavalishin – Ivanova, Volobuev).

K.D. Glinka /18, 19, 20/ distinguishes the following soil groups: exodynamomorphic (formed under the influence of external factors of soil formation; stress is laid on the importance of the climate in the process of the decomposition of organic matter; classes were isolated according to the moisture regime), and endodynamomorphic (in which the role of the soil-forming rocks is stressed).

D.G. Vilenskii /4, 5/ distinguishes analogous series A B C, etc, of zonal soils and sections: thermogenic, phytogenic, hydrogenic, halogenic, and a number of complex soils. Each section has analogous series in each zone (belt). Within each zone – the development stages of the soils are given,with the fulvo-acid stage of soil podzolization being the final stage in each of the zones.

S.A. Zakharov /30/ divides the zonal soil complex into the following sections: climatogenic, orogenic, hydrogenic, hydrohalogenic, fluviogenic and lithogenic. Within each section the division according to the climatic belts or to the ecological conditions is given.

G.N. Vysotskii /8/ subdivides the soils into classes of zonal, intrazonal and undeveloped soils, which are subdivided further according to their reliefs and parent rocks.

Gerasimov – Lavalishin – Ivanova /11/ – subdivide the genetic soil types and subtypes into series according to the extent of the moisture regime: floodland-alluvial series of subsurface moistening, the eluvial-hydromorphic series, the eluvial series, and the eluvial-xeromorphic series. The soils within each series were not examined separately,but as if being the probable stages in soil formation dependent on the development of the relief.

Finally,there appeared recently the classification by V.R. Volobuev /7/ in which the soils are classified also according to factors. Soil associations have been isolated according to the climatic data, and are subdivided into two orders and into intrazonal types. The first order includes the automorphic soils, the second order – the hydromorphic; the marshy soils and the solonchak soils are referred to the intrazonal soils. A further subdivision into soil types is carried out according to the plant formations.

P.S. Kossovich /34, 36/ was the first to apply, while elaborating on Dokuchaev's ideas, the genetic principle in soil classification. He considered that "the genetic classification must be based on the inner properties and characteristics of the soils themselves" /36/*.

Kossovich /36/ subdivides the soils into two classes: the class of the genetically independent – the eluvial soil formations, and the class of the illuvial, or genetically dependent soil formations. The class of the genetically independent soil formations includes the following types of soil formation: desert, desert-steppe (solonetsous), steppe (chernozem), podzolized, tundra, laterite and marshy-mossy soils. The class of the genetically dependent soils includes the following types of soil formation: the dry steppe soils of subsurface moistening, soils of subsurface moistening pertaining to the chernozem belt, the swamp and semi-swamp soils of the podzolic region, the slightly swampy soils of the humid tropical and subtropical regions. To each type of soil formation there are similar soil types with respect to their processes.

The classification is based on the processes of transformation of the mineral soil mass as well as on the processes of decomposition and of organic matter accumulation. The author maintains that clarifying the nature of the humus substances in accordance with the types of soil formation is extremely important and much attention should be given to their study /36/.

The classification by Tumin /51/ was published almost simultaneously with that of Kossovich. Their classifications are rather similar in respect to the principles on which they are based. Tumin isolates types of soil formation according to the type of humic acid: lateritic crenoacid (podzolized), humic acid (chernozem), humic alkali (solonetsous), and classes of soils are isolated within each type according to the moisture regime (normally-moistened and excessively moistened).

Thus we observe that in Tumin's classification the genetically dependent soils of Kossovich and the intrazonal soils of Sibirtsev are both dependent on the basic processes.

The classes are subdivided into genera according to the qualitative expression of the process. Some genera are subdivided into species:

----

* Professor Slezkin suggested somewhat earlier (1902) (according to Tumin – 51) that for the soil groupings the predominating soil formation process should be used, according to the following scheme:
    1. The process of physical destruction (the polar areas and deserts).
    2. The process of laterite formation.
    3. The process of biochemical accumulation of neutral humus.
    4. The process of salinization.
    5. The process of leaching or podzolization.

1) the podzolic – into podzols, podzolized and slightly podzolized; the chernozem – into ordinary and southern. Groups are isolated according to the parent rocks of the soils.

K.D. Glinka /21, 22, 23/ in 1922–1924 compiled a soil classification based on the soil formation processes. The author substantiates the soil formation types established by him by the following considerations: "The main properties of the soils arise during the very process of soil formation . . . . These properties originate due to the changes in the organic matter on the one hand and to the disintegration of minerals on the other, and to the migration of the final products of these two processes . . . /21/. So far as we know the soil processes . . . we may now indicate five principal types of soil formation" (stressed by the author).

The types are as follows: 1) laterite, 2) podzolized, 3) steppe, 4) solonetsous, 5) swampy.

The subsequent genetic classification was that of K.K. Gedroits, based on the soil adsorbing complex. Its main groups coincide with the soil formation types of Glinka. Gedroits formulates the principles of his classification as follows: "Such a basis, which represents the most essential component of the soil, is most closely connected with the rest of the soil's solid phase and with the soil solution, and which moreover . . . indicates the past history of the soil (and to a certain extent, also, its future), is, in my opinion, the one I designated as the soil adsorbing complex" /9/.

In the classification by S.S. Neustruev /40/ the soils were also grouped according to the types of soil formation, which, in respect to the soil processes, completely coincide with the types of Glinka and Gedroits. The soil processes are isolated according to the combination of the processes of mineral transformation, organic matter accumulation, the properties of the adsorbing process and the nature of the migration of soil formation products. All the types of soil formation form two groups: the automorphic and the hydromorphic soil formations. The evolution relations between the soil types are also pointed out in this classification.

L.I. Prasolov /42/, while preparing the legend for the summary soil map of the USSR, wrote the following: "If soil classification be based on the main and common elements of their genesis, then all the soils will fall into three series:

I. The eluvial series (corresponding to the series of the "climatic types" or to the soils of atmospheric moistening, called also the automorphic soils);

II. The salinization and desalinization series (resulting from an eluvial process taking place on saline soils subjected to subsurface moistening or on primary saline soils;

III. The swampy-meadow series (of hydromorphic soil formation) . . The first . . . falls into three groups according to the intensity of the aluvial process and its various combinations with the process of biological accumulation: 1) the soils of the humid regions . . . 2) the soils of the transitional (sub-humid and sub-arid) regions . . . 3) the soils of the dry (arid and extremely arid) regions . . . These three groups reflect the climatic zones and facies, but are not analogous of them, being based independently on the soil properties . . . Genetic soil types are isolated in the above groups, which are used in publications and which have been established according to the combination of their biodynamic components (first of all, of vegetation) . . . with the parent rocks".

The dynamic-genetic classifications (Kossovich, B.B. Polynov, V.V. Gemmerling) examine the development stages of the eluvial process, starting with the alkaline soil formation and ending with the acidic.

Kossovich bases his scheme on the fundamental concept of the soil formation process. Already in 1903 Kossovich /34/ stated the necessity to establish soil-weathering types and indicated three basic types: laterite, the alkaline and the podzolic. In his subsequent work /35/ the soil classification was based on these ideas: "all soils . . . belong either . . . to the alkaline type of weathering or to the acidic type of weathering". Each type of soil is considered by the author to represent one of the development stages of this process: "if we take a highly developed soil and try to indicate the individual stages of its past history, we shall see that for each of these stages we can indicate corresponding soil formations existing at present . . . the soil may pass in the course of its gradual development from one soil type to another" . . . and, furthermore, analogies are drawn with the phylogenetic and autogenetic development of plants and animals.

All the plakor* soils known at that time were subdivided into two groups.

The author refers the aeolian-loess soils (as, for instance, the desert soils), the brown and the chestnut (desert-steppe) soils, the chernozems (the steppe soils) and the rendzinas [humus carbonated soil] to the soil group of the alkaline type of weathering and arranges them according to the degree of development of the soil formation process. The northern and leached chernozems and rendzina soils are considered to be soils transitional between one group and the other. The gray forest soils, podzolic soils and the tundra soils are referred to the soils of acidic weathering.

The author thus distinguishes two principal types of soil formation, which include groups of soil types, and considers all the soils as a single development series. Later V.V. Gemmerling /10/ accepted these ideas.

As noted before, in 1910 Kossovich completed the soil classification, but based it on genetic principles. However, the various types of soil formation were also treated in the classification of 1910, not individually, but within the general development series of the eluvial process /36/

In his classification B.B. Polynov /41/ distinguishes two series of soil formations: 1) the first major series of the eluvial soil formations, and 2) the second major series of the lacustrine-marshy-alkaline soil formations.

The eluvial series is subdivided into two groups of soil formations according to the parent material: one on a calcareous and the second on an allophanic non-calcareous weathering crust. A transition of soil from one group to another takes place in the course of the soil-formation process. Types of soil formations which correspond to the phases of leaching are isolated. The following types are distinguished within the alkaline soil-formation groups: primitive alkaline, pre-chernozem and

----

* [Russian neologism from Greek. Designates a flat area with deep water table, e.g., plateau or tableland].

chernozem soil. The latter passes into the group of acidic soil formations:
the pre-podzolic soils and podzolic soils,changing with the further increase
of moisture into the group of marshy soil formations.

The soil-formation process of the lacustrine-marshy-alkaline series
consists of phases which correspond to the transition from the sulfate-
chloride alkaline group through the calcareous to the marshy group on the
allophanite weathering crust. Thus, the final form of the first series (the
marshy phase) passes into the final form of the second series. The author
has not elaborated this scheme to its final classification stage and it
mainly represents the criteria for a classification. Examples of soil types
referring to the individual stages in the soil formation are given,as well
as a scheme for further subdivisions. In general, these are the same zonal
types of the former classifications, but are now not considered as separate
units, but as stages of a single process.

V.R. Vil'yams, in accordance with his notions of the single soil
formation process,considers the soil development as pertaining to time,
together with the general development of the planet. Vil'yams maintains,
while treating the synthesis and decomposition of organic matter as being
the essence of the soil formation process, that life is the principal moving
force in this process . . . "from the point of view of soil-science, life is
a continuous succession of the process of creation and destruction of the
organic matter". It brings about changes in the environment. "These
changes, on becoming numerous, will inevitably bring about profound
qualitative differences, which we perceive as evolution . . . Soil formation is one of the indicators of this continuous process of evolution on the
earth's surface. This is one general process,colossal in scale and duration, which began when . . . life started to appear; the development
of this process later determined the superceding geological epochs"

The changeability of soil with time has been repeatedly emphasized
from the very beginning of the genetical approach to their study. As
early as 1898 Dokuchaev wrote the following, while summing up the main
stages in soil formation: "The law of soil progress and regress,or their
eternal changeability (the life of the soils),as pertaining to time and space,
states that the soil, like any other plant or animal organism,
lives and is continually subject to changes, either developing or being
destroyed, progressing or regressing /29/.

Similar ideas can be found in earlier works of Dokuchaev /25/,as
well as in the works of Sibirtsev /47/. Their ideas were elaborated later
by Kossovich, Vil'yams, Polynov and others.

The soils are peculiar in that not being organisms, they possess no
inherited characteristics by which it would have been possible to trace
the development of each type through the lengthy geological periods.
Consequently, the soil types show no true phylogenetic development;
nevertheless,however, some elements of this process are present in soils,
expressed in the eternal changes which take place both in the earth's crust
and in the living organisms in the course of their common geological
development and which are firstly expressed in the development of soil
fertility.

Since soils, in their development, are closely connected with the
development of plant types, it might appear possible, on the basis of
Dokuchaev's law concerning the "constancy of relations between the soil

(its physics and chemistry) on the one hand, and the plant and animal organisms inhabiting it on the other" /29/, to reconstruct from paleontological data the major stages in the development of soils in each of the geological periods.

The history of vegetational development in the USSR indicates [N. M. Strakhov /50/, I. P. Gerasimov /13/] that the first developed soils might have been found in the swampy lowlands covered by bushy and tree-like spore species. Forests of tree-like spore species and of coniferous trees appeared first in the Middle and Upper Carboniferous. Moss, lichen, and mushrooms were found in the forests. It was then, apparently, that the development of the vegetational-continental soils began. In the middle of the Cretaceous period the first angiospermae made their appearance. Towards the end of the Cretaceous period began the division of the vegetative cover into zones and regions. During the Paleogene the grasses appeared but the shaded forests were still predominant. Then also appeared the first indications to differentiate the forests into types: the tropical ever-green (of the Poltava flora) and those of the temperate regions consisting of deciduous varieties (of the Turgai flora). The savannas appeared in the Eocene, and the true steppes were formed during the Miocene, and the deserts — in the Pliocene. At approximately the same time the taiga zone (Siberia) came into being. The tundras appeared during the Quaternary period.

The attempt to outline the history of the development of soil from the data provided by the analysis of the history of the development of vegetation encounters many difficulties due to the fact that vegetation, like any other organism, adapts itself to its environmental conditions, and consequently one cannot guarantee that the plant types, which at present correspond to definite soil types, corresponded to the same soil types of former epochs, i.e., it cannot be certain whether it is possible to trace the present-day genetic connections between the soils and the plants of all the past geological periods.

Moreover, according to observations carried out on the current relationships between soils and plants, it is evident that the range of types of soil types to which a certain plant may be referred is much larger than originally assumed.

The relationships between soils and plants are determined, under the current conditions, not by individual species, but by ecological plant groups. However, the paleobotanical finds are, as a rule, insufficiently complete and only describe larger ecological groups than those corresponding to soil types. Therefore, no sequence showing the historico-genetic development of particular soils can be established from the paleobotanical data. Only the sequence concerning the time of appearance of the soil formation processes may be established.

The paleobotanical method is not sufficient to establish the historico-genetic relations in the geological development of individual soils nor for the purpose of devising an evolutional classification of soil types.

Vil'yams erred in the practical application of his concept. This was already pointed out in publications /13, 16, 43/. He expressed the essence of his system in the following manner: . . . "the entire extensive territory of the USSR successively passed through all the stages of soil formation beginning with the tundra" /6/. This interpretation of the zonal distribution of soils as pertaining to space is based on the assumption that the present-day soils are all members of a single evolutional series and that their evolution in all regions was similar in time,so that the contemporary zonal distinctions are but a function of time.

Paleobotanical data, however, prove that there was no common tundra landscape after the retreat of the glaciers; that regions which had differing resources of radiation and energy, and which lost the glaciers at different periods, were colonized by different plant types. The first development stages of the present soil cover of various discontinuous surfaces were apparently very ancient soils; moreover, their subsequent development in various geographic environments continued in accordance with the environmental conditions.

K.K. Markov /37/ maintains that the development of the northern landscape during the Quaternary period has undergone substantial changes, while that of the south remained more stable. There is no reason, therefore, if only the spatial distribution of soils and their absolute age be taken into consideration, to arrange them in a single evolutional series and to use this series as a basis for a classification.

The classification of M.S. Tsyganov /52/ may serve as an example of such a classification. It is based on the " . . . progressive-evolutional process of soil formation, determined by the progressive development of the continent caused by the increase in its geological age".

The author lists, on pages 58–60, a basic evolutional-progressive series of soil types,from the marshy soil,through the turfy-meadow, turfy-podzolic, turfy-steppe,to the steppe and desert-serozem soil. All in all, 16 major soil types were distinguished.

It cannot be proved that this series is an historico-genetic one.

In his scheme the author also deals with the intrazonal series of soil development in connection with the relief development. The latter relations undoubtedly exist and were justly taken into consideration in the suggested classification.

A survey of the existing classifications shows that their respective basic principles are not contradictory and rather supplement each other, thus enabling us to clarify more profoundly and systematize the conformities pertaining to the genesis and development of soils.

An Attempt at Establishing a Classification Scheme

In postwar years, extensive cartographical works were undertaken for the purpose of summarizing the large amount of material, which was accumulated by various institutes for soil-geographic research,carried out during the last twenty-five years. The data which were collected according to various scales are often not related to each other neither by the methods of the field investigations, nor by the soil classification – or nomenclature. It was only natural that this task drew the attention of many to the urgent need of introducing order in the classification of soils. The soil-geographical section of the Soil Institute of the Academy of Sciences of USSR carried on with this work for a number of years. Here, under the instruction of I.P. Gerasimov /12, 15, 16/ the following system of taxonomic units was elaborated: type, subtype, genus, species and variety.

The genetic meaning of these concepts was established as well as their taxonomic value; detailed systematic lists were compiled, which enabled us to combine both the already accumulated and the newly acquired data provided by the soil and geographical investigations. These lists proved to be of practical value and much attention should also be devoted in future to further elaborate and unify them. The introduction into soil science of the ideas concerning the biological nature of soil formation is characteristic of this period (Vernadskii, Vil'yams). These ideas are reflected in the definition of the principal taxonomic unit of the classification – "the soil type".

At the same time a search for new methods to elaborate upon the soil classification continued. Using Vil'yams' concept as a starting point, the soil scientists began searching for evolutional relations between the various soil types for the establishment of an evolutional classification /14, 16, 53/.

An examination of the concept of a single soil formation process, as proposed by Vil'yams, has proved that it is impossible to base a soil classification exclusively on the principle of evolution. However, evolutional relations between the various soils do exist and these must be taken into consideration. Therefore, the establishment of a complex classification based on geographical-genetic, genetic and evolutional principles deserves the greatest possible attention.

Much material has been accumulated in soil science since the days of Dokuchaev (1882) and Sibirtsev (1898),enabling us to elaborate upon the evolutional series of the various groups of soil types.

One of the main laws governing the development of life is the unity of environment and organism, which, applied to soils, means that any change in the environment brings about a change in the properties of the soils, and the soil itself affects the environment. These interactions of soil and environment bring about an accumulation of characteristics within the soil which affect the course of the processes which take place within it, and subsequently also the structure and properties of it. In some cases these changes may be so great that the soil will be transferred from one soil type to another.

We know of the following forms of soil development which are accompanied by a transformation of the soil profile and which culminate in their transformation to another soil type:

1) the influence of the evolution of the vegetative cover, e.g., when a solonets becomes a chestnut soil due to a long turfy process or when a turfy-carbonated soil becomes podzolic; in these cases, the possibility is established of a transition of some soil types into others, due to the influence of vegetation within the boundaries of a single zone,and without bringing about any drastic changes in the vegetative cover.

2) the relief development; for example, the transition from floodland to the watershed area brings about either the transition of the turfy-floodland soil to the meadow-chernozem or chernozem soil, or the evolution of the saline soils: solonchak-solonets-solod; this is also an example of the transition of certain types of soils into others occurring mainly within the boundaries of a single zone;

3) changes in the natural conditions brought about by man, as for instance, when some turfy-podzolic soils turn into gray forest soils owing

179

to the felling of forests in extensive areas of the southern extremity of the forest zone and the progradation of soils.

4) profound changes in the climatic conditions, and the displacement of bioclimatic zones (and facies). A striking example of such a transition is the secondary-podzolized soils, which, while having the profile of a podzolic soil, have, beneath the podzolized horizon, a residual humus horizon which belongs to a former stage in the soil's development. With slow changes in the environment, the profile of the soil changes gradually and its horizons are transformed into the profile of the new soil. With abrupt changes in the environment, the new soil develops on the former as if on parent material and preserves the residual horizons of the former soil (as in secondary podzolized soil). The relation between the old and the new soils in this case is expressed rather in the historical sequence of soil changes than in their evolutional development.

However, it is not always possible to exactly establish the type of the former soil from the residual characteristics. Even in the case of the secondary podzolized soils it is difficult to establish, according to the second (lower residual) humus horizon, which soil has been podzolized, whether it was a chernozem, a dark-gray or a turfy-carbonated soil — while the establishment of the soil development series depends on the correct solution to this problem. We see, however, that this evolution does not include extensive soil groups, but affects only soil types which are geographically close to each other: from the chernozem to the podzolic soils.

Transition showing residual characteristics, was not observed in all the zonal soil types, which means that, on the one hand, there is no reason to maintain that all of the zonal boundaries have undergone considerable territorial displacements, and, on the other hand, that residual characteristics may be preserved in the soils only upon abrupt zonal displacements.

The majority of soils either have no residual characteristics or we do not know how to detect them, and it is therefore not always possible to correctly retrace the history of a soil's development from its profile. In such cases we can only tentatively speak of probable connections. This is also determined by the fact that the development of soils is polygenetic, i.e., one and the same type of soils can be formed of different preceding types of soils. Moreover, soils which at present belong to a single type were frequently developed in directly opposite ways, which means that the processes may change their succession. The gray forest type of soils may serve as an example. Some of the soils belonging to this type developed from soils which had a higher humus content due to their podzolization, other developed from podzolic soils owing to their progradation, and still a third group have no residual characteristics, neither of the preceding podzolic soil stage, nor of the steppe soils, and are the "primary gray forest soils". Thus, in the case of the gray forest soils there are characteristics from which the history of their development can be retraced. There are soils of which the complex development may be assumed (for instance, from floodland soils to watershed soils), but the characteristics of their former development stages are difficult to determine; however, upon a more profound study of the soil properties and using the comparative geographical method, we are more and more able to recognize the evolutional connections between the soils.

The above cited examples show that soils may pass, in the course of their development, from one soil type to another; this has been proved in respect to intrazonal groups and to some of the adjacent zonal types.

This does not mean that the development of all the soils included these transitions, this being especially true with respect to changes in soils caused by displacement of zones and facies. Insofar as the displacements were not universal and occurred in limited local areas, they affected only the geographically adjacent groups of soil types and could supercede each other in various successions; consequently, no continuous, successive evolutional connections exist between the soils and there is no possibility to retrace any single series of soil development.

V.A. Kovda /32/,while treating the possibilities of applying the evolutional principle in soil classification, pointed out that " . . . the possible stages of the soil formation process and the connections between them are not obligatory for all cases in the first place, and, in the second, they may skip a number of stages or begin from any of them " (p 20).

All the aforesaid indicates that the evolutional principle, taken by itself – cannot be used as the basis for a classification.

However, the establishment of the actual evolutional relations enables us to more correctly determine the genesis of the various soils, to clarify their essential properties and to outline the ways of their future development. Therefore,a classification must take into consideration the evolution of soils based on "their genesis (the manner of their origin) and development (the life of the soil)" /29/.

The soil type is accepted as the basic unit in soil classification, as Dokuchaev long ago suggested, maintaining that " . . . the numerous variations in the course of the soil formation processes themselves . . " result in the appearance of ". . . various soil types, which, while being related to each other through a number of transitional forms, are,however, sufficiently differentiated and specific if only by their external properties" /29/.

Dokuchaev has elaborated also the basic nomenclature of the soil types, which was later followed by all Russian scientists.

Prasolov defined, in 1937, the concept of "type" more precisely,and in the years 1952, and 1954 it was made even more specific by Gerasimov. At present a soil type includes the soils in which:

1. The processes of transformation and the migration of substances are similar;

2. The water and thermal regimes are of a similar nature;

3. The structure of the soil profile belongs to a single type in respect to the genetic horizons (being the result of the development process);

4. The extent and quality of natural fertility are similar;

5. The development of all of the soils referred to a single type takes place beneath ecologically similar types of vegetation.

Sibirtsev combined all the soil types into large groups of zonal and intrazonal (or semizonal) soils. The latter soils occur in different zones (turfy soils, solonetses, marshy soil, meadow soils, solonchaks). Neustruev /39/ introduced the concept concerning " . . . the soil systems of a given zone . . .",taking into account that the effect of zonal conditions creates, in all the soils of the zone, the features characteristic of it. Afanas'ev, Prasalov and others agreed with this conclusion. However, it was never applied in subsequent classifications.

Recent investigations have proved still more convincingly that the soil formation processes which take place in the intrazonal soils are determined mainly by the same transformations of the organic and mineral substances, as in the case of the zonal soils, and also that the specific features of the intrazonal soils (as, for instance, the marshy process expressed in marshy soils, the development of the solonets process in the solonetses, etc) are determined by the bioclimatic conditions to such an extent that these soils develop entirely different formations in different zones. It would be more correct, therefore, to isolate several intrazonal soil types (rather than isolating single types) which will correspond to their particular common processes of transformation of organic and mineral substances. We suggest, for instance, to subdivide the intrazonal solonets soil type into the following types: solonets soils of the steppe, the solonets-desert soils, and the solonets permafrost soils, etc. This pertains also to the other intrazonal types: to the solonchak, to the meadow and to the marshy soil types. It is worth-while noticing that the solonetses of the deserts and of the dry steppes and also the solonetses of the chernozem region were isolated by Kossovich /36/ as independent soil types. Sibirtsev /48/ isolated subtypes of the solonetses: the desert-steppe subtype and the subtype of the chernozem regions.

Table I

Soil	Dry organic matter, ts /ha		Annual dry litter ts /ha	Annual introduction kg /ha		Humus, %
	in the grass stand	in the roots		minerals	nitrogen	
Meadow sero-zem soil: vetch-foxtail grass mea-dow	96	355	327	1676	545	3
Meadow-chernozem soil: mea-dow-motley grass – steppe	60	130	103	833	121	15

The comparison of the meadow-chernozem and the meadow-serozem soils has been cited as an example (Table I)*.

----

* Data by N. I. Bazilevich and L. E. Rodin /3, 44/.

Both soils develop under the influence of the meadow vegetation under conditions of increased atmospheric and ground moisture. However, the meadow-serozem soil is apparently close to the meadow soil in respect to their moisture regime. The soils are characterized by the different ways the substances enter into the biological cycle and by their different ways of transformation and fixation.

The meadow-serozem soil receives a great amount of organic matter and nitrogen, but only a negligible quantity of these becomes fixed.

The above-cited facts show that in spite of the fact that some of the soil formation elements (in this case it will be the increased moisture content) remain in some of their parts essentially unchanged in the various zones, the soil formation process on the whole is dependent on the general bioclimatic conditions.

Our isolation of soil types was based on the list /13/ which had been discussed at the All-Union Conference of Soil Scientists of 1950, and we now suggest a further subdivision of the intrazonal types. This list is supplemented by new soil types which were established subsequent to the latest investigations of the poorly investigated territories. Our classification includes 77 soil types.

The soil classification suggested by us is based on their genesis, i.e., on all their essential properties conditioned by their origin and development (i.e., by the processes taking place in these soils), on the determination of the regularities governing their geographical distribution, and on the establishment of the evolutional and historico-genetic relations between the soils, both the existing and the probable relations.

In a genetic classification the soils should be grouped according to the similarity of their soil-forming processes: 1) the processes of the transformation and migration of substances, and 2) the nature of the moisture and thermal regimes. The soils referred to a single group of soil formations develop beneath an ecologically similar vegetation. In respect to these groups of similar soil types the concept "soil-formation type" is usually applied. We suggest the term "soil formation groups" instead of the former. In our scheme (column 2) there are 40 of such groups.

The inclusion of former intrazonal soils in these groups should be explained. They may be grouped in two ways: firstly, according to the nature of the transformation of substances and the general ecological conditions of the soil formation, and secondly, according to those elements of the soil formation process which have formerly grouped these soils into the intrazonal types. It is clear that these principles contradict each other. From the point of view of our concept of the soil formation groups, we naturally prefer the first principle. The genetic relations existing between these types expressed in their specific properties may be taken into consideration in the particular soil series. We suggest the isolation of the following soil series:

1) biogenic (the former zonal) soils;
2) biolothogenic (turfy lithogenic soils: taiga soils, humid-forest soils, sub-tropical soils, tropical soils);
3) biohydrogenic (all the meadow and marshy soils);
4) biohalogenic (all the solonetses, solonchaks and solods).

The following is a brief description of the biogenic soil-formation

183

groups which are widespread in the USSR (see diagram).

I_1. Soils of the tundra-arctic soil formation are characterized by a low humus content, by a crysgenic (due to low temperatures) water regime, with winter thermal (from the warm horizons to the cooler ones) migrations of the soil-formation products rising to the surface (sesquioxides, carbonates, easily soluble salts),and with negligible quantities of substances being introduced into the biological cycle.

The processes of transformation of the mineral and organic substances are accomplished in these soils under low temperatures and mainly under anaerobic conditions. The clay minerals are still poorly investigated. The processes of bulging and freezing are the characteristic features of these soil formations.

I_2. The subarctic turfy soils develop on the northern border of the forest, beneath the long-standing grass vegetation. The soils are poorly investigated. They are widespread in the oceanic regions of the boreal belt.

II. Soils of the taiga-permafrost soil formation with a low humus content in the 1 m layer (apparently about 100 t/ha). Their mineralogical composition as well as the composition of their organic matter have not been studied. The soils are characterized by the presence of eternal freezing and by a cryogenic water regime with seasonal ascending and descending migrations of the soil-formation products (sesquioxides, carbonates, easily soluble salts). The soils are characterized by a high aluminic acidity and are therefore weakly podzolized or not podzolized at all.

III. Soils of a podzolic-forest soil formation (the podzolic and the gray forest soils), containing the beidellite-amorphous group of secondary minerals* with an average adsorption capacity in the loamy and clayey varieties (20–30 m. equiv.) and rather active introduction of substances into the biological cycle: annually from 60-100 to 300 kg/ha of minerals, 20–72 kg/ha of nitrogen and 80–140 ts**/ha of dry organic matter in the forest litter, 2000–2500 ts/ha in the tree stand,and about 600 ts/ha in the roots***. Soils with an intensive decomposition of organic matter and forming medium resources of humus (in the 1 m layer 100–200 t/ha of humus and 7–12 t/ha of nitrogen – M.M. Kononova /33/), with a mixed fulvo-humic-acidic organic matter with a predominant acidic reaction, with a flushing water regime,and with a predominantly descending migration

----

* Data by N. I. Gorbunov. Pochvovedenie, No 2, 1956.

** [A Russian centner ("tsentner") is equal to 100 kg].

*** The figures cited were taken from Bazilevich's work /3/. The new investigations carried out by V.N. Min /38/ were published recently, giving smaller absolute quantities of the substances entering into the biological cycle,while the orders of the numerical quantities are close to the above (in the broad-leaved forests the general quantity of the minerals in the litter is about 200 kg/ha and of nitrogen about 50 kg/ha).

of the soil-formation products (sesquioxides, carbonates, easily soluble salts).

IV. Soils of humid forest burozem (brown soil) soil formation, with a predominance of the beidellite-amorphous group of secondary minerals, having a slightly reduced (as compared to the podzolic and gray forest soils) adsorption capacity with respect to cations, "an increased clay accumulation and podzolization the nature of which is not clear", with medium humus reserves, with a sharply expressed fulvo-acidic composition (the soils are highly acidic, non-podzolized or weakly podzolized), with a flushing water regime, with a predominance of descending migrations of soil-formation products (carbonates, easily soluble salts, and partly of sesquioxides) with a rather active introduction of substances into the biological cycle and their average biological fixation.

V. Soils of the sub-boreal-steppe soil formation, containing the beidellite group of secondary minerals (having, in comparison with the podzolic soils, a lower content of substances of the amorphous group of minerals), highly active in introducing substances into the biological cycle (annually – 800–1100 kg/ha of minerals, 120 kg/ha of nitrogen, 100–180 ts/ha of dry organic matter in the litter, 60–100 ts/ha in the grass stand and 100–200 ts/ha in the roots), forming large humus resources (200–700 t/ha of humus and 13–36 t/ha of nitrogen annually), with the predominance of humic acid organic matter showing a neutral reaction, with a high adsorption capacity of the organic-mineral complex. The soils of the non-flushing or periodically flushing water regime, with a descending migration of carbonates and easily soluble salts (which are fixed in the lower horizons of the soil and in the upper horizons of the parent material).

VI_1. Soils of the sub-boreal-steppe soil formation, having a high content of primary minerals, while the formation of the secondary minerals is limited. The latter are represented mainly by the hydromica group, having a rather low exchange capacity, with a weakened introduction of substances into the biological cycle: annually – about 475 kg/ha of minerals, 90 kg/ha of nitrogen, about 100 ts/ha of organic matter in the litter, 25 ts/ha in the grass stand and 90 ts/ha in the roots.

The plant residues decompose rapidly with the organic matter reaching almost complete mineralization; there is an insignificant humus accumulation in the 1 m layer (humus content, 60–100 t/ha, and nitrogen, 6–7 t/ha); the clearly expressed nonflushing water regime with weakened descending migrations of the biogenic easily soluble salts, which determine the part taken by the latter in the soil formation upon the development of the solonetsization processes.

The introduction of substances is weak and their fixation is more intensive, followed by a carbonation of the surface horizons, due to the biologically accumulated carbonates (with a formation of carbonate [carbonate-containing] crusts in the soils most approaching those of desert soils).

VI_2. The takyz process of the sub-boreal-desert soil formation is characterized by a slight weathering of the primary minerals of the parent rock; of the secondary minerals there are the beidellit hydromicas and beidellite.

The peculiar water regime is a characteristic property of these soils:

it is nonflushing, being rather abundantly surface flooded and with a very slight leaching of the takyr crust, because of its low water permeability. The stagnant surface waters exclude the possibility of desert brushwood development and yet they do not provide enough moisture for the development of meadow plants; therefore, the takyr soils develop mainly under the influence of the lower plant groups: algae and lichen. The soils are distinguished by the negligible humus content, which is no more than its content in the parent material (alluvium). These soils are further characterized by weak migrations of soil formation substances (of the easily soluble salts and, to a certain extent, of carbonates) and an insignificant introduction of substances into the biological cycle (annually – 7–20 kg/ha of nitrogen, 1–3 ts/ha of litter and 1–2 ts/ha of organic matter in the grass stand and roots, respectively).

Soils which are carbonated along the entire profile, in some cases express an increasing carbonate content towards the surface. The easily soluble salts are detected on the surface or (this being more often the case) they are washed into the subsurface horizon. The specific properties of the soils are mostly expressed in the heavy-clayey parent rocks. With a lighter mechanical composition the crust is less typical and the easily soluble salts are more deeply washed in (the solonchakic takyrs).

VII$_1$. The soils of the sub-tropical humid zheltozem [yellow-earth]

soil formation are considered by M.N. Sabashvili /45/ in his recent work as being soils in which the krasnozem [red-earth] formation process is less expressed: " . . . the zheltozems and the yellow loams are the preliminary stage of the red soils and they originated in every place where the iron content in the parent material is not sufficient to cause a brighter red hue". These soils are still poorly investigated and there are not sufficient data to more accurately establish their genesis and properties and to compare them with the krasnozems.

VII$_2$. Soils of the sub-tropical humid krasnozem soil formation,

having a flushing water regime and descending migrations of the soil-formation products (of the easily soluble salts and carbonates containing all the bases, silicic acid and, to a certain extent, the sesquioxides) under conditions of intensive weathering, of intensive introduction of substances into the biological cycle and of an intensive loss of products (total loss of rock substances up to 50–60 %), and a relative accumulation in the weathering products of aluminum and iron hydroxides (molecular ratio – $\frac{SiO_2}{Al_2O_3}$ 1.5-2.0).

A thick profile having a rather constant composition of mineral elements is characteristic of the typical krasnozems. Marked clay accumulation and deep weathering are also characteristic of them. The silt fraction with respect to its total composition approaches the total composition of the entire soil mass and is characterized by the goethite-hydroargillite-beidellite-kaolinite group of secondary minerals, having a relatively low adsorption capacity with respect to cations and a high content of exchangeable aluminum.

The humus content of these soils is on the average 6–7 %, the resources being about 300 t/ha of humus and 10 t/ha of nitrogen. In respect to composition, the humus is fulvo-acidic, represented by fractions connected with the sesquioxides. The soils often express signs of podzolization, the nature of which has not been sufficiently clarified.

VIII. The soils of the subtropical brown-earth soil formation are distinguished by their marked clayeyness (in the middle and lower parts of the profile especially, with the ratio $\dfrac{SiO_2}{Al_2O_3}$ being constant) and by deep weathering of the minerals, but which does not,however, reach the allitic stages. Their secondary minerals are not adequately investigated. Ferribeidellite, beidellite hydromicas and,sometimes − the hydroargillite occur. The soils show a neutral or a weakly alkaline reaction in their upper horizons are deeply humified with a relatively low humus content. The humus composition, in comparison with the sub-boreal steppe soil formation, has a low content of humic acids and a higher content of stable (to acids and alkalis) humic substances and a lower C : N ratio. The adsorption capacity is high due to their increased clayeyness.

The soils have a nonflushing water regime, seasonally ascending (early summer period) and descending (winter and early spring) migrations of soil-formation substances (carbonates) and practically completely devoid of easily soluble salts in the profile. The introduction of the substances into the biological cycle is very active.

The combination of apparently contradicting features which is characteristic of these soils (of deep weathering and the alkaline reaction) is explained by Gerasimov /17/ by the fact that the contradicting processes are of a seasonal nature: the intensive development of the weathering and soil-formation processes − during the cold and moderately cold half of the year, and the fact that these soils are only slightly leached, as expressed in the formation of the carbonated [carbonate-containing] − illuvial horizon and in the neutral or alkaline reaction, pertaining to the dry phase of soil formation during the warm half of the year.

IX. The soils of the serozem- [gray soil] desert- sub-tropical soil formation are characterized by the high content of primary minerals and by a limited formation of secondary minerals which include nontronite, ferribeidellite and hydromica. The secondary minerals are stable and determine the clayization of the soil profile, being only slight in comparison with the parent material, and they show a low adsorption capacity. Also characteristic of them is the relatively limited introduction of minerals into the biological cycle (in the case of serozems it is annually about 400 kg/ha of minerals, 92 kg/ha of nitrogen, about 100 ts/ha of dry mass in the litter,and 20 and 92 ts/ha,respectively,in the grass stand and roots), the rapid decomposition of organic residue and a poor accumulation of humus (the resources being about 70 t/ha of humus and 7 t/ha of nitrogen) with a predominant concentration of it in the upper layer. The peculiar rather highly dispersed forms of humic substances (of the fulvoacidic type) are predominant in its composition. However, their effect on the mineral part of the soil is limited owing to an extreme deficiency in moisture,as well as to the predominance in them of the compounds with calcium. Their water regime is a nonflushing one with seasonal ascending (early summer) and descending (winter − early spring) migrations of the soil-formation products,thus bringing about, in the serozems, a decarbonation of the profile at the expense of the biogenic carbonates (with the content of the latter

becoming less towards the surface), the removal of easily soluble salts
from the soil horizons and a limited development of the solonets process.

The types of soils combined into soil formation groups constitute
the soil classes (see diagram, column 1). The classes combine the soil
of similar soil-formation groups, which are determined by a similar trend
of mineral weathering and by the transformation of the organic matter, and
they make-up zonal complexes and combinations in accordance with those
indicated in Neustruev's work /39/. The close relationship of the weathering
crusts is characteristic of the soil classes.

All the soil types included in a class may be interconnected by
evolutional ties in the soil development processes along with the development of vegetation and relief. The transition of soil from one class to
another takes place upon displacement of bioclimatic zones and facies.

All in all, 12 classes of soils are distinguished (see diagram).

The soil classes are formed into international groups of soil classes:
the boreal, sub-tropical and tropical groups, which differ in their thermal
regimes and in the intensity of the processes of mobilization and transformation of substances.

In the grouping of soils which we suggested, the types are mainly
unified into the same groups as in the former classification (Kossovich,
Glinka, Neustruev), excepting for the application of new investigation
methods which enabled us to characterize the soil processes more profoundly. The extensive geographical investigations in new territories
enabled us to clarify new soil formation processes, to form new groupings
of soil types, and to clarify the evolutional connections between the soils
(see the diagram in the article of N. N. Rozov).

All the genetic classifications are based, as we have already seen,
on the processes of transformation, both of the mineral and the organic
substances, which demonstrate most fully the essence of soil formation.
This method of studying the inner properties of the soils was already outlined by Dokuchaev in the year 1886, who maintained that "the center,
so to say, the focus of all the soil components (and, consequently, of their
properties) is the clay, and the most obvious expression of their common
nature are the organic substances" /26/.

The soil genesis is interrelated with its geographical location, as
the soil formation process is determined by the soil-forming conditions
(the factors) which are distributed on the earth's surface according to
certain regularities. Hence, a correct genetic soil classification not
only clarifies the laws of the genesis but also the regularities of the geographical distribution and development of soils, and thus turns naturally
into an evolutional, geographical-genetic and genetic classification.

The basic property of a soil is its fertility and, consequently, the
soil classification must also reflect this property of the soil.

We know that the first scientific soil classification was that
established by Dokuchaev for the evaluation of land in the Nizhnii Novgorod
gubernaya. "The natural classification and the relative evaluation of the
soils, on the basis of all their natural properties taken as a whole, provided
the criteria for the distribution of land plots and property into territorial
regions and their arrangement into evaluation categories to a certain
gradation. The computation of the data concerning the yielding capacity
(productivity) of arable land and meadow is carried out according to the

An Attempt at a General Soil Classification

Classification scheme of soils*

Class of soil	Groups of soil formations (subclasses of soils)	Types of soil automorphic	Types of soil automorpho-hydromorphic	hydromorphic
1	2	3	4	5

1. Global group of classes of boreal soil formation.

Class of soil	Groups of soil formations (subclasses of soils)	automorphic	automorpho-hydromorphic	hydromorphic
Class I Tundra-arctic soils	1. Tundra-arctic	Arctic soil Tundra soils	– –	– –
	2. Subarctic-turf	Turf meadow soils	–	–
	3. Tundra-marshy	–	–	Marshy-tundra soils
	4. Arctic saline (solonchak)	–	Arctic saline (solonchak)	–
Class II Boreal-permafrost taiga soils	1. Permafrost-taiga soils	Taiga ferruginous soils	–	–
		Pale yellow taiga	Pale yellow gleyey soils	–
	2. Permafrost-marshy		–	Marshy-permafrost soils
	3. Permafrost-solonets soils	Permafrost solods	Permafrost gleyey solods	–
Class III Boreal-taiga and forest soils	1. Podzolic forest soils	Podzolic soils	Podzolic marshy soils	–
		Gray-forest soils	Gray-forest gleyey soils	
	2. Turfy taiga soils	Turfy-taiga (including the turfy-carbonated)	Turfy-gleyey soils	–
	3. Marshy soils	–	–	Marshy soils

---

* Compiled in collaboration with N.N. Rozov.

189

Class of soil 1	Groups of soil formations (sub-classes of soils) 2	automorphic 3	Types of soil automorpho-hydromorphic 4	hydromorphic 5
Class IV Sub-boreal humid-forest and meadow soils	1. Burozem (brown earth)	Humid-forest nonpodzolic acid soils	Humid-forest nonpodzolic acid gleyey soils	–
		Brown forest soils	Brown-forest gleyey soils	–
	2. Meadow-burozem	?	Meadow-burozem soils (of the prairies)	?
	3. Turfy humid-forest soil	Humus-carbonated*	Humus-carbonated gleyey soils	–
	4. Marshy soil	–	–	Marshy soils
Class V Sub-boreal steppe soils	1. Steppe soils	Chernozems	Meadow-chernozem soils	–
		Chestnut soils	Meadow-chestnut soils	–
	2. Meadow soils	–	–	Meadow soils
	3. Meadow-marshy soils	–	–	Meadow-marshy soils
	4. Solonetses	Steppe solonetses	Meadow-solonets solods	–
	5. Solonchak soils	–	–	Steppe solonchaks
Class VI Sub-boreal desert soils	1. Desert soils	Brown semi-desert soils	Brown meadow-desert soils	–
		Gray-brown desert soils	–	–
	2. Takyr soil	–	Takyrs	–

----

* Soil on carbonated parent material amidst the brown forest soils.

Class of soil	Groups of soil formations (sub-classes of soils)	automorphic	Types of soil automorpho-hydromorphic	hydromorphic
1	2	3	4	5
	3. Solonets desert soil	Desert solonetses	Meadow-desert solonetses	–
	4. Solonchak desert soil	–	–	Desert solonchaks

II. Global group of classes of subtropical soil formations

Class of soil	Groups of soil formations (sub-classes of soils)	automorphic	Types of soil automorpho-hydromorphic	hydromorphic
Class VII Subtropical humid forest soils	1. Zheltozem (yellow soils)	Zheltozems	Gleyey zheltozems	–
	2. Krasnozem (red earth)	Krasnozems	Gleyey krasnozems	–
	3. Marshy-subtropical	–	–	Marshy subtropical soils
Class VIII Sub-tropical dry forest, savanna and steppe soils	1. Cinnamon earth	Cinnamon soils	Cinnamon-meadow-forest soils	–
		Gray-cinnamon soils	Gray-cinnamon meadow-steppe soils	–
	2. Turfy-subtropical ?	–		–
	3. Subtropical-meadow	–	–	Subtropical meadow soils
	4. Solonets sub-tropical	Subtropical solonetses	Meadow-solonets subtropical soils	
Class IX Sub-tropical desert soils	1. Serozem-desert	Serozems	Meadow serozem soils	–
		Desert sub-tropical soils	–	–

Class of soil	Groups of soil formations (sub-classes of soils)	automorphicic	Types of soil automorpho-hydromorphic	hydromorphic
1	2	3	4	5
	2. Solonchak sub-tropical	–	–	Subtropical solonchaks

III. Global group of classes of tropical soil formations

Class of soil	Groups of soil formations	automorphicic	automorpho-hydromorphic	hydromorphic
Class X Tropical humid savanna and forest soils	1. Laterite soils	Laterite soils Red soils of tall-grass savannas	Laterite gleyey soils Red gleyey soils of tall-grass savannas	– –
	2. Marshy tropical	–	–	Marshy tropical soils
Class XI Tropical dry forest savanna soils	1. Red-colored tropical	Red-cinnamon soils of the dry forests Red-brown soils of the dry savannas	Red-cinnamon meadow-forest soils Red-brown meadow-savanna soils	– –
	2. Turfy tropical	Black soils of the dry savannas (regurs)	Black gleyey savanna soils	–
	3. Meadow tropical	–	–	Meadow tropical soils
	4. Solonets tropical	Tropical savanna solonetses	Meadow-solonets tropical soils	–
Class XII Tropical desert soils	1. Desert tropical	Red-brown soils of the savannas turned to desert Desert-tropical soils	Red-brown meadow-desert soils –	– –
	2. Tropical solonchak	–	–	Tropical solonchak

grouping of the land plots with respect to the type and quality of the soils. It cannot be denied that a close interdependence exists between the soil types on the one hand,and the yielding capacity, the cereal varieties, and the methods of cultivation on the other . . ." /29/.

Considering the described soil groupings from this point of view, we observe that the world soil groups,isolated according to the quantity of radiation energy, determine the most common ways of utilizing the natural resources in the State economy.

The classes and the groups of the soil formations determine the types of management and cultivation of certain groups of cultures (hothouse, selective vegetables of fallow land, unreclaimed land, slightly cultivated, intensively cultivated, drained, irrigated, irrigation-cultivated soils, etc).

The soil types, regional groups, genera, subtypes and species of soils enable us to establish with greater accuracy the regionally suitable cultures and the general agricultural methods, and help determine the necessary melioration methods and the degree of arability of the soils used for agricultural purposes, etc.

The soil varieties enable us to improve the agricultural measures and to establish them according to the different regions with respect to their soils.

BIBLIOGRAPHY

1. Afanas'ev, Ya.N., Zonal'nye sistemy pochv (Zonal soil systems). – Zapiski Goretskogo sel'sko-khozyaistvennogo in-ta, 1922.
2. Afanas'ev, Ya.N., Klassifikatsionnaya problema v russkom pochvovedenii (The problem of classification in Russian soil science). – Uspekhi pochvovedeniya (Achievements in soil science). – Byuro upolnomoch. pochvovedov SSSR. Doklady delegatov na I-m kongresse Mezhdunar. ob-va pochvovedov (Reports by delegates at the 1st Congress of the International Society of Soil Scientists), 1927.
3. Bazilevich, N.I., Osobennosti krugovorota zol'nykh elementov i azota v nekotorykh pochvenno-rastitel'nykh zonakh SSSR (The peculiarities of the minerals and nitrogen cycle in some soil vegetation zones of the USSR). – Pochvovedenie, No. 4: 1955.
4. Vilenskii, D.G., Analogichnye ryady v pochvoobrazovanii i ikh znachenie dlya postroeniya geneticheskoi klassifikatsii pochv (Analogous series in soil formation and their significance for the construction of a genetic soil classification). – (Tiflis) 1924.
5. Vilenskii, D.G., Russkaya pochvenno-kartograficheskaya shkola (The Russian soil-cartographic school). – SOPS [Sovet proizvoditel'-nykh sil Severa] Akademiya Nauk SSSR, 1945.
6. Vil'yams, V.R., Pochvovedenie. Zemledelie s osnovami pochvovedeniya (Soil science. Agriculture based on soil science). – Sel'khozgiz, 1939.
7. Volobuev, V.R., Nekotorye voprosy ucheniya o geneticheskom tipe pochv (Some problems concerning the genetic type of soils). – Pochvovedenie, No. 11: 1955.

8. Vysotskii, G.N., Ob oroklimatologicheskikh osnovakh klassifikatsii pochv (Concerning the oroclimatological principles of soil
   classification). – Pochvovedenie, No. 1–4: 1906.
9. Gedroits, K.K., Pochvennyi pogloshchayushchii kompleks i
   pochvennye pogloshchennye kationy, kak osnova geneticheskoi pochvennoi klassifikatsii (The soil adsorption complex and the adsorbed
   cations of the soils as a basis of the genetic soil classification).–
   Izd. Nosovskoi sel'sko-khozyaistvennoi opytnoi stantsii, 1927.
10. Gemmerling, V.V., O metamorfoze pochvennykh obrazavanii
    (Concerning the metamorphosis of formations). – Dnevnik 12-ogo
    S''ezda russkikh estestvoispytatelei i vrachei v 1910 g. (Diary of the
    12th conference of the Russian naturalists and physicians in the year
    1910). – (Moskva) 1910.
11. Gerasimov, I.P., A.A. Zavalishin and E.N. Ivanova, Novaya
    skhema obshchei klassifikatsii pochv SSSR (A new scheme of a
    general classification of the soils of the USSR). – Pochvovedenie,
    No. 7: 1939.
12. Gerasimov, I.P., Gosudarstvennaya pochvennaya karta SSSR
    (The State soil map of the USSR). – Pochvovedenie, No. 1: 1947.
13. Gerasimov, I.P., Proiskhozhdenie prirody sovremennykh geo;
    geograficheskikh zon na territorii SSSR (The origin of the natural
    conditions prevailing in the present-day geographical zones of the
    USSR). – Izv. AN SSSR, seriya geograficheskaya, No. 2: 1951.
14. Gerasimov, I.P., Paleogeograficheskoe znachenie ucheniya
    V.R. Vil'yamsa o edinom pochvoobrazovatel'nom protsesse (The
    paleogeographic significance of V.R. Vil'yams' teaching concerning
    the single soil-formation process). – Problemy fizicheskoi geografii,
    Vol. XVI: 1951.
15. Gerasimov, I.P., Nauchnye osnovy sistematiki pochv (The
    scientific fundamentals of soil classification). – Pochvovedenie,
    No. 11: 1952.
16. Gerasimov, I.P., Nauchnye osnovy sistematiki i klassifikatsii
    pochv (The scientific basis of soil systematics and classification). –
    Pochvovedenie, No. 8: 1954.
17. Gerasimov, I.P., Korichnevye pochvy sredizemnomorskikh
    oblastei (The cinnamon soils of the Mediterranean regions). –
    Doklad na V Mezhdunarodnom kongresse pochvovedov AN SSSR
    (Report at the 5th International Congress of Soil Scientists of
    the Ac. of Sc. USSR), 1954.
18. Glinka, K.D., Posletretichnye obrazovaniya i pochvy Pskovskoi,
    Novgorodskoi i Smolenskoi gubernii (Klassifikatsiya pochv) [The
    post-Tertiary formations and soils of Pskov, Novgorod and Smolensk
    (soil classification)]. – Ezhegodnik po geologii i mineralogii. Vol.
    5: No. 4–5: 1902.
19. Glinka, K.D., Pochvovedenie (Soil science). – Kurs lektsii
    (Series of lectures) 1–oe izd. (1st edition), St. Petersburg, 1908.
20. Glinka, K.D., Pochvovedenie (Soil science). – Kurs lektsii,
    2-oe izd. (2nd edition), 1915.
21. Glinka, K.D., Pochva, ee svoistva i zakony rasprostraneniya
    (The soil, its properties and the regularities of its distribution). –
    Izd. N. Derevnya (Moskva),1922.

22. G l i n k a , K.D., Dispersnye systemy v pochve (The dispersion systems in the soil). – Izd.,1924.

23. G l i n k a , K.D., Differents types d'après lesquels se forment les sols et la classification de ces derniers. Comité intern. de pédologie, IV Commis., No. 20: 1924.

24. D o k u c h a e v , V.V., Kratkii istoricheskii ocherk i kriticheskii razbor vazhneishikh sushchestvuyushchikh pochevennykh klassifikatsii (A brief historical survey and a critical analysis of the most important extant soil classifications). – Trudy SPb., obshchestva estestvoispytatelei (Works of the St. Petersburg society of naturalists), 1879, Vol. X: pp 64–67, Protokoly. Sochineniya. Vol. II: 1950.

25. D o k u c h a e v , V.V., Po voprosu o sibirskom chernozeme (Concerning the problem of the Siberian chernozem). – Trudy V.E.O., 1882. Sochineniya. Vol. II: No. 3: Vol. II: 1950.

26. D o k u c h a e v , V.V., Materialy k otsenke zemel' Nizhegorodskoi gubernii (Data for the evaluation of the soils of the Nizhnii Novgorod). – Sochineniya. Vol. IV: izd AN SSSR.

27. D o k u c h a e v , V.V., Materialy k otsenke zemel' Nizhegorodskoi gubernii. Pochvy, rastitel'nost' i klimat Nizhegorodskoi gubernii (Data for the evaluation of the soils of the Nizhnii Novgorod. Soils, vegetation and climate of the Nizhnii Novgorod). – Gl. I (Chapter I), 1886, Sochineniya. Vol. V: 1950.

28. D o k u c h a e v , V.V., Katalog pochvennoi kollektsii prof. V.V. Dokuchaev i ego uchenikov (The catalogue of the collection of Prof. V.V. Dokuchaev and his students). – (St. Petersburg) 1896, Sochineniya, Vol. VII: 1953.

29. D o k u c h a e v , V.V., K voprosu o pereotsenke zemel' Evropeiskoi i Aziatskoi Rossii s klassifikatsiei pochv (Concerning the problem of revaluation of the lands of both European and Asiatic Russia upon soil classification). – (Moskva) 1898, Sochineniya. Vol. VI: 1951.

30. Z a k h a r o v , S.A., Kurs pochvovedeniya (Soil science manual). – 1927.

31. I v a n o v a , E.N., Sistmatika pochv severnoi chasti Evropeiskoi territorii SSSR (Soil classification of the northern part of the European USSR). – Pochvovedenie, No. 1: 1956.

32. K o v d a , V.A., Printsipy klassifikatsii pochv (The principles of the soil classification). – Sbornik (collection), Zadachi i metody pochvennykh issledovanii. Tr. Sov. sekts. MAP, Vol. II: Kom. V (genezis pochv), No. 1: 1933.

33. K o n o n o v a , M.M., Problema pochvennogo gumusa i sovremennye zadachi ego izucheniya (The soil humus problem and the present-day purposes of their investigation). – AN SSSR, 1951.

34. K o s s o v i c h , P.S., Solontsy, otnoshenie k nim rastenii i metody opredeleniya solontsevatosti pochv (The solonetses, the relations of the plants towards them and the methods for establishing the degree of salinity of the soils). – Zhurnal opytnoi agronomii, Book I, pp. 30–31: 1903.

35. K o s s o v i c h , P.S., K voprosu o genezise pochv i ob osnovakh dlya geneticheskoi pochvennoi klassifikatsii (Concerning the problem of the soil genesis and the basis for a genetic soil classification). – Zhurnal opytnoi agronomii, Book IV, Vol. VII: 1906.

36. K o s s o v i c h , V.S., Pochvoobrazovatel'nye protsessy,kak osnova geneticheskoi pochvennoi klassifikatsii (The soil formation processes as a basis for a genetic soil classification). – Zhurnal opytnoi agronomii, Book 5, Vol. XI: 1910.
37. M a r k o v , K.K., Vzaimootnosheniya lesa i stepi v istoricheskom osveshchenii (The historical interrelations of the forest and steppe). – Voprosy geografii, Sbornik (collection) 23, (Moskva) 1950.
38. M i n a , V.N., Krugovorot azota i zol'nykh elementov v dubrasvakh lesostepi (The nitrogen and mineral cycle in the forests of the forest-steppe). – Pochvovedenie, No. 6: 1955.
39. N e u s t r u e v , S.S., O pochvennykh kombinatsiyakh ravninnykh i gornykh stran (Concerning the soil combinations in the plain and mountain countries). – Pochvovedenie, No. 1: 1915.
40. N e u s t r u e v , S.S., Opyt klassifikatsii pochvoobrazovatel'nykh protsessov v svyazi s genezisom pochv (An attempt at a soil classification of soil formation processes concerning the genesis of soils). – Izvestiya Geograficheskogo komiteta, No. 6: 1926.
41. P o l y n o v , B.B., Osnovy postroeniya geneticheskoi klassifikatsii pochv (predvaritel'nye tezisy) [The basis for the construction of a genetic soil classification (Preliminary theses). – Sbornik. Zadachi i metody pochvennykh issledovanii. Trudy sovetskoi sektsii MAP, Vol. II: kom. V (genezis pochv), No. 1: 1933.
42. P r a s o l o v , L.I., K voprosu o klassifikatsii i nomenklature pochv (Concerning the problem of the classification and nomenclature of soils). – Trudy Pochvennogo instituta im. V.V. Dokuchaeva, Vol. VIII: 1934.
43. R o d e , A.A., Pochvoobrazovatel'nyi protsess i evolutsiya pochv (The soil formation process and the evolution of soils). – 1947.
44. R o d i n , L.E., and N.I. B a z i l e v i c h , O krugovorote zol'nykh elementov i azota v nekotorykh pustynnykh biogeotsenozakh (Concerning the mineral and nitrogen cycle in some desert biogeocoenoses). – Botanicheskii zhurnal. Vol. XL: No. 1: 1955.
45. S a b a s h v i l i , M.N., Subtropicheskie krasnozemy SSSR (Subtropical krasnozems of the USSR). – Doklad na V Mezhdunarodnom kongresse pochvovedov (Lecture at the 5th International Congress of Soil Scientists). AN SSSR, 1954.
46. S i b i r t s e v , N.M., Ob osnovaniyakh geneticheskoi klassifikatsii pochv (Concerning the fundamentals of genetic soil classification). – 1895. Zapiski Novo-Aleksandrivskogo instituta khozyaistva i lesovodstva, Vol. IX: No. 7: Izbrannye sochineniya (Selected works). Vol. II: 1953.
47. S i b i r t s e v , N.M., Kratkii obzor glaveneishikh pochvennykh tipov Rossii (so skhemoi) [A brief survey of the main soil types of Russia (supplemented by a diagram)]. – Zapiski Novo-Aleksandrovskogo instituta sel'skogo khozyaistva i lesovodstva, Vol. XI: No. 3: 1898. Izbrannye proizvedeniya (Selected works), Vol. II: 1951.
48. S i b i r t s e v , N.M., Pochvovedenie (Soil science). – Izbrannye proizvedeniya (Selected works), Vol. I: 1951.
49. S i b i r t s e v , N.M., Klassifikatsiya pochv (Soil classification). – Polnaya entsiklopediya russkogo sel'skogo khozyaistva (Complete encyclopedia of Russian agriculture). – Vol. IV: (St. Petersburg)

pages 174–192: 1901. Izbrannye proizvedeniya, Vol. II: 1951.

50. S t r a k h o v , N. M. , Istoricheskaya geologiya (Historical geology). – 1938.

51. T u m i n , G. M. , Printsipy klassifikatsionnykh skhem Dokuchaeva, Ruspolozhenskogo, Sibirtseva, Glinki, Slezkina i Kossovicha (The principles of classification schemes of Dokuchaev, Rispolozhenskii, Sibirtsev, Glinka, Slezkin and Kossovich). – Ezhegodnik po geologii i mineralogii Rossii, Vol. VIII: No. 10: 1907.

52. T s y g a n o v , M. S. , Osnovnye printsipy geneticheskoi klassifikatsii i nomenklatury pochv (The basic principles of a genetic classification and nomenclature of soils). – Pochvovedenie, No. 12: 1955.

53. S h i l o v a , E. I. , Ob uchenii akademika Vil'yamsa o edinom pochvoobrazovatel'nom protsesse (With reference to the teachings of Academician Vil'yams concerning the single soil formation process). – Vestnik Leningradskogo universiteta. seriya biologii i geologii, No. 10: 1953.

Soil institute
im. V. V. Dokuchaev
AN SSSR

Date of submission
28 April 1956

14

Reprinted from *Soil Taxonomy: A Basic System for Making and Interpreting Soil Surveys*, U.S. Government Printing Office, Washington, D. C., 1975, pp. 71–81

The Categories of the System

Soil Survey Staff

A category of this system is a set of classes that are defined approximately at the same level of generalization or abstraction and that include all soils. There are six categories in the system. In order of decreasing rank and increasing number of differentiae and classes, the categories are order, suborder, great group, subgroup, family, and series.

In one sense, taxonomy is a sorting process. In the highest category of this system, one sorts all kinds of soil into a small number of classes. The number of classes is small enough to permit one to comprehend and remember them and to understand the distinctions between them. The sorting must make distinctions that are meaningful for our purposes. Obviously, when all soils are sorted into a very few classes, such as the ten orders, each order is very heterogeneous with respect to properties that are not used for the sorting and that are not accessory to the properties that are used.

To reduce the heterogeneity requires another sorting in the next lower category, the suborder. Again, the sorting must be meaningful, but the sorting is made in one order at a time and properties relevant to the sorting in one order may have little meaning in another order. In this taxonomy there are now 47 suborders, a number larger than can be remembered conveniently along with all their properties; however, if we focus on the suborders of a single order we have, at the most, seven suborders to understand and remember. Each of these suborders in an order has the properties common to the order plus the properties used for sorting into that suborder. In each of the 47 suborders, there is still great heterogeneity, so we must sort again to obtain, at the next lower level, a set of meaningful great groups. There are about 225 great groups, more than one can remember. We need focus, however, on only one suborder at a time. We sort each suborder into a few great groups, few enough that we can remember the properties that distinguish each of them.

The sorting process continues in the remaining categories down to the soil series. The soils in any one series are nearly homogeneous in that their range of properties is small and can be readily understood.

Collectively, the thousands of soil series are far beyond our powers of comprehension, but we can sort them category by category, and we seldom need to try to comprehend more than a few of them at any one time.

Orders

There are 10 orders. They are differentiated by the presence or absence of diagnostic horizons or features that are marks in the soil of differences in the degree and kind of the dominant sets of soil-forming processes that have gone on. If the soils in a given taxon are thought to have had significantly different genesis, the intent has been to sort out the differences in the next lower category.

We know that soil properties must be the consequences of a variety of processes acting on parent materials over time. At this highest category we have looked for distinctions among orders that would be as meaningful as possible for understanding soils and remembering them on a grand scale. We have also looked for distinctions that appear in an orderly fashion in nature. The processes that go on in soils must be orderly in relation to the soil-forming factors, which are climate and living organisms acting on parent materials over time as conditioned by relief. These factors, in turn, have geographic order. We can see in the soil the marks of the soil-forming processes, but we can only infer the details of the processes. The distinctions made in classifying soils cannot be based on the processes themselves because new knowledge is certain to change our ideas about the processes, but the marks of the processes are facts that can be observed and measured and used as a basis for distinctions. Thus, the distinctions between orders are based on the marks of sets of processes that experience indicates are dominant forces in shaping the character of the soil. In this framework, the lack of marks or the zero degree is also a logical criterion.

The 10 orders and the major properties that differentiate them are discussed next to illustrate the nature of this category. Complete definitions are given later.

[Editor's Note: None of the Plates mentioned in this article have been reproduced here.]

These orders are not the only possible orders in the taxonomy. The hierarchy is flexible, and other *ad hoc* orders may be defined to emphasize properties not considered in these. The method of doing this is discussed later in connection with the nomenclature.

Alfisols

Soils in the order of Alfisols have marks of processes that translocate silicate clays without excessive depletion of bases and without dominance of the processes that lead to formation of a mollic epipedon. The unique properties of Alfisols are a combination of an ochric or an umbric epidedon, an argillic horizon, a medium to high supply of bases in the soil, and water available to mesophytic plants more than half the year or more than 3 consecutive months during a warm season. Because these soils have water and bases, they are, as a whole, intensively used, but problems with tilth are very common. Figure 26 shows the distribution of organic carbon and clay and the degree of base saturation in a typical Alfisol (pedon 12). Note the shallow penetration of organic carbon, the distinct accumulation of clay at a depth of about 60 cm, and the high degree of base saturation throughout the soil. The horizon of clay accumulation may be thick or, as in this example, relatively thin. Plates 2D, 3A, 4B, 5C, 6A, 7D, 10B, and 10C show representative Alfisols.

Aridisols

The unique properties common to Aridisols (plates 5D and 6C) are a combination of a lack of water available to mesophytic plants for very extended periods, one or more pedogenic horizons, a surface horizon or horizons not significantly darkened by humus, and absence of deep wide cracks (see Vertisols). The Aridisols have no "available" water (held at tension <15 bars) during most of the time that the soil is warm enough for plant growth (warmer than 5° C or

41° F), and they never have water continuously available for as long as 90 days when the soil temperature is above 8° C (47° F). Figure 27 shows the distribution of organic carbon, clay, and calcium carbonate in typical Aridisols (pedons 36 and 59). Note the very small amount of organic carbon near the surface and the large amount of accumulated carbonates and clay.

Aridisols are primarily soils of arid areas. They are in places that preclude much entry of water into the soil at present, either under extremely scanty rainfall or under slight rainfall that for one reason or another does not enter the soil. Vegetation, if present, consists of scattered plants, ephemeral grasses and forbs, cacti, and xerophytic shrubs. Some Aridisols furnish limited grazing. If irrigated, many of them are suitable for a wide variety of crops.

Entisols

The absence of marks in the soil of any major set of soil-forming processes is itself an important distinction. There can be no accessory characteristics. The unique properties common to Entisols are dominance of mineral soil materials and absence of distinct pedogenic horizons. Entisols are soils in the sense that they support plants, but they may be in any climate and under any vegetation. The absence of pedogenic horizons may be the result of an inert parent material such as quartz sand, in which horizons do not readily form; of formation from a hard, slowly soluble rock such as limestone, which leaves little residue; of the lack of time for horizons to form, as in recent deposits of ash or alluvium; of occurrence on slopes where the rate of erosion exceeds the rate of formation of pedogenic horizons; or of recent mixing of horizons by animals or by plowing to a depth of 1 or 2 m.

Figure 28 shows the distribution of organic carbon, clay, and calcium carbonate as functions of depth in two kinds of Entisols (pedons 67 and 68). Note the lack of evidences of translocation of materials or of alteration other than accumulation of small amounts of organic carbon that have not appreciably darkened the soils. The variability in the percentages of clay and of organic carbon is a reflection of stratified alluvium (fig. 28, left). In this soil, even the fine stratification within the more sandy layers has not been disturbed by biologic activity. Plate 4A shows an Entisol in sand that has some properties of an Alfisol. Plate 9A shows an Entisol in recent alluvium (pedon 67).

Histosols

The unique property of Histosols (plate 8D) is a very high content of organic matter in the upper 80 cm (32 in.) of soil. The amount of organic matter is at least 20 to 30 percent in more than half of this thickness, or the horizon that is rich in organic matter rests on rock or rock rubble. Most Histosols are peats or mucks, which consist of more or less decomposed plant remains that accumulated in water, but some have formed from forest litter or moss, or both, in a perhumid environment and are freely drained. The freely drained Histosols are discussed in chapters 4 and 11.

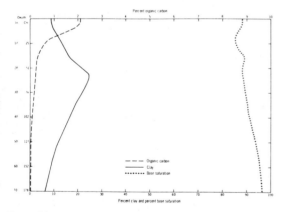

Figure 26.—Percentages of clay, organic carbon, and base saturation as functions of depth in a representative Alfisol.

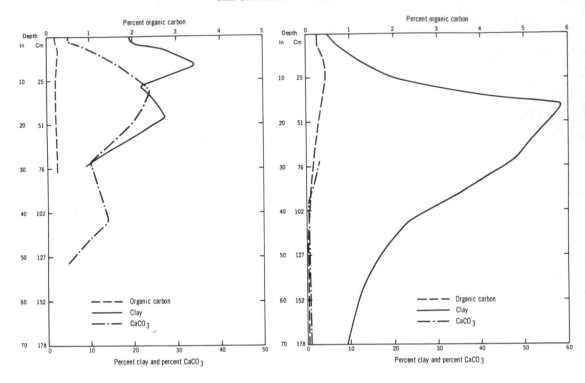

Figure 27.—**Percentage of clay, organic carbon, and calcium carbonate as functions of depth in two contrasting Aridisols.**

Inceptisols

The unique properties of Inceptisols are a combination of water available to plants during more than half the year or more than 3 consecutive months during a warm season; one or more pedogenic horizons of alteration or concentration with little accumulation of translocated materials other than carbonates or amorphous silica; texture finer than loamy sand; some weatherable minerals; and moderate to high capacity of the clay fraction to retain cations. In addition, Inceptisols do not have one or more of the unique properties of the order of Mollisols, which are the thick, dark surface horizon, the high calcium supply, and the crystalline clays. Figure 29 shows the distribution of organic carbon and the bulk density in two Inceptisols as functions of depth. Clay distribution cannot be measured for the soil in figure 29 (left), which was derived from volcanic ash, so is not given for either of the soils. Pedon 29 is an Inceptisol. Note that there are wide ranges within Inceptisols in the content of organic carbon and in the cation-exchange capacity. Similar variability exists in the degree of base saturation.

Inceptisols can form in almost any but an arid environment, from polar to tropical, and the comparable differences in vegetation are great. Very commonly, Inceptisols have rock at a shallow depth. Most Inceptisols are on relatively young geomorphic surfaces, late Pleistocene or Holocene for the most part. Plates 9B and 9C show Inceptisols that are similar to pedons 74 and 29. Other Inceptisols are shown in plates 2C, 11A, 11B, and 11C.

Mollisols

The unique properties of Mollisols (plates 1D, 2B, 8B, 11D, 12A, and 12B) are a combination of a very dark brown to black surface horizon (mollic epipedon) that makes up more than one-third of the combined thickness of the A and B horizons or that is >25 cm (10 in.) thick and that has structure or soft consistence when dry; a dominance of calcium among the extractable cations in the A1 horizon and the B horizons; a dominance of crystalline clay minerals of moderate or high cation-exchange capacity; and <30 percent clay in some horizon above 50 cm if the soil has deep wide cracks (≥1 cm wide) above this depth at some season. Figure 30 shows the distribution of organic carbon, calcium carbonate, clay, cation-exchange capacity, and degree of base saturation as functions of depth in a typical Mollisol. Note the high de-

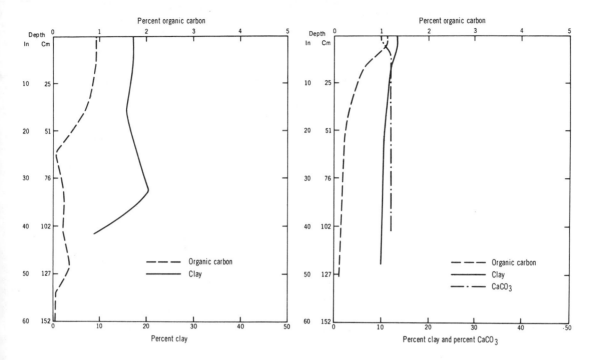

*Figure 28.—*Percentages of clay, organic carbon, and calcium carbonate as functions of depth in two contrasting Entisols, left, on a floodplain, and right, on a hillside.

gree of base saturation and the relatively deep penetration of organic carbon. One or both of the horizons of clay and of carbonate accumulation may be absent in Mollisols.

Mollisols characteristically form under grass in climates that have a moderate to pronounced seasonal moisture deficit. A few form in marshes or on marls in humid climates. Mollisols are the extensive soils of the steppes of Europe, Asia, North America, and South America.

Oxisols

The unique properties of Oxisols are extreme weathering of most minerals other than quartz to kaolin and free oxides; very low activity of the clay fraction; and a loamy or clayey texture (sandy loam or finer). Figure 31 shows as functions of depth the organic carbon, cation-exchange capacity, total clay, and water-dispersible clay in a typical Oxisol (pedon 32). Note the high content of organic carbon, the low cation-exchange capacity, and the decreasing clay content with depth. The decrease of cation-exchange capacity with depth is due to the decreasing amount of organic matter. Pedon 33 is the Oxisol that is shown in plate 4D.

Oxisols characteristically occur in tropical or subtropical regions on land surfaces that have been stable for a long time. Mostly the surfaces are early Pleistocene or much older. Oxisols develop in a humid climate, but because climates change, some are now in an arid environment.

Spodosols

Spodosols (plates 4C, 7B, and 8C) have marks in at least an upper sequum of dominant processes that translocate humus and aluminum, or humus, aluminum, and iron as amorphous materials. The unique property of Spodosols is a B horizon of accumulation of black or reddish amorphous materials that have high cation-exchange capacity, the spodic horizon. In most undisturbed soils, an albic horizon overlies the B horizon. The spodic horizon has accessory characteristics of being moist or wet, loamy or sandy texture, high pH-dependent exchange capacity, and few bases. Figure 32 shows the distribution of clay and organic carbon, base saturation, cation-exchange capacity, and extractable iron and aluminum as functions of depth in a typical Spodosol (pedon 25). Note that the cation-exchange capacity is related to the amount of organic carbon rather than to the clay.

201

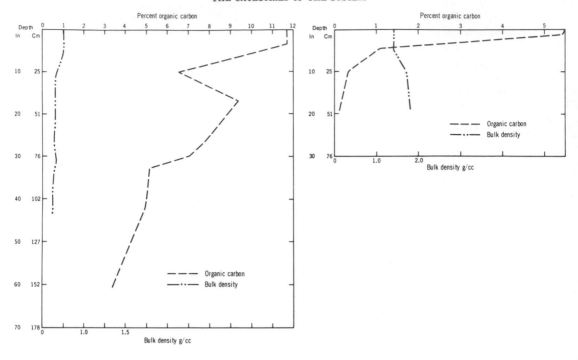

Figure 29.—Percentage of organic carbon and bulk densities of two contrasting Inceptisols, left, from volcanic ash, and right, from old alluvium.

Ultisols

Ultisols (plate 9D), like Alfisols, have marks of clay translocation, but they also have marks of intense leaching that are absent in Alfisols. The unique properties common to Ultisols are an argillic horizon, a low supply of bases, particularly in the lower horizons, and a mean annual soil temperature higher than 8° C (47° F).

Figure 33 shows the distribution of clay, cation-exchange capacity, and base saturation as functions of depth in two Ultisols (pedons 21 and 15). Note that the clay content increases and then decreases with increasing depth, but the horizon of clay accumulation may be thin or thick. The cation-exchange capacity in Ultisols is mostly moderate or low. The decrease in base saturation with depth reflects cycling of bases by plants or additions in fertilizers. In soils that have not been cultivated, the highest base saturation is normally in the few centimeters just beneath the surface. Like Alfisols, the Ultisols have water, but they have few bases. Without fertilizers, they can be used for shifting cultivation, but because they are relatively warm and moist, they can be made highly productive if fertilizers are used.

Vertisols

These soils (plates 10A, 12C, and 12D) have marks of processes that mix the soil regularly and prevent development of diagnostic horizons that one might otherwise expect to find. Because the soil moves, the diagnostic properties have many accessory properties. Among them are high bulk density when dry, very slow hydraulic conductivity when moist, an appreciable rise and fall of the soil surface as the soil becomes moist and then dries, and rapid drying as a result of open cracks. The unique properties common to Vertisols are a high content of clay; pronounced changes in volume with changes in moisture; deep wide cracks (≥ 1 cm wide at a depth of 50 cm) at some season; and evidences of soil movement in the form of slickensides, gilgai microrelief, and wedge-shaped structural aggregates that are tilted at an angle from the horizontal. Figure 34 shows the distribution of clay, organic carbon, and cation-exchange capacity in a typical Vertisol (pedon 127). Note the relatively low content of organic carbon, the high content of clay, the high cation-exchange capacity, and the lack of marked changes in these properties with depth. At some season, rains come when the cracks are open,

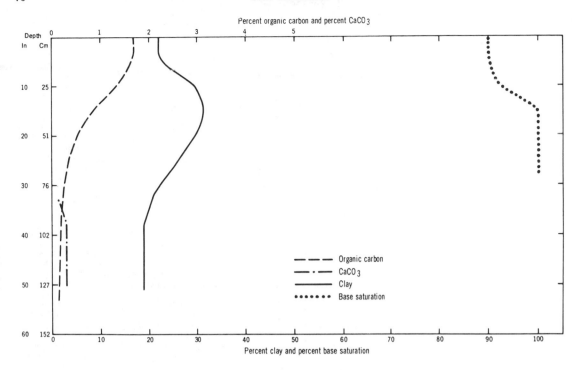

Percent organic carbon and percent CaCO₃

Figure 30.—Percentages of clay, organic carbon, base saturation, and calcium carbonate as functions of depth in a representative Mollisol.

Examples of differences in orders

and some of the surface soil falls or washes into the cracks before the whole soil becomes moist and the cracks close. Because of this movement, these soils have been said to swallow themselves. Plate 7C shows slickensides from a Vertisol.

The preceding discussion illustrates the kinds of marks that have been chosen to differentiate the orders and the sets of processes that they indicate. The Entisols lack distinctive horizons, although they may have stratified layers. Too little has happened in them to provide distinct pedogenic marks in the soil. The Entisol on a hillside has accumulated a little organic carbon, but not enough to darken it appreciably (fig. 28, right). The upper 10 cm of soil have lost a little calcium carbonate, but all of these alterations would be obliterated by the first plowing.

The Vertisol (fig. 34) shows few changes with depth. The organic carbon has penetrated deeply but the amounts are small. There are deep wide cracks, gilgai, abundant slickensides below a depth of 50 cm, and a change in structure but little change in other properties at increasing depth. These are marks of the dominance of self swallowing over most other processes.

The Alfisol (fig. 26) has a distinct horizon of clay accumulation and shallow penetration of organic carbon. Base saturation is high. Too little water has gone through the soil to remove the bulk of the bases. Release of bases from primary minerals or their retention has so far been dominant over leaching and removal of the bases from the soil.

Although Alfisols are defined as having an argillic horizon, such a definition also implies that an eluvial horizon is present or has been present. The eluvial horizon may have been removed by erosion, but the classification of the soil is not changed as a result of truncation until the B horizon has also been lost.

In contrast to Alfisols, the Ultisols (fig. 33) have few exchangeable bases. The few bases that are present in the soil shown in figure 33 have been held by plants against leaching. The soil has virtually no weatherable minerals. The processes of eluviation and illuviation seem the same in Ultisols and Alfisols.

The Spodosol (fig. 32) has negligible evidences of eluviation and illuviation of clay but rather has evidences of eluviation and illuviation of amorphous mixtures or compounds of organic carbon, iron, and aluminum. Retention of bases by the soil or by plants has been negligible, and the effects of leaching of bases are dominant. In contrast to the Ultisols, the cation-ex-

203

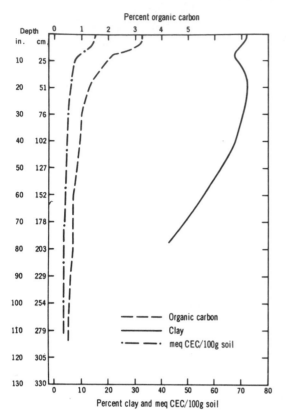

Percent organic carbon

Percent clay and meq CEC/100g soil

- - - - - Organic carbon
——— Clay
·—·—· meq CEC/100g soil

Figure 31.—**Percentages of organic carbon, total and water-dispersible clay, and cation exchange capacity as functions of depth in a representative Oxisol.**

change capacity is a function of the amount of organic carbon rather than the amount of clay.

Suborders

Forty-seven suborders currently are recognized. The differentiae for the suborders vary with the order but can be illustrated by examples from two orders. The Entisol order has five suborders that distinguish the major reasons for absence of horizon differentiation. One suborder includes soils that have an aquic moisture regime. They consist of marshy recent alluvium and the soils of coastal marshes that are saturated with water and have a blue or green hue close to the surface. This segregates the wet varieties. A second suborder includes soils that are not wet and that consist of recent alluvium, which generally is stratified

(fig. 28, left). This segregates the very young soils that do not have horizons because there is continuing deposition of new sediments. A third suborder provides for soils on recently eroded slopes (fig. 28, right). This segregates soils that are kept young by removal of soil materials at a rate that is more rapid than that of horizon differentiation. A fourth suborder includes sands that may range from recent to old. If old, either they lack the building blocks for pedogenic horizons or they do not have enough moisture. Although the reasons for absence of horizons in the sands vary, the sands have many common physical properties such as low capacity for moisture retention, rapid hydraulic conductivity, susceptibility to blowing, and so on. The sorting of these differences is continued in the lower categories. The fifth suborder of Entisols provides for soils in which horizons have been mixed, as by deep plowing, where man has destroyed the pedogenic horizons as such but not the fragments of the horizons.

The order of Alfisols also has five suborders. As in Entisols, one suborder provides for wet soils in which the colors are dominantly gray. A second suborder provides for the Alfisols that are not wet but are cold or cool enough that the low temperature of the soil restricts biologic activity during much of the year. A third suborder provides for Alfisols that are warm and that have a udic moisture regime, so that they rarely lack water available to plants. A fourth suborder has an ustic moisture regime, is warm or hot, and has extended or frequent periods when the soil does not have water that is available to mesophytic plants in some or all horizons. The fifth suborder includes the Alfisols that have a xeric moisture regime. These soils are cool and moist in winter, but they are dry in summer for extended periods.

The differentiae used in defining suborders of Alfisols include important properties that influence genesis and that are extremely important to plant growth. The differentiae in four of the other orders parallel those of Alfisols. In the remaining orders differentiae were selected to reflect what seemed to be the most important variables within the orders.

Great groups

About 185 great groups are currently known to occur in the United States.

At as high a categoric level as possible, it is desirable to consider all the horizons and their nature collectively as well as the moisture and temperature regimes. The moisture and temperature regimes are causes of properties, and they also are properties of the whole soil rather than of specific horizons. At the levels of the order and suborder, only a few of the most important horizons could be considered because there are few taxa in those categories. At the great group level, therefore, we try to consider the whole soil, the assemblage of horizons, and the most significant properties of the whole soil, selected on the basis of the numbers and importance of accessory properties. Although the definition of a great group may involve only a few differentiae, the accessory properties

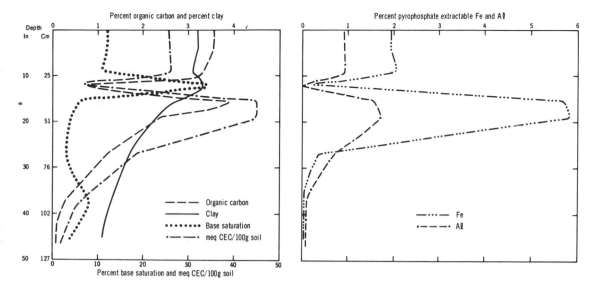

Figure 32.—Percentages of clay, organic carbon, base saturation, and pyrophosphate-extractable iron and aluminum, and cation exchange capacity as functions of depth in a representative Spodosol.

are many times that number. In a few taxa, soil moisture and soil temperature regimes are not defined but are accessory properties to some that are defined.

Differentiae in the great group category place soils together that have in common the following properties:

Close similarities in kind, arrangement, and degree of expression of horizons.—Exceptions are made for some thin surface horizons that would be mixed by plowing or lost by erosion and for horizons that indicate transitions to other great groups. For example, an argillic horizon that underlies the spodic horizon is permitted in the Spodosol order because that combination is considered to represent a kind of transition between the Spodosols on the one hand and Alfisols and Ultisols on the other. Emphasis is placed on the upper sequum in the great group category because it is thought to reflect the current processes and is more critical to plant growth than the deeper horizons.

Close similarities in soil moisture and temperature regimes.—Wider ranges in the moisture regime are permitted in soils that are very cold than in soils that are warm. Irrigation and drainage can modify the moisture regime, but there is no practical way to raise the temperature of a cold soil on a large scale.

Similarities in base status.—If the base status varies widely within a suborder, the range is narrowed at the great group level.

Examples.—In the order of Alfisols, the suborders were defined on moisture and temperature regimes. In

addition to the argillic horizon that is common to all Alfisols, other kinds of horizons may be present. A fragipan or a duripan restricts root development and water movement, which in turn affect current processes of soil formation. These horizons are used as one basis for defining great groups. The argillic horizon may have a fine texture and may be abruptly separated from an overlying albic horizon. This combination also affects root development and water movement, inducing shallow perched ground water and intermittent reducing conditions in the soil. The horizons may be thick as a result of very long periods of development. They may be undergoing destruction and may have deep tongues of an eluvial horizon penetrating them. These marks are also used as differentiae for great groups.

In contrast to Alfisols, emphasis in the Entisols is placed on soil moisture and temperature regimes to differentiate the great groups. Because the various suborders occur in all parts of the world, they have extreme ranges in moisture and temperature regimes, and those regimes are the main factors in determining what can happen in the soils and how we can use them.

A few suborders have no great groups. The reasons vary. In some suborders, we lack information about the soils. The soils of other suborders are very homogeneous in their properties. In still other suborders, it seems more appropriate to subdivide the soils at a lower category, as in the situation with those Entisols that have resulted from mixing of horizons by the activities of men.

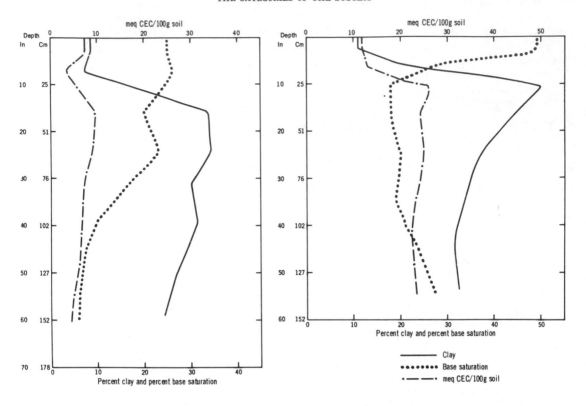

Figure 33.—Percentages of clay and base saturation and cation exchange capacity as functions of depth in two contrasting Ultisols.

Subgroups

About 970 subgroups are currently recognized in the United States.

Through the categories of order, suborder, and great group, emphasis has been placed on marks or causes of sets of processes that appear to dominate the course or degree of soil development. In addition to these dominant marks, many soils have properties that, although apparently subordinate, are still marks of important sets of processes. Some of these appear to be marks of sets of processes that are dominant in some other great group, suborder, or order, but in a particular soil they only modify the marks of other processes. For example, some soils have an aquic moisture regime and have throughout their depth gray colors that are mottled with shades of brown, red, or yellow. Other soils have an aquic moisture regime only in their lower horizons, and in those horizons the dominant colors may be shades of brown, red, or yellow with some gray mottles. The effects of ground water are apparent in both sets of soils, but they have less importance in the latter set.

Other properties are marks of sets of processes that are not used as criteria of any taxon at a higher level. For example, a Mollisol at the foot of a slope, where there has been a slow accumulation of materials washed from higher parts of the slope, may have a greatly overthickened mollic epipedon.

Thus, we have three kinds of subgroups:

The central concept of the great group.—This is not necessarily the most extensive subgroup.

The intergrades or transitional forms to other orders, suborders, or great groups.—The properties may be the result of sets of processes that cause one kind of soil to develop from or toward another kind of soil, or otherwise to have intermediate properties between those of two or three great groups. The properties used to define the intergrades may be:

1. Horizons in addition to those definitive of the great group, including an argillic horizon that underlies a spodic horizon as discussed under great groups and a buried horizon such as a thick layer of organic materials that is buried by a thin mineral soil;

2. Intermittent horizons such as those discussed in the section of chapter 1 that deals with the pedon; or

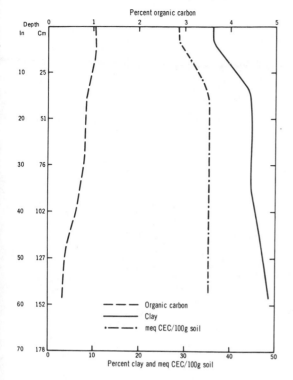

Figure 34.—**Percentage of clay and organic carbon and cation exchange capacity as functions of depth in a representative Vertisol.**

3. Properties of one or more other great groups that are expressed in a part of the soil but are subordinate to the properties of the great group of which the subgroup is a member. One example of different levels of ground water was given earlier. Another example might be that of an Alfisol that has an ochric epipedon a little too thin or a little too light in color to be a mollic epipedon. This could result from an invasion of grassland by forest or the reverse. Or, it could be the effect of both grass and forest coexisting on the same polypedon.

Extragrades.—These subgroups have some properties that are not representative of the great group but that do not indicate transitions to any other known kind of soil. One example of an overthickened mollic epipedon was given earlier. Other examples would be soils that are very shallow over rock or that have permafrost.

Families

In this category, the intent has been to group the soils within a subgroup having similar physical and chemical properties that affect their responses to management and manipulation for use. The responses of comparable phases of all soils in a family are nearly enough the same to meet most of our needs for practical interpretations of such responses. Soil properties are used in this category without regard to their significance as marks of processes or lack of them. About 4,500 families are currently recognized in the United States.

Families are defined primarily to provide groupings of soils with restricted ranges in

1. Particle-size distribution in horizons of major biologic activity below plow depth;
2. Mineralogy of the same horizons that are considered in naming particle-size classes;
3. Temperature regime;
4. Thickness of the soil penetrable by roots; and
5. A few other properties that are used in defining some families to produce the needed homogeneity.

These properties are important to the movement and retention of water and to aeration, both of which affect soil use for production of plants or for engineering purposes. The differentiae are discussed in more detail in chapter 18.

Series

The series is the lowest category in this system. About 10,500 series have been recognized in the United States, although a number of them have been divided or replaced and do not appear in the present taxonomy. The taxa and many of the differentiae in the series category have been carried over from earlier classifications. Modifications of the original concept of the series category, as well as of the series differentiae, have been made repeatedly over the years (Simonson 1964).

The differentiae used for series are mostly the same as those used for classes in other categories, but the range permitted in one or more properties is less than is permitted in a family or in some other higher category. A series may have virtually the full range that is permitted in a family in several properties, but in one or more properties the range is restricted. The purpose of the series category, like that of the family, is mainly pragmatic, and the taxa in the series category are closely allied to interpretive uses of the system.

Two kinds of distinctions, therefore, are made between series. First, the distinctions between families and between classes of all higher categories are also distinctions between series; a series cannot range across the limits between two families or between two classes of any higher category. Second, distinctions between similar series within a family are restrictions in one or more but not necessarily all of the ranges in properties of the family. Taken collectively, the number of the latter kind of distinctions is too large to be comprehended readily. We can only state the basis for decisions to separate individual series. We must keep in mind the purpose of the taxonomy. It is not an end in itself. Attention is centered on the genetic horizons below the depth of normal plowing or, if they are thin, faint, or absent, on the zone of major biologic activity below depth of normal plowing.

The differentiae for series in the same family are expected to meet three tests. The first is that properties serving as differentiae can be observed or can be inferred with reasonable assurance. The second is that they must create soil series whose unique range of properties is significantly greater than the normal errors of measurement, observation, or estimate by qualified men. The third is that the differentiae have some relation to horizon differentiation if horizons are present. This significance can be reflected in the nature or degree of expression of one or more horizons. By nature of horizons we mean the composition, including mineralogy, structure, consistence, texture of the subhorizons, moisture and temperature regimes, and so on. If color is accessory to some other property, it too is included. Degree of horizon expression includes thickness, contrast between horizons or subhorizons, and nature of boundaries. If horizons are absent, the nature of the whole zone of major biologic activity below plow depth is considered. The limits of this zone are given in chapter 18.

Important differences, as shown by experience or research to condition or influence the nature of the statements that we can make about the behavior of the soil, should be considered as series differentiae.

A number of soil properties condition the statements we may make about a soil or its use but are not series differentiae. A steep slope or stones on the surface may be very important to the use of a soil in mechanized farming, but they may have virtually no importance to the growth of a forest. If we assume that these soil characteristics are not reflected in the nature of the soil, or in the nature or degree of expression of horizons explained earlier,[1] then they may be used as one of the bases of phases. The phases provide for a utilitarian classification that can be superimposed on the taxonomy at any categoric level to permit more precise interpretations and predictions of the consequence of the various alternative uses of the soil that can be foreseen.

The primary use of soil series in the classification system is to relate the polypedons represented on de-

[1] The loss of surface horizons as a result of erosion is reflected in the nature of the soil but is used as one of the bases for phases to facilitate the purposes of the taxonomy. If the series is changed by the loss of a horizon that is not diagnostic in a higher category, relations between the eroded and uneroded soils become unnecessarily complicated for the users of the soil surveys.

tailed soil maps to the taxa and to the interpretations that may follow. The dominant kinds of polypedons that are delineated on maps are given the names of series. Polypedons are real things, but series are conceptual. The Miami series, for example, cannot be seen or touched, but the polypedons that we identify as physical entities within the concept of the Miami series can be seen and touched.

Soil series names are used with several meanings that must be kept in mind. We may speak of the Miami series as a taxonomic class, a concept of a narrowly defined kind of soil. Or, we may examine a pedon and say, "This is Miami," meaning that the properties we find in the pedon are those we ascribe to the Miami series and that the pedon is a proper example. We also use Miami as part of the name for an area shown on a soil map if the Miami series is dominant in that area, for example, Miami silt loam, 2 to 6 percent slopes. These are three common meanings of Miami, or of any series name, and all are proper. It is essential, however, to keep in mind that a series, as used in this taxonomy, is conceptual; the meaning is not identical with the meaning intended on soil maps, for an area of Miami soil has inclusions of soils of other series. Inclusions of Miami soil are also permitted in areas named for other series. The application of the taxonomy to soil surveys is discussed in chapter 19.

Soil types

The soil type has been the lowest category in all previous classification systems that we have used in the United States. Types have been distinguished within series on the basis of texture, a single characteristic. At first, the distinction involved the texture of the soil when texture meant a combination of particle size, structure, and consistence. In recent years, when texture has been defined in terms of particle size distribution alone, the texture of the plow layer or of its equivalent in virgin soils was used to distinguish types within a series. Because the significance of the texture or particle size distribution of the plow layer is mainly pragmatic, and the texture classes can be made more useful if adjusted to fit circumstances at a given time, the soil type has not been retained as a category of this system. This mention is made here to explain its disappearance as a category. The texture of the plow layer will commonly be shown in the soil name in published soil surveys as heretofore, but will be considered as a part of the phase name rather than the name of a taxon.

Part IV

LIMITED OR REGIONAL APPROACHES

Editor's Comments
on Papers 15 Through 23

Limited or regional approaches to soil classification are, in general, based on more detailed considerations than otherwise possible in comprehensive world sytems. The examples collected here represent attempts to subdivide the soil mantle within a range of limited environmental conditions. Spanning the gamut from tropical rainforests (Paper 18) to polar landscapes (Paper 23), the territories where rather uniform environmental conditions prevail are by no means inconsequential in terms of global eco-systems. Papers 15 through 23 are based on different approaches that range from morphogenetic classifications (Paper 15) to alpha-numeric taxonomies (Paper 22). Aspects of the various systems are briefly considered in the following remarks.

Classifiers in Western Europe, drawing on U.S., Russian, and local taxonomic experience, have developed some distinctive systems suited to the European region. In the French example (Paper 15), based mainly on experience in African colonies, differentiation of the soil mantle is expressed in terms of an evolutionary sequence of soils from poorly developed skeletal soils to highly weathered sesquioxide-rich formations. Higher categories of the system (classes, subclasses, groups, and subgroups) are based on intensity of profile development and mode of mineral alteration, type of organic matter present, and certain other factors such as poor drainage or salt accumulation that influence the course of soil development, as well as morphogenetic impacts of the pedo-climate. The British classification (Paper 16) carries genetic over-tones for pedogenically altered recent alluvial deposits but is largely based on soil characteristics. Classes of higher categories (major group, group, subgroup) are defined partly by the compo-sition of soil material, presence or absence of diagnostic horizons, or evidence of recent alluviation. The system of soil classification for The Netherlands (Paper 19) was developed for the unusual Dutch conditions, where much of the land has been reclaimed from the sea. Soils of the polders and other wetlands are developed from marine- and fresh-water alluvium and from eolian sediments. Hardly any soils in this region are derived from bedrock. The Dutch system thus recognizes two different universes of alluvial and eolian soils that are differentiated on the basis of texture and organic matter content. The current German effort (Paper 21), a genetic classification that reflects much of the Kubiëna (1953, 1958) system, is still another European approach. The main cate-gories are differentiated according to interpretations of profile characteristics that can be directly attributed to soil genesis, soil dynamics, interstitial structure, and the degree and direction of translocated dissolved and colloidal particles. Soil forms, the lowest category, are differentiated on the basis of parent materials.

The Canadian system (Paper 17) applies to a vast area of North America and retains many features in common with the present U.S. system. Influenced by Marbut's translation of Glinka's (1914) book, ideas outlined in the 1938 U.S. system and international developments in conceptions of soil, the system mainly reflects increased understanding of Canadian soils that dates back to pioneer soil surveys (for example, Joel, Paper 3; Ellis, 1932). Soils are classified on the basis of their characteristics, not on concepts of genesis or interpretations of use. The system does, however, have a genetic bias because properties that reflect dynamic soil processes are favored as differentiae in higher categories. The categorical levels are, from highest to lowest: order, great group, subgroup, family, and series.

The example from South America is drawn from the classification of tropical soils in Brazil (Paper 18). Although the system is based on morphological, physical, chemical, and mineralogical properties of tropical soils, a dichotomy is effected at a high level producing two broad groups of soils, one with a latosolic B horizon (approximately equivalent to an oxic horizon in the U.S. system) and another with a textural B horizon (approximately equivalent to an argillic horizon). Soils are further set apart using chemical criteria such as cation exchange capacity and percent base saturation. Subdivision at lower levels considers soil characters such as iron content, color, parent material, texture, and concretions. Plinthite is used as an additional differentiating criterion for soils with a textural B horizon.

In Japan, where much of the arable land is presently under intensive agriculture and where many soils have been in continuous use for centuries, the classification scheme (Paper 20) emphasizes, perhaps not surprisingly, the productivity and amelioration of soils. Consideration of these important attributes led Japanese pedologists to a system based on: (1) profile characters such as humus content, color, clay pans, mottles and concretions, structure, peat or muck horizons, gleying, and thickness of profile; (2) soil acidity; (3) mode of sedimentation; and (4) parent material. Soil series having the same diagnostic horizons and a similar genetic history are organized in terms of soil groups, the main classification unit.

The Australian system (Paper 22) is based on relatively few soil properties as compared to the U.S. and Canadian systems and the FAO legend. The system is reasonably free of judgments about factors and processes of soil formation. One important advantage of such a scheme is that it provides names for soils when limited information about them is available. The Australian key also permits recognition of soil profiles that otherwise might have been overlooked because they did not support extant models of soil genesis. All soils are divided into four divisions or primary profile forms designed as organic (0), uniform (U), gradational (G), and duplex (D). Names of the last three classes refer to textural trends within the profile. Further subdivision is based on pedological organization and nature and color of material below A1 and A2 horizons. Northcote's classification, which religiously follows Leeper's (1943) suggestion that soils be classified according to their properties, stands in marked contrast to the earlier ecological approaches of Jensen (1914), Prescott (1926, 1931), Teakle (1937), and Stephens (1953).

The soils of polar landscapes contain many features not seen in contemporary soils of other regions, except possibly those occurring at higher elevations in mountainous terrains. The effects of soil freezing, frost heaving and thrusting, up-freezing and tilting of stones, frost sorting, thermal expansion and contraction, cryostatic pressure, and thermokarst processes give a distinctive character to the soil mantle in cold regions. Tedrow's review of polar soil classifications (Paper 23) pretty well summarizes most important developments over the last half century. In addition to the official taxonomic systems of Canada, Russia, and the United States, which take in soils of cold regions in their special categories, Tedrow discusses the role of individual efforts toward developing workable soil classifications in this unique bioclimatic zone. Tables outlining Tedrow's soil classifications for tundra, polar desert, subpolar desert, and cold desert soil zones are appended to the main article rounding out the topic.

REFERENCES

Ellis, J. H., 1932, A Field Classification of Soils for Use in Soil Survey, *Sci. Agric.* **12**:338–345.

Glinka, K. D., 1914, *The Great Soil Groups of the World and Their Development*, C. F. Marbut, trans., Edward Bros., Ann Arbor, Mich., 235 p.

Jensen, H. I., 1914, *The Soils of New South Wales*, Department of Agriculture of New South Wales, Sydney, Aust., 107 p.

Kubiëna, W. L., 1953, *Bestimmungsbuch und Systematik der Böden Europas*, Enke, Stuttgart, 392 p.

Kubiëna, W. L., 1958, The Classification of Soils, *J. Soil Sci.* **9**:9–19.

Leeper, G. W., 1943, The Classification and Nomenclature of Soils, *Aust. J. Sci.* **6**:48–51.

Prescott, J. A., 1926, Soil Classification and Survey, *Aust. & New Zealand Assoc. Adv. Sci. Proc.* **18**:724 *et seq.*

Prescott, J. A., 1931. The Soils of Australia in Relation to Vegetation and Climate, *C.S.I.R. (Australia) Bull.* 52.

Stephens, C. G., 1953, *A Manual of Australian Soils*, Commonwealth Scientific and Industrial Research Organization, Melbourne, 48 p.

Teakle, L. J. H., 1937, A Regional Classification of the Soils of Western Australia, *J. Royal Soc. West. Aust.* **16**:202–230.

15

Reprinted from *Classification des Sols*, INRA Commission de Pédologie et de Cartographie des Sols, Laboratoire de Géologie-Pédologie de l'ENSA, Paris, 1967, pp. 1, 5–13, 15

CLASSIFICATION DES SOLS

EDITION 1967

D'après les travaux de

Messieurs G. AUBERT, R. BETREMIEUX, P. BONFILS, M. BONNEAU, J. BOULAINE, J. DEJOU, J. DELMAS, G. DROUINEAU, P. DUCHAUFOUR, J. DUPUIS, P. DUTIL, H. FLON, F. FOURNIER, J. GELPE, B. GEZE, J. HEBERT, S. HENIN, M. HOREMANS, F. JACQUIN, M. JAMAGNE, R. MAIGNIEN, Madame S. MERIAUX, Messieurs G. PEDRO, J. PORTIER, R. SEGALEN, E. SERVAT, et J. VIGNERON.

Commission de Pédologie et de Cartographie des sols.✶

✶ - La Commission de Pédologie et de Cartographie des sols est un groupe de personnalités réunies à l'initiative de Monsieur l'Inspecteur
Général de l'INRA, G. DROUINEAU, afin d'étudier les modalités d'une
mise en ordre des travaux concernant la Pédologie en France, et par-
ticulièrement les règles de réalisation des cartes des sols ainsi
que la classification des sols.

INTRODUCTION

La classification des sols que nous présentons dans ce texte
a été mise au point, de 1964 à 1967, par la Commission de Pédologie et
de Cartographie des sols.

Elle est le fruit des libres discussions et des travaux des
membres de cette commission qui se sont réunis plusieurs fois par an
pour discuter les propositions des groupes de travail animés par des
responsables.

Dès le départ, une option fondamentale devait être prise :
devait-on tenter une construction entièrement originale, comme l'avaient
fait les pédologues du Soil Survey Staff du soil conservation service
de l'USDA dans leur travail connu sous le nom de septième approximation
(1960), ou devait-on au contraire conserver la cadre déjà élaboré par
G. AUBERT et P. DUCHAUFOUR et publié sous des formes diverses notamment
en 1956 (5ème congrès international de la science du sol).

Cette dernière méthode de travail a été choisie.

Elle est justifiée par le fait que tous les travaux des pédologues français effectués depuis vingt ans ont utilisé des notions qui
correspondent à l'esprit de cette classification.

D'autre part, elle permet de mieux exprimer les conceptions
fondamentales de la pédologie française qui est à la fois morphologique
et génétique.

I - LES REGLES DE CLASSIFICATION

A - OBJET DE LA CLASSIFICATION

La classification est conçue comme un système de référence qui
permet d'ordonner les sols décrits dans une étude et les unités d'une
carte suivant un plan commun. C'est en même temps un ensemble définis-
sant un langage.

Les termes utilisés doivent permettre de désigner avec un minimum de mots les sols qui ont fait l'objet de descriptions ou d'analyses suffisantes.

Mais cette classification n'a pas l'ambition de prévoir tous
les cas possibles et d'englober tous les sols existant à la surface du
globe. D'ailleurs, elle s'arrête le plus souvent au niveau du sous-groupe,
c'est-à-dire à un niveau de généralisation suffisamment élevé pour qu'il
soit nécessaire de distinguer encore des familles, des séries, des types
et des phases pour exprimer à peu près correctement les variations consi-
dérables que présente le "phénomène sol".

En outre, la Commission de Pédologie a prévu la création de trois sortes d'unités de classement qui permettent de tenir compte à la fois du caractère provisoire, en perpétuel devenir, de la classification pédologique et des difficultés inhérentes à l'objet même de celle-ci. En effet, le sol est un phénomène continu dont le découpage en unités homogènes n'est possible et justifié que pour un certain pourcentage des cas.

B - UNITES DE CLASSIFICATION

La classification distingue donc les types d'unités suivantes :

1) Unités simples :

Unités_génétiques : les unités de la classification de référence.

Unité_d'apparentement : les unités dont la détermination est encore hypothétique à cause d'un manque d'informations suffisantes : ces unités sont provisoirement classées, sous toutes réserves, à côté d'une unité génétique en attendant que des études plus poussées, ou un diagnostic mieux affirmé, permettent de les placer définitivement dans la classification.

Unités_intergrades : les unités intermédiaires entre deux unités de classification et qui possèdent des caractères communs à l'une et à l'autre.

Dans le cas d'un inventaire cartographique des sols, ces unités sont appelées unités simples par opposition aux unités complexes.

2) Unités complexes :

Les unités complexes ne correspondent pas à des unités de classification, mais permettent de rendre compte de certains aspects de la distribution des sols. Elles ne doivent être employées comme catégorie d'une légende de carte pédologique, que lorsque l'échelle utilisée ne permet pas de faire apparaître à sa place chacune des unités simples.

Une_juxtaposition_de_sols : est un ensemble de sols dont chacun d'eux ne comporte qu'une surface petite à l'échelle de la carte et dont la coexistence ne paraît dépendre d'aucune règle de répartition précise.

Une_séquence_de_sols : est un ensemble de sols dont la succession se retrouve constamment dans un ordre déterminé, sans qu'il y ait lien génétique apparent entre eux. La raison de leur juxtaposition régulière est l'influence prépondérante, et régulièrement répétée, d'un de leur facteur de formation.

Une chaîne de sols : est un ensemble de sols liés génétiquement, chacun d'eux ayant reçu des autres, ou cédé aux autres, certains de ses éléments constituants.

Dans la pratique, on distingue aussi les unités majeures et les unités mineures. Les premières sont les unités de niveau supérieur (classe, sous-classe, groupe et sous-groupe) et sont utilisées dans les travaux généraux et pour les cartes à petite échelle.

Les secondes permettent une expression plus fine du phénomène sol et sont utilisées dans les cartes à moyenne et grande échelle, ainsi que pour définir les sols dans les travaux de recherches.

3) Les unités Majeures de classification du sol :

La classification française utilise les unités majeures suivantes :

CLASSE

SOUS-CLASSE

Groupe

Sous-groupe

Les CLASSES : au niveau le plus élevé les sols sont répartis en classes. On range dans la même classe des sols qui ont en commun certains caractères majeurs qui sont les suivants :

1 - Un certain degré de développement du profil ou d'évolution du sol. Les matériaux qui n'ont subi aucune évolution constituent la classe des sols minéraux bruts, ceux dont l'évolution est faible constituent la classe des sols peu évolués. Les autres sols possèdent au moins un horizon dans lequel le matériau primitif a été altéré et modifié dans sa nature chimique.

2 - Un mode d'altération des minéraux en relation avec les caractères généraux des conditions physico-chimiques régant dans la partie supérieure du sol, caractérisé par la nature des sesquioxydes libérés et qui se maintiennent individualisés, ou constituent des complexes caractéristiques. L'importance relative de cette libération, ainsi que la dominance de certains types d'argile traduisent ces divers types d'altération. Elles s'expriment dans le profil soit par des couleurs, soit par des propriétés physiques (structure) de certains horizons, soit par des propriétés chimiques (saturation du complexe).

3 - Une composition et une répartition typique de la matière organique susceptible d'influencer l'évolution du sol et la différenciation des horizons du profil. Par exemple : concentration dans les horizons supérieurs ou répartition dans tout le profil ; humus évolué calcique, humus évolué de type mull, apte à favoriser la migration des colloïdes minéraux ; humus grossier de type mor, capable de dégrader le complexe minéral des sols.

4 - Certains facteurs fondamentaux d'évolution du sol qui deviennent prédominants. C'est le cas de la présence d'eau (hydromorphie) ou de sels très solubles (halomorphie).

Ces deux types de processus diffèrent des autres en ce qu'ils sont, non pas vraiment "transitoires", comme il a été dit parfois mais beaucoup plus rapides. Cependant ils peuvent être si intenses qu'il modifient entièrement le mode d'évolution du sol et son profil. En ce cas seulement, ils sont pris comme caractéristiques de classes de sols ; moins intenses, ou moins intensément exprimés, ils ne définissent que des groupes et surtout des sous-groupes, séries ou phases de sols.

Les *SOUS-CLASSES* : dans la mesure du possible, la différenciation des sous-classes repose sur des critères résultant des conditions de pédo-climat.

En effet, les sols peuvent avoir acquis certains caractères majeurs par le jeu de combinaisons variables des facteurs climatiques. Par exemple, la faible évolution des sols peu évolués peut être due à un climat très froid ou à un climat sec et chaud.

Les sols dans lesquels la matière organique, d'origine principalement racinaire, s'accumule sur une forte épaisseur peuvent se former dans les climats à hiver froid tempéré ou chaud et la position de la saison sèche au cours de l'année est un facteur essentiel de la nature du sol.

Parfois, c'est aussi l'absence de relations avec le climat qui est pris comme critère au niveau de la sous-classe. C'est le cas pour les sols minéraux bruts et les sols peu évolués qui ne doivent leurs caractères de classe qu'au fait que leur durée d'existence est faible et que les processus majeurs n'ont pas encore eu le temps de jouer.

Nous avons donc donné à la notion de pédo-climat, un contenu très élargi dans lequel entrent non seulement les notions de température et de plus ou moins grande humidité du sol (liées ou non au climat local), mais aussi la notion de concentration des solutions du sol en certains ions, que ce soient des cations comme le sodium ou le calcium. L'ion hydrogène, lui aussi intervient à ce niveau au moins indirectement par l'intermédiaire de la notion de milieu réducteur ou oxydant.

C'est donc l'ambiance physico-chimique qui est utilisée pour définir les sous-classes. Les éléments principaux sont la température, l'humidité, l'état réduit ou oxydé, la concentration des solutions du sol en tel ou tel cation (tous ces facteurs ayant des variations annuelles, saisonnières, ou même journalières).

Dans certaines classes (sols hydromorphes, sols sodiques) les relations avec le climat du sol sont modifiées du fait même de l'existence d'un processus dominant et les critères choisis sont une traduction indirecte du complexe des caractères pédoclimatiques.

Les *groupes* : Les groupes sont définis :

par des caractères morphologiques du profil correspondant à des processus d'évolution de ces sols : différenciation de certains horizons, lessivage du calcaire, des éléments colloïdaux, etc... Parfois deux groupes voisins peuvent être caractérisés par un même processus pédologique général ; ils sont alors différenciés par une forte variation de son intensité correspondant à des profils nettement distincts. Tels sont les cas des Groupes Podzoliques et des Podzols, dans la classe des sols à humus grossier et hydroxydes ; ou des groupes châtains, bruns, etc... dans la classe isohumique.

Les *sous-groupes* : Les groupes comprennent en général plusieurs sous-groupes dont les caractères essentiels des profils sont les mêmes, mais qui sont différenciés soit par une intensité, variable d'une catégorie à l'autre, du processus fondamental d'évolution caractéristique du groupe, soit par la manifestation d'un processus secondaire, indiqué par certains éléments nouveaux du profil (concrétionnement, induration, tache d'hydromorphie, élargissement de la structure, etc...).

4) Les unités Mineures de Classification des sols :

Dans des études détaillées, on divise les sous-groupes en utilisant les notions de Famille, de Série de Type et de Phase.

La Famille : à l'intérieur d'un même sous-groupe, toutes les séries formées à partir du même matériau pétrographique constituent une famille. Réciproquement, on peut diviser les sols d'un même sous-groupe en tenant compte du matériau originel.

Un classement, dans un but cartographique, des matériaux, pourrait être dressé pour servir de base à la classification des familles. Après plusieurs tentatives, il semble préférable dans une région donnée, de tenir compte des faciès locaux des roches. Celles-ci ont en général des caractères très importants pour les sols et la référence à un système conventionnel ne pourrait introduire qu'une perte de précision.

La Série : est une unité (cartographique et de classification) dont les caractères édaphiques sont suffisamment homogènes pour que les variations typologiques n'aient pas d'influence notable sur la croissance des plantes spontanées ou cultivées.

La définition d'une série devrait s'appuyer sur des critères d'homogénéité statistique des caractères des sols de la dite série. Le relativement petit nombre d'études de ce genre faites à ce jour a conduit à la définition suivante :

Définition : Une série de sols est l'ensemble des sols qui présentent sur un matériau originel de composition lithologique définie, et dans des positions comparables dans le paysage, le même type de profil.

Les profils des sols d'une série sont semblables non seulement par la succession, l'aspect et la constitution générale de leurs divers horizons, mais aussi par l'ordre de grandeur de l'épaisseur de chacun de ces derniers. Cet ordre de grandeur est envisagé en fonction de l'influence possible de la présence de chacun d'eux sur les propriétés générales des sols. La série est dénommée d'après le lieu où elle a été caractérisée.

La définition de la série constitue l'étape préalable à son interprétation génétique.

Le type : A l'intérieur d'une série les sols ayant la même texture de l'horizon superficiel appartenaient aux mêmes types : exemple type limono-argileux et type argileux de la série X.

En fait, la dénomination des textures est purement arbitraire et d'autre part la nouvelle définition de la série est telle qu'elle implique des variations de texture plus faibles par rapport à une texture modale que celles des zones des triangles des textures en usage. La notion de type perd donc beaucoup de sa valeur à l'intérieur d'une série définie de façon rigoureuse.

La phase : Lorsque des phénomènes naturels (érosions, colluvionnements, action des animaux et des végétaux ou l'action de l'homme), modifiant de façon éventuellement temporaire la nature, l'organisation et la dynamique des horizons superficiels d'une série, on peut en tenir compte en divisant cette série en phases.

Dans tout ce qui suit, la classification s'arrête au niveau du sous-groupe ; elle ne concerne donc que les unités majeures. Nous avons cependant défini les unités mineures dans les pages précédentes pour orienter les divisions éventuelles lorsque cela s'avèrera nécessaire.

C - LES HORIZONS

Afin d'alléger le texte concernant la classification de chaque classe, on a souvent désigné les horizons des sols par des groupes de lettres et de chiffres qui sont en usage chez les pédologues.

On a jugé utile de rappeler ici la signification de ces lettres et de ces chiffres et de définir les horizons principaux.

1) - *HORIZON A.*

Les horizons A_{00}, A_0 d'une part, les horizons A d'autre part, se superposent dans l'ordre indiqué quand ils sont présents simultanément dans le profil.

A_{00} - Horizon de surface, formé de débris végétaux facilement identifiables (feuilles, brindilles, et autres) et non reliés ensemble par du mycélium. Cet horizon correspondant à ce que divers auteurs désignent par la lettre L.

A_0 - Horizon constitué principalement de débris végétaux partiellement décomposés et pratiquement non reconnaissables sur le terrain. Les horizons peuvent être subdivisés en F et H. La couche H se distingue de F par l'absence complète de structure végétale.

A l'analyse ils contiennent en général plus de 30 % de matière organique totale. Ils sont mesurés de bas en haut à partir du sommet de A_1. Certains de ces horizons peuvent manquer.

L'horizon A est un horizon majeur occupant la partie supérieure ou l'ensemble du profil du sol et présentant l'un ou l'autre des caractères suivants ou les deux en même temps :

a) Présence de matière organique.
b) Appauvrissement en constituants tels que argile, fer, alumine, etc...

L'horizon A_1 est un horizon minéral présentant en général moins de 30 % de matière organique bien mélangée à la partie minérale et de couleur généralement sombre. Il peut être ou non un horizon éluvial.

L'horizon A_2 est un horizon de couleur plus claire que l'horizon susjacent ; il est appauvri en fer, en argile, en alumine avec concentration corrélative de minéraux résistants. C'est un horizon d'éluviation par lessivage de matériaux en solution ou suspension. Les éléments se déplacent généralement à l'état dissous ou dispersés vers l'horizon B et / ou hors du profil.

L'horizon A_3 est un horizon de transition entre A et B mais il est plus proche de A que de B. Si l'horizon de transition ne peut être valablement attribué à l'un ou l'autre, on écrira AB.

2) - <u>*HORIZON B*</u>.

Horizon majeur situé au-dessous de A et caractérisé par des teneurs en argile, en fer, en humus, plus élevées qu'en A ou C. Cet enrichissement peut être dû, soit à des transformations sur place des minéraux préexistants, soit à des apports illuviaux. On désigne cet horizon par B.

Si la variation de teneur est très faible et que la différenciation avec A ou C ne porte que sur la consistance, la structure, ou la couleur, on désignera cet horizon par (B).

Une lettre minuscule, placée après B, précisera la nature de l'enrichissement ou de la différenciation. L'horizon est divisé en :

B_1 Horizon de transition avec A, mais plus proche de B que de A.

B_2 Horizon constituant la partie essentielle de B, correspondant soit à l'accumulation principale, soit au développement maximum de la différenciation.

B_3 Horizon de transition avec C, mais plus proche de B que de C.

> N.B. On peut affecter les horizons d'un nouveau chiffre secondaire (tel que B_{21}, B_{22}, etc...) sans autre signification que d'avoir introduit une subdivision.

3) - <u>*HORIZON C*</u>.

Horizon minéral, autre que la roche brute, placé sous B (ou sous A s'il n'y a pas de B), analogue ou différent du matériau dont dérive le couple AB et relativement peu affecté par les processus pédogénétiques ayant conduit à l'individualisation des horizons A et B sus-jacents et ne présentant pas leurs caractéristiques.

4) - <u>*HORIZON R*</u>.

Roche brute sous-jacente.

5) - <u>*Hétérogénéité des matériaux*</u>.

En cas de discontinuité lithologique, on désignera chaque matériau originel par un chiffre romain qui précèdera l'horizon. S'il n'y a qu'un seul matériau, on omettra le chiffre romain. Dans le cas de plusieurs matériaux, celui du dessus (I), pourra être omis.

Exemple : A_1 - A_2 - B_1 - B_{21} - II B_3 - II C_1 - III C_2 - IV R.

6) - *Transition et mélanges d'Horizons.*

Les horizons qui assurent une transition entre deux horizons majeurs sont indiqués par deux lettres majuscules désignant les deux horizons majeurs concernés. L'ordre des lettres indique les caractères dominants de l'horizon de transition (par exemple : AB ou BA). Les horizons de mélange sont indiqués par les deux lettres majuscules désignant les deux horizons majeurs concernés, mais séparés par un trait oblique, par exemple : A / B.

[*Editor's Note:* Material has been omitted at this point.]

T A B L E A U   D E S   C L A S S E S

- I -   Classe des sols minéraux bruts.

- II -   Classe des sols peu évolués.

- III -   Classe des vertisols.

- IV -   Classe des andosols.

- V -   Classe des sols calcimagnésiques.

- VI -   Classe des sols isohumiques.

- VII -   Classe des sols brunifiés.

- VIII -   Classe des sols podzolisés.

- IX -   Classe des sols à sesquioxydes de fer et manganèse.

- X -   Classe des sols ferrallitiques.

- XI -   Classe des sols hydromorphes.

- XII -   Classe des sols sodiques.

Dans ce qui suit le n° de la classe est rappelé dans les tableaux de classification jusqu'au niveau du groupe. Les sous-classes, groupes et sous-groupes sont désignés par un code numérique.

Chaque chapitre, concernant chacune des classes, comporte une définition générale des sols de la classe, les règles de classification à l'intérieur de la classe et la classification jusqu'au niveau du sous-groupe.

En annexe, on trouvera un tableau synoptique de tous les niveaux de classification jusqu'au sous-groupe.

[*Editor's Note:* In the original, material follows this excerpt.]

16

SOIL CLASSIFICATION IN THE SOIL SURVEY OF ENGLAND AND WALES

B. W. AVERY

(*Soil Survey of England and Wales, Rothamsted Experimental Station, Harpenden, Herts.*)

Summary

The development of soil classification as a basis for soil mapping in England and Wales is briefly reviewed, and a system for future use is described. The things classified are soil profiles, and classes are defined by relatively permanent characteristics that can be observed or measured in the field, or inferred within limits from field examination by comparison with analysed samples. Profile classes are defined at four categorical levels by progressive division, and are termed major groups, groups, subgroups, and soil series respectively.

Classes in the three higher categories are defined partly by the composition of the soil material and partly by the presence or absence of particular diagnostic horizons, or evidence of recent alluvial origin, within specified depths. Soil series are distinguished by other characteristics, chiefly lithologic, not differentiating in higher categories. Most of the soil groups, regarded as the principal category above the soil series, are closely paralleled in other European systems, in the U.S.D.A. system (7th Approximation with subsequent amendments), or in both. Compared with the system used hitherto, the main innovations are the use of specific soil properties to define classes at all categorical levels, and the separation at group level of classes based primarily on inherited lithologic characteristics.

The soil-profile classification provides a uniform basis for identifying soil map units, considered as classes of delineated soil bodies. When a map unit is identified by the name of a profile class, it is implied that most of the soil in each delineation conforms to that class, and that unconforming inclusions belong to one or more closely related classes or occupy an insignificant proportionate area. Map units identified by land attributes not differentiating in the profile classification are termed phases.

Introduction

THE classification used as a basis for soil mapping in England and Wales originated in the 1930s with the introduction of the soil series, described by Robinson (1943) as a 'group of soils similar in the character and arrangement of the horizons of the profile, and developed under similar conditions from one type of parent material'. To systematize differentiation of soil series and to show their relationships, each is now conceived as a profile class (Beckett and Burrough, 1971) with particular lithologic characteristics, within a broader group based primarily on character or arrangement of horizons. Definitions of major soil groups and subgroups were first compiled by a small committee of soil surveyors (Clarke, 1940) and subsequently modified (Avery, 1956, 1965; Mackney and Burnham, 1964) to give a system in general accord with those evolved in France (Commission de Pédologie et de Cartographie des Sols, 1967; Duchaufour, 1970) and Germany (Mückenhausen, 1965; Ehwald *et al.*, 1966) over the same period.

During the last 10 years, the extension and wider use of soil surveys have directed attention to the need for an improved classification using specific soil properties to define class limits, as in the systems developed for soil-survey purposes in the United States (Soil Survey Staff, 1960, 1967) and the Netherlands (De Bakker and Schelling, 1966). In attempting to meet this need, the need to maintain continuity with previous work was also considered important. Accordingly, several modifications of the currently used system were compiled and tested by trial and error as described by Macvicar (1969) and Schelling (1970).

This paper outlines the latest scheme, which will now be used in surveys. A monograph explaining it in more detail is in preparation.

Aims and basic concepts

The main purpose is to provide a nationally uniform, systematic basis for legends of soil maps made to aid land use in England and Wales. Soil is accordingly considered to include any unconsolidated material directly below a ground surface, and classification is based on horizons within the upper 1·50 m. The things classified are soil profiles, considered for the purpose as three-dimensional, with lateral dimensions large enough to evaluate diagnostic properties of soil horizons at a particular place. This unit is similar in function to the *pedon* (Soil Survey Staff, 1960), but is purposely limited to a volume (about 1 m³) that in practice can be adequately described and sampled as a single entity.

As the classification is intended for use in general-purpose surveys of cultivated and uncultivated land, classes are, in principle, defined by properties that can be observed or measured in the field, or inferred within limits from field examination by comparison with analysed samples, and that are relatively permanent. Properties of thin surface layers that are quickly destroyed or obscured by normal cultivation, and chemical properties that are readily altered or that cannot be assessed in the field are therefore excluded as differentiating characteristics.

Class structure and nomenclature

Profile classes are defined at four levels of abstraction by progressive division. Table 1 lists classes in the three higher categories, termed major groups, groups, and subgroups respectively, the fourth category being the soil series. Differentiating characteristics are summarized below, and those of major groups and groups are given in the table. Major groups 1 and 2 have not been formally subdivided below group level.

Classes in higher categories are differentiated primarily by combinations of the following factors:

1. Composition of the soil material within specified depths.
2. Presence or absence of diagnostic horizons generally reflecting degree or kind of alteration of the original material.

In addition, soils in recent alluvium are distinguished from otherwise similar soils. These separations, involving an inference about the origin

Table 1

*Classes in higher categories**

Major group	Group	Subgroup
1 Terrestrial raw soils Mineral soils with no diagnostic pedogenic horizons or disturbed fragments of such horizons, unless buried beneath a recent deposit more than 30 cm thick	**1.1 Raw sands** Non-alluvial, sandy (mainly dune sands) **1.2 Raw alluvial soils** In recent alluvium, normally coarse textured **1.3 Raw skeletal soils** With bedrock or non-alluvial fragmental material at 30 cm or less **1.4 Raw earths** In naturally occurring, unconsolidated, non-alluvial loamy, clayey or marly material **1.5 Man-made raw soils** In artificially disturbed material, e.g. mining spoil	
2 Hydric raw soils (Raw gley soils) Gleyed mineral soils, normally in very recent marine or estuarine alluvium, with no distinct topsoil, and/or ripened no deeper than 20 cm	**2.1 Raw sandy gley soils** In sandy material **2.2 Unripened gley soils** In loamy or clayey alluvium, with a ripened topsoil less than 20 cm thick	
3 Lithomorphic (A/C) soils With distinct, humose or organic topsoil over C horizon or bedrock at 40 cm or less, and no diagnostic B or gleyed horizon within that depth	**3.1 Rankers** With non-calcareous topsoil over bedrock (including massive limestone) or non-calcareous, non-alluvial C horizon (excluding sands) **3.2 Sand-rankers** With non-calcareous, non-alluvial sandy C horizon **3.3 Ranker-like alluvial soils** In non-calcareous recent alluvium (usually coarse textured) **3.4 Rendzinas** Over extremely calcareous non-alluvial C horizon fragmentary limestone or chalk **3.5 Pararendzinas** With moderately calcareous non-alluvial C horizon (excluding sands) **3.6 Sand-pararendzinas** With calcareous sandy C horizon **3.7 Rendzina-like alluvial soils** In recent alluvium	3.11 humic ranker 3.12 grey (non-humic) ranker 3.13 brown (non-humic) ranker 3.14 podzolic ranker (with greyish E) 3.15 stagnogleyic (fragic) ranker 3.21 typical sand-ranker 3.22 podzolic sand-ranker 3.23 gleyic sand-ranker 3.31 typical ranker-like alluvial soil 3.32 gleyic ranker-like alluvial soil 3.41 humic rendzina 3.42 grey (non-humic) rendzina 3.43 brown (non-humic) rendzina 3.44 colluvial (non-humic) rendzina 3.45 gleyic rendzina 3.46 humic gleyic rendzina 3.51 typical (non-humic) pararendzina 3.52 humic pararendzina 3.53 colluvial pararendzina 3.54 stagnogleyic pararendzina 3.55 gleyic pararendzina 3.61 typical sand-pararendzina 3.71 typical rendzina-like alluvial soil 3.72 gleyic rendzina-like alluvial soil
4 Pelosols Slowly permeable (when wet), non-alluvial clayey soils with B or BC horizon showing vertic features and no E, non-calcareous Bg or paleo-argillic horizon	**4.1 Calcareous pelosols** Without argillic horizon **4.2 Non-calcareous pelosols** Without argillic horizon **4.3 Argillic pelosols** With argillic horizon	4.11 typical (stagnogleyic) calcareous pelosol 4.21 typical (stagnogleyic) non-calcareous pelosol 4.31 typical (stagnogleyic) argillic pelosol

* Names in parenthesis are alternative or explanatory.

TABLE I (cont.)

Major group	Group	Subgroup
5 *Brown soils* Soils, excluding pelosols, with weathered, argillic or paleo-argillic B and no diagnostic gleyed horizon at 40 cm or less	**5.1** *Brown calcareous earths* Non-alluvial, loamy or clayey, without argillic horizon	**5.11** typical brown calcareous earth **5.12** gleyic brown calcareous earth **5.13** stagnogleyic brown calcareous earth
	5.2 *Brown calcareous sands* Non-alluvial, sandy, without argillic horizon	**5.21** typical brown calcareous sand **5.22** gleyic brown calcareous sand
	5.3 *Brown calcareous alluvial soils* In recent alluvium	**5.31** typical brown calcareous alluvial soil **5.32** gleyic brown calcareous alluvial soil
	5.4 *Brown earths (sensu stricto)* Non-alluvial, non-calcareous, loamy or clayey, without argillic horizon	**5.41** typical brown earth **5.42** stagnogleyic brown earth **5.43** gleyic brown earth **5.44** ferritic brown earth **5.45** stagnogleyic ferritic brown earth
	5.5 *Brown sands* Non-alluvial, sandy or sandy gravelly	**5.51** typical brown sand **5.52** gleyic brown sand **5.53** stagnogleyic brown sand **5.54** argillic brown sand **5.55** gleyic argillic brown sand
	5.6 *Brown alluvial soils* Non-calcareous, in recent alluvium	**5.61** typical brown alluvial soil **5.62** gleyic brown alluvial soil
	5.7 *Argillic brown earths* Loamy or clayey, with ordinary argillic B	**5.71** typical argillic brown earth **5.72** stagnogleyic argillic brown earth **5.73** gleyic argillic brown earth
	5.8 *Paleo-argillic brown earths* Loamy or clayey, with paleo-argillic B	**5.81** typical paleoargillic brown earth **5.82** stagnogleyic paleo-argillic brown earth
6 *Podzolic soils* With podzolic B	**6.1** *Brown podzolic soils* (podzolic brown earths) With Bs below an Ap or 15 cm, and no continuous albic E, thin ironpan, distinct Bhs with coated grains, or gleyed horizon at 40 cm or less	**6.11** typical (non-humic) brown podzolic soil **6.12** humic brown podzolic soil **6.13** paleo-argillic brown podzolic soil **6.14** stagnogleyic brown podzolic soil **6.15** gleyic brown podzolic soil
	6.2 *Humic cryptopodzols* (Humic podzolic rankers) With very dark humose Bhs more than 10 cm thick and no peaty topsoil, thin ironpan, continuous albic E, Bs, or gleyed horizon	**6.21** typical humic crypto-podzol
	6.3 *Podzols (sensu stricto)* With continuous albic E and/or distinct Bh or Bhs with coated grains and no peaty topsoil, bleached hardpan or gleyed horizon above, in or directly below the podzolic B or at less than 50 cm	**6.31** typical (humo-ferric) podzol **6.32** humus podzol **6.33** ferric podzol **6.34** paleo-argillic (humo-ferric) podzol **6.35** ferri-humic podzol
	6.4 *Gley-podzols* With continuous albic E and/or distinct Bh or Bhs, gleyed horizon directly below the podzolic B or at less than 50 cm, and no continuous thin ironpan or bleached hardpan	**6.41** typical (humus) gley-podzol **6.42** humo-ferric gley-podzol **6.43** stagnogley-podzol **6.44** humic (peaty) gley-podzol
	6.5 *Stagnopodzols* With peaty topsoil and/or gleyed E or bleached hardpan over thin ironpan or Bs horizon (wet above a podzolic B)	**6.51** ironpan stagnopodzol **6.52** humus-ironpan stagno-podzol **6.53** hardpan stagnopodzol **6.54** ferric stagnopodzol

TABLE 1 (*cont.*)

Major group	Group	Subgroup
7 *Surface-water gley soils* (Stagnogley *sensu lato*) Non-alluvial soils with distinct, humose or peaty topsoil, non-calcareous Eg and/or Bg or Btg horizon, and no G or relatively pervious Cg horizon affected by free groundwater	7.1 *Stagnogley soils* (*sensu stricto* ≈ *Pseudogley*) With distinct topsoil	7.11 typical (argillic) stagnogley soil 7.12 pelo-stagnogley soil 7.13 cambic stagnogley soil 7.14 paleo-argillic stagnogley soil 7.15 sandy stagnogley soil
	7.2 *Stagnohumic gley soils* With humose or peaty topsoil	7.21 cambic stagnohumic gley soil 7.22 argillic stagnohumic gley soil 7.23 paleo-argillic stagnohumic gley soil 7.24 sandy stagnohumic gley soil
8 *Ground-water gley soils* With distinct, humose or peaty topsoil and diagnostic gleyed horizon at less than 40 cm, in recent alluvium ripened to more than 20 cm, and/or with G or relatively pervious Cg horizon affected by free ground water	8.1 *Alluvial gley soils* With distinct topsoil, in loamy or clayey recent alluvium	8.11 typical (non-calcareous) alluvial gley soil 8.12 calcareous alluvial gley soil 8.13 pelo-(vertic)alluvial gley soil 8.14 pelo-calcareous alluvial gley soil 8.15 sulphuric alluvial gley soil
	8.2 *Sandy gley soils* Sandy, with distinct topsoil and without argillic horizon	8.21 typical (non-calcareous) sandy gley soil 8.22 calcareous sandy gley soil
	8.3 *Cambic gley soils* Non-alluvial, with distinct topsoil, loamy or clayey Bg horizon and relatively pervious Cg or G horizon	8.31 typical (non-calcareous) cambic gley soil 8.32 calcaro-cambic gley soil 8.33 pelo-(vertic) cambic gley soil
	8.4 *Argillic gley soils* With distinct topsoil and argillic (Btg) horizon over relatively pervious Cg	8.41 typical argillic gley soil 8.42 sandy-argillic gley soil
	8.5 *Humic-alluvial gley soils* With humose or peaty topsoil, in loamy or clayey recent alluvium	8.51 typical (non-calcareous) humic-alluvial gley soil 8.52 calcareous humic-alluvial gley soil 8.53 sulphuric humic-alluvial gley soil
	8.6 *Humic-sandy gley soils* Sandy, with humose or peaty topsoil and no argillic horizon	8.61 typical humic-sandy gley soil
	8.7 *Humic gley soils* (*sensu stricto*) Non-alluvial, loamy or clayey, with humose or peaty topsoil	8.71 typical (non-calcareous) humic gley soil 8.72 calcareous humic gley soil 8.73 argillic humic gley soil
9 *Man-made soils* With thick man-made A horizon or disturbed soil (including material recognizably derived from pedogenic horizons) more than 40 cm thick	9.1 *Man-made humus soils* With thick man-made A horizon, including Plaggen soils	9.11 sandy man-made humus soil 9.12 earthy man-made humus soil
	9.2 *Disturbed soils* Without thick man-made A horizon	
10 *Peat* (*organic*) *soils*	10.1 *Raw peat soils* Without earthy topsoil or ripened mineral surface layer	10.11 raw oligo-fibrous peat soil 10.12 raw eu-fibrous peat soil 10.13 raw (unripened) oligo-amorphous peat soil 10.14 raw (unripened) eutro-amorphous peat soil
	10.2 *Earthy peat soils* With earthy topsoil or ripened mineral surface layer	10.21 earthy oligo-fibrous peat soil 10.22 earthy eu-fibrous peat soil 10.23 earthy oligo-amorphous peat soil 10.24 earthy eutro-amorphous peat soil 10.25 earthy sulphuric peat soil

of the soil material, are made because the resultant classes usually characterize natural soil landscape units (Schelling, 1970) of practical significance. The soils classed as alluvial are required to have properties defining Fluvents and fluventic subgroups in the U.S.D.A. system (Soil Survey Staff, 1967).

Several other differentiating criteria are derived from that system, with or without minor modifications, but some given emphasis therein, notably the mollic epipedon and the argillic horizon, are not used or are used at a lower categorical level, either because of problems of identification, or because the resultant groupings seemed less significant in Britain than others already in use (Ragg and Clayden, 1973).

Major groups 5, 6, 7, 8, and 10 correspond closely to classes originally defined by Clarke (1940). The main innovations are major groups to accommodate raw and ranker-like (non-calcareous A–C) soils and soils profoundly modified by human activity, and the separation at group level of very sandy soils and cracking clay soils classed as Pelosols by Mückenhausen (1965). There are two reasons for the latter change, firstly to create a manageable number of classes more homogeneous for practical objectives than those based on pedogenic horizons alone, and secondly because extreme lithologies tend to modify or obscure the expression of such horizons.

To facilitate use of the system by non-pedologists in Britain, classes in higher categories are named using English words, or terms made familiar by previous usage, whenever it was judged that no undue confusion of meaning or sacrifice of brevity was entailed. Other classes are named by terms derived from current European classification schemes or from the U.S.D.A. system. To conform with British usage the term *gley soil* is applied broadly to soils characterized by morphological evidence of reduction and segregation of iron, and those to which this term is restricted in the French and West German systems are identified as ground-water gley soils. Hydromorphic or semi-hydromorphic soils with impeded drainage or excess surface wetness, distinguished in these systems by the terms *Pseudogley* and *Stagnogley*, are identified at group or subgroup level by the prefix *stagno-*, used with the broader connotation of *Staunässeböden* (Mückenhausen, 1965) or *Staugley* (Ehwald et al., 1966).

Soil series are differentiated by profile characteristics, chiefly lithologic, that are not differentiating at subgroup level, and identified in one of two ways:

1. As hitherto by a geographic name referring to the location of a type profile.
2. By appending terms denoting the series-differentiating characteristics to the appropriate subgroup name.

To apply the system, established soil series concepts are now being reviewed: most are transferable with or without minor changes in definitions, but some will be abandoned and others introduced to conform with the subgroup limits.

Differentiating criteria

Kinds of soil material

Organic soil materials are either:

1. Seldom saturated with water for more than a month at a time and have more than 20 per cent organic carbon (35 per cent O.M.).
2. Saturated with water for longer periods or artificially drained, and have more than 18 per cent organic carbon (30 per cent O.M.) if the < 2 mm inorganic fraction is 50 per cent or more clay, more than 12 per cent organic carbon (20 per cent O.M.) if there is no clay, and proportionate organic-carbon contents with intermediate clay values.

The first criterion applies to well-drained superficial organic accumulations (L, F, H) and the second to materials commonly described as peat. The variable limits for the latter are derived from the U.S.D.A. system, and accord with field experience that a given proportion of organic matter modifies physical properties of a sand more than it does those of a clay.

Peaty materials are further categorized as *fibrous* (fibric), *semi-fibrous* (mesic or hemic) or *amorphous* (humic or sapric) according to the degree of decomposition of plant remains as determined by solubility in sodium pyrophosphate and proportions and durability of 'fibres' retained by an 0·20 mm sieve (National Soil Survey Committee of Canada, 1968; Soil Survey Staff, 1968). Limmic materials, including sedimentary peat and detritus muds, are also distinguished. This classification is not yet well tested in Britain, and so is considered provisional.

Mineral soil materials, with less organic carbon, are further differentiated as humose and non-humose, and into particle-size (textural) classes based on proportions of stones (> 2 mm), sand (2000–60 μm), silt (60–2 μm) and clay (< 2 μm) in the inorganic fraction.

Humose mineral materials have more than 4·5–7·0 per cent organic carbon (8–12 per cent O.M.), depending on clay content as above.

The basic particle-size classification (Fig. 1), which replaces the U.S.D.A. (Soil Survey Staff, 1951) system now in use, accords with those used by engineers (British Standards Institution, 1967; Road Research Laboratory, 1970) in so far as class limits are based on Massachusetts Institute of Technology size grades and similar separations are made. The limits at 18 and 35 per cent clay are common to U.S.D.A. (Soil Survey Staff, 1967) and Netherlands (De Bakker and Schelling, 1966) systems. Sandy classes are subdivided into coarse, medium, and fine sub-classes according to proportions of coarse (2–0·6 mm), medium (0·6–0·2 mm), and fine (0·2–0·06 mm) sand fractions.

Clay-size carbonates are treated as silt when placing soil materials in particle-size classes. Materials containing more than 40 per cent $CaCO_3$ (< 2 mm) are considered *extremely calcareous*, and those in which more than one-third of the clay-size fraction is $CaCO_3$ are classed as marls rather than as loams or clays.

Materials in which dithionite-extractable Fe_2O_3 (< 2 mm basis) amounts to more than 4 per cent and more than half the measured clay percentage are considered *ferritic*.

Materials containing more than 35 per cent stones by volume (50 per cent by weight with fine earth of bulk density 1·5) are classed as gravel (with stones dominantly < 60 mm) or fragmental, and those with 35–70 per cent (*c*. 80 per cent by weight) are further categorized as sandy, loamy, or clayey (e.g. sandy gravel, loamy-fragmental).

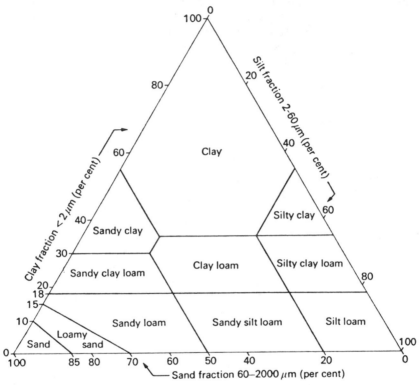

Fig. 1. Limiting percentages of sand, silt, and clay fractions in particle-size classes.

Horizon designations

Soil horizons are designated by letters and numbers generally according with international usage (International Society of Soil Science, 1967). Superficial organic horizons that are seldom wet are designated L, F, and H, peat or peaty horizons O, and mineral horizons A, E (equivalent to A_2 or A_e), B, and C. Specific differentiating characteristics are denoted by lower-case suffixes (e.g. Ap, Bt), and lithologic discontinuities by numerical prefixes.

Organic and mineral soils

Organic (peat) soils (major group 10) are required to have O horizons at least 40 cm thick, starting at the surface or at less than 30 cm depth. Other soils, including those with a thinner organic layer resting more or less directly on hard rock or fragmental material, are considered *mineral soils*.

Differentiating characteristics of mineral soil groups and subgroups

Distinct topsoil. A cultivated soil with an appreciably darkened Ap containing at least 0·6 per cent organic carbon (1 per cent O.M.) in the upper 15 cm, or an uncultivated soil with as much or more organic matter and continuous O or H and Ah horizons (whichever is present) together more than 7·5 cm thick, is considered to have a distinct topsoil.

Humose topsoil. An A horizon that is humose over at least 15 cm depth, or 10–15 cm if directly over bedrock or fragmental material.

Peaty topsoil. An O horizon 7·5–40 cm thick, over a mineral sub-surface horizon. Soils with humose and peaty topsoils are classed together in *humic* subgroups, and distinguished when feasible as phases in mapping.

Thick man-made A horizon. Dark A horizon more than 40 cm thick, resulting from addition of earth-containing manure or otherwise attributable to human occupation.

Podzolic B horizon. B horizon or horizons (cf. spodic horizon in the U.S.D.A. system) in which organic matter and aluminium and/or iron have accumulated in amorphous forms, as evidenced by morphology and distribution of pyrophosphate-extractable iron, aluminium and carbon (Bascomb, 1968; McKeague *et al.*, 1971). It usually underlies a bleached (albic) E horizon or a dark Ah, H, or O in which mineral grains are uncoated, and is required to extend to at least 15 cm depth, excluding surface litter (L, F). In cultivated soils, it occurs below the ploughed layer or as recognizable fragments within it.

The following horizons, diagnostic at group or subgroup level in major group 6, can form all or part of a podzolic B.

Bh   Normally dark coloured, with coated grains but little ($<$ 0·3 per cent dithionite-extractable) iron (*humus*-subgroups).

Bhs Dark coloured, more than 1 cm thick, with proportionately more iron (*humo-ferric* subgroups).

Bs   Brown or ochreous, with coated grains or 'pellety' microfabric (*humo-ferric* subgroups). It is required to contain more than 0·3 per cent pyrophosphate-extractable Fe+Al, amounting to at least 5 per cent of the measured clay percentage.

Bf   Thin ironpan (placic horizon in the U.S.D.A. system) (ironpan subgroups).

Humic and *ferri-humic* podzols have Bhs horizons that are humose over at least 10 cm, and generally lack an E horizon (crypto-podzols).

Soils with a Bs horizon and no continuous albic E or distinct Bhs, classed in major group 6 as brown podzolic soils, often show little evidence of illuviation and may qualify as Dystrochrepts rather than as Spodosols (podzols) in the U.S.D.A. system.

Argillic (Bt) horizon (Soil Survey Staff, 1967). Textural B horizon, normally containing translocated silicate clay as evidenced by argillans (clay skins) and/or intra-ped concentrations of strongly orientated clay in some part. As a horizon diagnostic of argillic groups and subgroups, it is required to contain significantly more clay than all overlying horizons (except in evidently eroded soils) and to meet thickness and depth limits that vary with texture.

A *paleo-argillic* horizon shows features generally attributable to pedological reorganization before the last (Weichselian) glacial period. It normally has a dominant matrix chroma more than 4 in hues redder than 10YR and/or red (5YR or redder) mottles, especially in the lower part, and a complex sepic plasmic fabric. Overlying horizons are often in a lithologically distinct (more recent) deposit, and the soil material usually shows signs of disturbance by cryoturbation or solifluction.

Weathered B (Bw) horizon. Non-podzolic, non-argillic B horizon, usually brownish, differentiated by colour and/or structural features indicating pedological reorganization *in situ*. Included are ungleyed cambic horizons (Soil Survey Staff, 1967) and analogous sandy horizons. As a diagnostic horizon (major group 5), it is required to extend at least 10 cm below an Ap horizon or to more than 30 cm depth, and extremely calcareous material is excluded.

A shallower brownish weathered horizon (Bw or A/Bw) characterizes brown rankers and brown rendzinas.

Gleyed horizons. These have dominantly greyish and/or ochreous colours attributable to alteration of the original material by reduction, or reduction and segregation, of iron, and are normally wet for significant periods or artificially drained.

Intensely gleyed horizons containing enough iron to redden on ignition are designated by the additional symbol G (e.g. CG). They have a dominant chroma of 1 or less in yellowish, greenish, or bluish hues that change on exposure to air. A *sulphuric* gleyed horizon (in sulphuric subgroups) is extremely acid as a result of oxidation of pyrites. It has a pH of 3·5 or less in 0·01 M $CaCl_2$ and normally contains jarosite.

Other gleyed horizons, with or without ferruginous mottles, are denoted by the suffix g (Eg, Bg, Cg). The diagnostic 'greyish' colours can have chromas up to 4, depending on colour value and the original colour and composition of the material, but chromas less than 3 are normally required. Horizons dominated by ferruginous segregations are identified by the additional suffix f (e.g. Bgf).

Gleyed horizons (normally CG) that have remained waterlogged since deposition, and consequently have a fluid consistence, are described as *unripened*. They have an *n*-value (Pons and Zonneveld, 1965) greater than 0·7, and a sample will pass through the fingers when squeezed in the hand.

Gley soils (major groups 2, 7, and 8) are required to have a gleyed horizon starting at less than 40 cm depth, and no podzolic B. Other soils with gleyed horizons, or with distinct mottling or other evidence of wetness within the upper 60–70 cm, are distinguished at group or subgroup level in other major groups.

Ripened or partially ripened gley soils with a distinct, humose or peaty topsoil are separated at major group level from those which are raw or unripened, and subdivided into 'groundwater' and 'surface-water' major groups. The latter soils, formed in Pleistocene and pre-Pleistocene deposits, are characterized by relatively impervious massive, platy, prismatic, or blocky sub-surface horizons (Bg, Btg, or BCg), normally with greyish colours associated with ped faces or cleavage planes, and ferruginous mottles or inherited colours in the matrix. The former major group includes soils in recent alluvium of medium to fine texture (alluvial and humic-alluvial groups), in sandy materials (sandy groups), and in non-alluvial loamy or clayey materials with relatively pervious substrata affected by fluctuating groundwater (cambic and argillic gley soils). These soils can have G horizons, and/or relatively pervious Cg, Bgf, or Cgf horizons in which ferruginous segregations, when present, occur wholly or partly as coats on skeletal grains or ped faces, or bordering voids.

Bleached hardpan. Grey E (albic) horizon that is very hard when moist and dry, possibly because of cementation by silica.

Sandy soils. For inclusion in sandy groups or subgroups, at least half the upper 80 cm of mineral soil (excluding bedrock and buried sola) must be sand or loamy sand (Fig. 1).

Vertic features (cf. Soil Survey Staff, 1967; Schlichting, 1968). The soils classed as Pelosols or in Pelo-subgroups are identified by the following characteristics:

1. More than 35 per cent clay over at least 30 cm, starting at the surface, directly below an Ap, or at less than 25 cm depth.
2. Blocky, prismatic, or wedge-shaped peds with glazed and/or slickensided faces, often inclined.
3. Cracks 5 mm or more wide between 25 and 50 cm depth in most years.
4. Potential linear extensibility (Soil Survey Staff, 1967) more than 4 cm in the upper 80 cm.

Calcareous soils. Calcareous groups and subgroups comprise soils that lack an argillic horizon and have a Bw, Bg, or C horizon containing at least 1 per cent $CaCO_3$ (< 2 mm fraction, excluding added lime) at less than 40 cm depth.

Alluvial soils. Alluvial groups comprise soils wholly or partly in a recent (Holocene or Flandrian) fluviatile, marine, or lacustrine deposit at least 30 cm thick. They may have O, A, Bw, and/or gleyed horizons, but not an argillic horizon except as part of a buried solum. The organic-carbon content of loamy or clayey alluvial layers is more than 0·2 per cent throughout, or decreases irregularly with depth. Soils in unsorted colluvial (slope) deposits of recent origin are excluded.

Disturbed soil. Artificially disturbed or transported material, including recognizable fragments of pedogenic horizons without discernible natural order, occurring below any Ah or Ap horizon. Deeply worked ground and soil replaced after mineral extraction are included.

Differentiating characteristics of organic (peat) soil groups and subgroups

Classification of peat soils (major group 10) is based primarily on the presence or absence of an *earthy topsoil* (peaty earthy layer in the

Netherlands system) at least 20 cm thick, consisting of ripened amorphous material, or a ripened mineral surface layer. Subgroups are identified by characteristics of sub-surface horizons, normally between 20 and 60 cm depth, as follows:

Fibrous. Fibrous or semi-fibrous material present.

Amorphous. Amorphous or mineral material to 60 cm.

Sulphuric. With extremely acid sulphuric horizon (see above), often containing gypsum.

Oligotrophic. Peat derived chiefly from *Sphagnum, Eriophorum, Calluna,* or *Trichophorum,* and/or with pH less than 4·5 (undried samples in 0·01 M $CaCl_2$) between 20 and 60 cm.

Eutrophic. pH more than 4·5 in some part, and no sulphuric horizon (usually peat derived from grasses or sedges, or limnic material).

Criteria for soil series

The main criteria differentiating soil series within subgroups of mineral soils are:

1. Dominant particle-size class of the soil material between specified depth limits, normally the upper 80 cm (basic particle-size classes are grouped for this purpose).
2. Presence and nature of texturally contrasting layers, bedrock or horizons (indurated layer or fragipan) impenetrable to roots within a specified depth.
3. Origin of the soil material (e.g. marine or river alluvium, glacial drift).
4. Mineralogical or related characteristics (e.g. colour) of the soil material.

At present, many soil series are partly defined by the stratigraphic age of a presumed parent rock, but the intention is to base their differentiation on intrinsic soil properties whenever possible.

Possible criteria for differentiating soil series within subgroups of organic soils are:

1. Botanical origin and degree of decomposition of fibrous sub-surface layers, and mineral-matter content of humic or limnic layers.
2. Presence and composition of mineral substrata within a specified depth.
3. Presence of a humilluvic layer (Soil Survey Staff, 1968).

Soil-profile classification in relation to soil mapping

Whereas a profile class groups profiles according to their similarity, a soil map must group them according to their contiguity. Profile classes are therefore distinguished from *soil map units,* considered as classes of *delineated soil bodies* (Knox, 1965) that are characterized by some degree of geographic homogeneity and that may or may not correspond to natural *soil landscape units* as defined by Schelling (1970).

Profile classification serves to guide the creation of map units and provides a uniform basis for describing and identifying them. Accordingly, a map unit is usually identified by the name of a profile class, implying that most of the soil in each delineation conforms to that class, and that inclusions of other kinds of soil conform to one or more closely related classes or occupy an insignificant proportionate area (Simonson, 1968). More heterogeneous units (complexes, associations, undifferentiated groups) are similarly identified by the names of two or more classes.

A classification based on the morphology and composition of soil profiles excludes from consideration other land attributes affecting its use, notably slope, which can characterize delineated soil bodies because these have shape as well as depth. Also, in constructing a profile classification for general use, characteristics (e.g. of topsoils) that may be significant for one type of use (e.g. forestry) but not for another (e.g. arable farming) are deliberately excluded as differentiating criteria. To meet these situations, mapping units termed *phases* (Smith, 1965) are defined in conjunction with, or independently of, the profile classification and identified by whichever internal or external attribute sets them apart.

Important attributes that are not specifically differentiating in the proposed profile classification include soil-moisture and soil-temperature regimes (Soil Survey Staff, 1967). Data on these are generally too few to serve as bases for map units in detailed surveys, but establishment of 'climatic phases' to define separations on small-scale maps is under consideration, with average maximum soil-moisture deficit, mean annual soil temperature and length of the growing season as possible criteria.

Acknowledgement

The author thanks all those colleagues, at home and abroad, who have contributed through discussion or provision of unpublished information to the preparation of this paper.

REFERENCES

AVERY, B. W. 1956. A classification of British soils. Rapp. 6th int. Congr. Soil Sci. Paris. E, 279–85.
—— 1965. Soil classification in Britain. Pédologie, Gand, numéro spécial 3, 75–90.
BASCOMB, C. L. 1968. Distribution of pyrophosphate-extractable iron and organic carbon in soils of various groups. J. Soil Sci. 19, 251–68.
BECKETT, P. H. T., and BURROUGH, P. A. 1971. The relation between cost and utility in soil survey. IV. Comparison of the utilities of soil maps produced by different survey procedures, to different scales. Ibid. 22, 466–80.
British Standards Institution. 1967. Methods of testing soils for engineering purposes. British Standard 1377. London.
CLARKE, G. R. 1940. Soil Survey of England and Wales. Field Handbook. University Press, Oxford.
Commission de Pédologie et de Cartographie des Sols. 1967. Classification des sols. École Nat. Sup. Agr. Grignon. (Mimeographed.)
DE BAKKER, H., and SCHELLING, J. 1966. Systeem van Bodemclassificatie voor Nederland. De Hogere Niveaus (with English summary). Pudoc, Wageningen.
DUCHAUFOUR, P. 1970. Précis de Pédologie, 3rd Edition. Masson, Paris.
EHWALD, E., LIEREROTH, I., and SCHWANECKE, W. 1966. Zur Systematik der Böden der Deutschen Demokratischen Republik, besonders im Hinblick auf die Bodenkartierung. Sitz. Ber. Deut. Akad. Landwirtschaftswiss. Berlin. 15 (18).
International Society of Soil Science, 1967. Proposal for a uniform system of soil horizon designations. Bull. int. Soc. Soil Sci. 31, 4–7.

Knox, E. G. 1965. Soil individuals and soil classification. Proc. Soil Sci. Soc. Amer. **29,** 79–84.

Mackney, D., and Burnham, C. P. 1964. The soils of the West Midlands. Bull. Soil Surv. Gt. Br. 2, Harpenden.

Macvicar, C. N. 1969. A basis for the classification of soil. J. Soil Sci. **20,** 141–52.

McKeague, J. A., Brydon, J. E., and Miles, N. M. 1971. Differentiation of forms of extractable iron and aluminium in soils. Proc. Soil Sci. Soc. Amer. **35,** 33–8.

Mückenhausen, E. 1965. The soil classification system of the Federal Republic of Germany. Pédologie, Gand, numéro spécial **3,** 57–89.

National Soil Survey Committee of Canada. 1968. Proceedings of the 7th meeting of the National Soil Survey Committee of Canada. Mimeographed rept. Univ. of Alberta, Edmonton.

Pons, L. J., and Zonneveld, I. S. 1965. Soil ripening and soil classification. Int. Inst. Land Reclamation and Improvement. Publ. 13. Veenman, Wageningen.

Road Research Laboratory, 1970. Soil classification. R.R.L. Lf. 208. Crowthorne (Road Research Laboratory).

Ragg, J. M., and Clayden, B. 1973. The classification of some British soils according to the Comprehensive System of the United States. Tech. Monogr. Soil Surv. Gt. Br. Harpenden (in press).

Robinson, G. W. 1943. Mapping the soil of Britain. Discovery **4,** 118–21.

Schelling, J. 1970. Soil genesis, soil classification and soil survey. Geoderma **4,** 165–93.

Schlichting, E. 1968. Bodenbildende Prozesse in Tongesteinen unter gemässigt-humidem Klima. Trans 9th int. Congr. Soil Sci. Adelaide **4,** 411–18.

Simonson, R. W. 1968. Concept of soil. Adv. Agron. **20,** 1–45.

Smith, G. D. 1965. Lectures on soil classification. Pédologie, Gand, numéro spécial **4.**

Soil Survey Staff. 1951. Soil Survey Manual. U.S. Dept. Agr. Handbook 18. Government printer. Washington, D.C.

Soil Survey Staff. 1960. Soil classification: A Comprehensive System. 7th Approximation. U.S. Dept. Agr. Washington, D.C.

Soil Survey Staff. 1967, 1968. Supplements to 7th approximation (mimeographed) U.S. Dept. Agr. Soil Conservation Service. Washington D.C.

Discussion

D. V. Crawford (Nottingham): (*a*) Can we now use your new classification system instead of (or as well as) the series names? (*b*) If so, where or when may the equivalent terminology be found?

B. W. Avery: (*a*) Yes. (*b*) As stated, the question is difficult to understand. Soil series will be defined or redefined as subdivisions of soil subgroups in future Soil Survey publications.

J. J. Reynders (Utrecht): You said that in soil nomenclature Latin and Greek words are used as in the American and F.A.O. systems. The Dutch nomenclature is part of our language, using familiar or old words. Mapping and classification are to be used increasingly by farmers, engineers, geographers, etc., for planning. Do you think they will use difficult names after 10 years? More familiar names may be more effective.

B. W. Avery: Confusion can arise from the use of familiar or old words when their meaning is imprecise and when they mean different things to different people. For local use, we have the soil series names, and experience has shown that after 10 years these are being used by agronomists, planners, and increasingly by farmers. I believe the more exotic soil group and subgroup names will also be accepted in time in so far as they connote useful concepts.

J. J. Reynders (Utrecht): Older land surfaces (e.g. in France around the Basin of Paris and in southern England) have many soils formed in two layers ('profils a deux couches'). Where erosion occurred many subsoils (second layer) have another texture and some older genesis. For example, you can distinguish paleo-argillic horizons. Do you think that, in general, the new classification covers these 'two-layer' soils? It is a much more general problem than people thought 20 years ago.

B. W. Avery: I agree that 'two-layer' profiles of the kind you describe are widespread. The intention in defining 'paleo-argillic' classes is to separate those in which the subsoil (second layer) retains characteristics of an argillic horizon formed in an interglacial period, but more work is needed to derive better criteria for identifying this condition.

J. Tinsley (Aberdeen): Where do 'red' soils with colour inherited from Red Sandstone parent materials fit into the proposed hierarchical system of soil classification?

B. W. Avery: The question presumably concerns the recognition of 'weathered B horizons' in red (haematitic) materials, comparable with the brownish weathered B horizons in originally greyish materials. Where such horizons cannot be identified by colour differences, they can usually be recognized by structural characteristics or by an increase in clay content, compared with the original material.

17

HISTORY OF SOIL CLASSIFICATION IN CANADA

Canada Soil Survey Committee

The early years, 1914–1940

The classification of soils in Canada began with the first soil survey in Ontario in 1914. When A. J. Galbraith set out to map the soils of southern Ontario, concepts of soil and methods of soil classification were rudimentary in North America. G. N. Coffey, formerly of the U.S. Bureau of Soils, advised Galbraith during the early stages of the survey and the system of classification used was that of the U.S. Bureau of Soils based largely upon geological material and texture (Ruhnke 1926). Nine "soil series" were mapped in all of Ontario south of Kingston by 1920. The broad scope of "series," which were somewhat analogous to geological formations at that time, has narrowed progressively to the present.

Changes in the system of classifying the soils of Canada have resulted from the combined effects of international developments in concepts of soils and increasing knowledge of Canadian soils. Canadian pedologists were influenced from the time of the first surveys by the concept of soil as a natural body integrating the accumulative effects of climate and vegetation acting on surficial materials. This concept was introduced by Dokuchaev about 1870, developed by other Russian soil scientists, and proclaimed to western Europe by Glinka in 1914 in a book published in German. Marbut's translation of this book made the Russian concept of soil as a natural body easily available to the English-speaking world (Glinka 1927). The concept is of paramount importance in soil science because it makes possible the classification of soils on the basis of properties of the soils themselves rather than on the basis of geology, climate, or other factors. Classification systems based on the inherent properties of the objects classified are called natural or taxonomic systems.

Recognition of the relationships between soil features and factors of climate and vegetation was not limited to Russian scientists. In the United States, Hilgard had noted this association in a book published in 1860 (Jenny 1961), and Coffey had recognized soils as natural bodies by 1912 (Kellogg 1941). However, the development of the concept of soils as natural bodies with horizons that reflect the influences of soil-forming factors, particularly climate and vegetation, may be credited to the Russians.

In the earliest soil surveys undertaken in the Prairie Provinces during the 1920s classification was on the basis of texture, but an increased awareness of soil zones and of the soil profile is evident in soil survey reports published during that decade. Preliminary soil zone maps of Alberta and Saskatchewan were presented by Wyatt and Joel at the first International Congress of Soil Science in 1927. They showed the broad belts of brown, black, and gray soils. The Congress and the associated field tours brought Canadian pedologists into close contact with world concepts of soil and systems of soil classification.

Developments in soil classification occurred independently in each province because surveys were carried out by university departments of soils or chemistry. For example, a numbering system indicating the soil zone, nature of parent material, mode of deposition, profile features, and texture was developed in Alberta. J. H. Ellis in Manitoba recognized the impossibility of developing a scientific soil taxonomy based on the limited knowledge of Canadian soils in the 1920s. Influenced by concepts of C. C. Nikiforoff in Minnesota, Ellis developed a field system of soil classification that was useful in soil mapping and endures in various revised forms to this day. The system identified "associations" of soils formed on similar parent materials and "associates" that differed according to topographic position within the association (Ellis 1932, 1971).

During the 1930s soil surveys proceeded in Ontario and the Prairie Provinces and were started in British Columbia in 1931, Quebec and Nova Scotia in 1934, and New Brunswick in 1938. Soil surveys began in Prince Edward Island in 1943, the Northwest Territories in 1944, and Newfoundland in 1949. A few soil surveyors were employed on a permanent basis by federal and provincial departments of agriculture in the 1930s and they worked cooperatively with personnel of university soils departments. By 1936 about 15 000 000 ha (1.7% of the land area of Canada) had been surveyed, mainly in Alberta, Saskatchewan, and Ontario. Soil classification was limited by the fragmentary knowledge of the soils of Canada.

Canadian pedologists were influenced in the 1930s by Marbut's developing ideas on soil classification, Ellis's system of field classification (Ellis 1932), and the classification system of the U.S. Department of Agriculture (USDA) described by Baldwin, Kellogg, and Thorp (1938). The latter system divided soils at the highest category among three orders:

1. Zonal soils, which are those with well-developed characteristics that reflect the influence of the active factors of soil genesis such as climate

and organisms, particularly vegetation, e.g. Podzol.

2. Intrazonal soils, which are soils having more or less well-defined characteristics that reflect the dominant influence of some local factor of relief or parent material over the normal effects of climate and vegetation, e.g. Humic Gley.
3. Azonal soils, which are soils without well-developed characteristics due either to their youth or to some condition of relief or parent material, e.g. Alluvial soils.

Zonal soils were divided at the suborder level on the basis of climatic factors, and suborders were subdivided into great groups more or less similar to the great groups of today.

Canadian experience showed that the concept of zonal soils was useful in the western plains, but was less applicable in Eastern Canada where parent material and relief factors had a dominant influence on soil properties and development in many areas. However, the 1938 USDA system was used in Canada and it influenced the subsequent development of the Canadian system.

From 1940 to 1976

The formation of the National Soil Survey Committee of Canada (NSSC) was a milestone in the development of soil classification and of pedology generally in Canada. The initial organization meeting was held in Winnipeg in 1940 by the Soils Section of the Canadian Society of Technical Agriculturists (Ellis 1971). Subcommittees were established to prepare reports on six major topics including soil classification. At the suggestion of E. S. Archibald, Director of the Experimental Farms Service, the NSSC became a committee of the National Advisory Committee on Agricultural Services. The first executive committee of the NSSC consisted of: A. Leahey, chairman; P. C. Stobbe, secretary; F. A. Wyatt, western representative; and G. N. Ruhnke, eastern representative. Terms of reference for the NSSC were developed by A. Leahey, G. N. Ruhnke, and C. L. Wrenshall. They were modified and restated in 1970 by the Canada Soil Survey Committee (CSSC) as follows:

To act as a coordinating body among the soil survey organizations in Canada supported by the Canada Department of Agriculture, provincial departments of agriculture, research councils, and departments of soil science at universities. Its functions include:

1. Improvement of the taxonomic classification system for Canadian soils and revision of this system because of new information.

2. Improvement of the identification of physical features and soil characteristics used in the description and mapping of soils.
3. Review of methods, techniques, and nomenclature used in soil surveys and recommending changes necessary for a greater measure of uniformity or for their improvement.
4. Recommending investigations of problems affecting soil classification, soil formation, and the interpretation of soil survey information.
5. Recommending and supporting investigations on interpretations of soil survey information for soil ratings, crop yield assessments, soil mechanics, and other purposes.
6. Cooperating with specialists in soil fertility, agronomy, agrometeorology, and other disciplines in assessing interrelated problems.

Much of the credit for the present degree of realization of these objectives is due to A. Leahey, chairman from 1940 until 1966, and P. C. Stobbe, secretary from 1940 until 1969. W. A. Ehrlich was chairman from 1966 to 1971 and was succeeded by J. S. Clark. In 1969 the name of the committee was changed to the Canada Soil Survey Committee (CSSC).

Developments in soil classification in Canada since 1940 are documented in the reports of the meetings of the NSSC held in 1945, 1948, 1955, 1960, 1963, 1965, 1968, and of the CSSC in 1970, 1973, and 1976. Soil classification was one of the main items on the agenda of the first meeting and a report by P. C. Stobbe provoked a prolonged and lively discussion. He and his committee recommended a system of field classification similar to that developed in Manitoba by Ellis (1932). The proposed system was a hierarchical one with seven categories as follows:

Soil Regions—tundra, woodland, and grassland soils.

Soil Zones—broad belts in which a dominant kind of soil occurs, such as podzol or black soil.

Soil Subzones—major subdivisions of soil zones, such as black and degraded black.

Soil Associations or Catenas—the group of soils that are associated together on the same parent material to form a land pattern.

Soil Series, Members, or Associates—the individual kinds of soils that are included in an association.

Soil Type or Soil Class—subdivisions of associations or of series based upon texture.

Soil Phase—subdivisions of mapping units based upon external soil characteristics such as stoniness and topography.

This was a field system of classification or a system of classifying the units of soils mapped at various scales. The classes at all levels, phase to region, were segments of the landscape that included all of the soil variability within the area designated. Thus a soil zone was a land area in which a "zonal great soil group occurred as a dominant soil." The system was not intended to be a scientific or taxonomic one in which the classes at all levels had clearly defined limits based on a reasonably thorough knowledge of the properties of the entire population of soils in Canada. Unfortunately, a degree of confusion about the distinction between the nature of a soil taxonomic system and a system of classifying and naming soil mapping units persists to this day. Several of the issues debated at that first NSSC meeting in 1945 remain as controversial issues today. The system proposed was accepted for trial by the committee, which represented all provinces. Thus, an important step was taken in the development of a national system of classifying the units of soil mapped in soil surveys.

The first Canadian taxonomic system of soil classification was presented by P. C. Stobbe at the NSSC meeting in 1955. This system was a marked departure from the mapping or field classification system proposed in 1945 and it probably resulted from the following circumstances:

The greater knowledge of Canadian soils.

The desire to classify soils, even at the highest categorical level, on the basis of properties of the soils themselves.

The need for a taxonomic system better than the old USDA system (Baldwin, Kellogg, and Thorp 1938) that focused unduly on "normal" soils. The Soil Conservation Service had begun the development of a new system in 1951, but the 4th approximation of that system was judged to be too complicated and too tentative for Canadian needs (NSSC Report 1955).

Unfortunately, formal discussion of field systems of classifying soils or soil mapping systems was dropped for several years at NSSC meetings, but the need for such systems was recognized by leaders in pedology. This need can be illustrated by an example of mapping soils at a particular scale and classifying the kinds of soil that occur.

If the map is at a scale of 1:100 000 and the smallest area delineated is a square measuring 1 cm on each side, that area represents 100 ha. Such an area commonly includes upland and lowland positions in the landscape and the associated kinds of soils. The kinds of soils within that area are identified by digging pits at different topographic positions in the landscape. At each of these points the profile exposed usually has a rather narrow range of properties that reflects the influence of soil-forming factors at that point. Thus the soils at each point of observation can be classified as a single class in a taxonomic system, but the area delineated on the map cannot be classified as a single class in such a system because it includes several kinds of soils. However, the area mapped can be classified as a kind of soil mapping unit such as a soil association in the system of Ellis (1932). Thus the need was evident for both a taxonomic system to permit the naming and the ordering of information about specific kinds of soils, and a mapping system to permit the ordering of information about the areas delineated on soil maps and the naming of them.

The taxonomic system outlined in 1955, which is the basis of the system used today, had six unnamed categorical levels corresponding to the order, great group, subgroup, family, series, and type. The seven taxa separated at the order level were: Chernozemic, Halomorphic, Podzolic, Forested Brown, Regosolic, Gleisolic, and Organic. Taxa were defined only in general terms down to the subgroup level. Although this was inevitable because of the lack of sufficient information, it led to differences of interpretation of the taxa in various provinces and some lack of uniformity in the use of the system. The need for correlation was clearly recognized by senior Canadian pedologists.

Progress in the development of the Canadian system of soil classification since 1955 has been toward more precise definitions of the taxa at all categorical levels and an increasing emphasis on soil properties as taxonomic criteria. This is evident from the reports of NSSC meetings held in 1960, 1963, 1965, and 1968, at which the main topic of discussion was soil classification. Some changes in taxa at the order, great group, and subgroup levels were made at these meetings. For example, in 1963 the Meadow and Dark Gray Gleisolic great groups were combined as Humic Gleysol; in 1965 a system of classifying soils of the Organic order was presented and accepted; in 1968 the former Podzolic order was divided into Luvisolic (clay translocation) and Podzolic (accumulation of Al and Fe organic complexes) orders, and the concept and classification of Brunisolic soils were revised. Criteria of classification involving morphological, chemical, and physical properties became increasingly specific through this period. The bases of classifying soils at the family level were outlined and the series and type categories were defined more specifically.

Following the publication of *The System of Soil Classification for Canada* in 1970, topics other than soil taxonomy were emphasized at CSSC meetings. However, in 1973 a Crysolic order was proposed to classify the soils with permafrost close to the surface, and some refinements were made in several orders.

Between 1945 and 1970 very little consideration was given at NSSC meetings to systems of naming and classifying soil mapping units. In 1970 a report was presented on the Biophysical Land Classification System, which is a hierarchical system with four levels for classifying the units of land mapped. The 1973 CSSC meeting included reports on mapping units in surveys of forest land and permafrost areas, and miscellaneous land types. The main topic at a CSSC meeting in 1976 was soil mapping units. With the achievement of a reasonably satisfactory taxonomic system, the emphasis in the CSSC has swung to the equally basic matter of a system of defining the nature and distribution of soils within mapping units and of naming these units.

RATIONALE OF SOIL TAXONOMY IN CANADA

During some 60 years of pedological work in Canada, concepts of soil and systems of classification have progressed as a result of new knowledge and new concepts developed in Canada and elsewhere. An attempt is made here to enunciate the current rationale of soil taxonomy based on the historical material outlined in the previous section and on recent publications on pedology in Canada.

The nature of soil

The concept of soil in Canada and elsewhere (Cline 1961; Knox 1965; Simonson 1968) has changed greatly since 1914 when the first soil survey was started in Ontario. No specific definition is available from that early work, but clearly soil was thought of as the uppermost geological material. Texture was apparently considered to be its most important attribute. Currently, soil is defined in general terms by pedologists as the naturally occurring, unconsolidated, mineral or organic material at the earth's surface that is capable of supporting plant growth. Its properties usually vary with depth and are determined by climatic factors and organisms, as conditioned by relief and hence water regime, acting on geological materials and producing genetic horizons that differ from the parent material. In the landscape soil merges into nonsoil entities such as exposed, consolidated rock or permanent bodies of water at arbitrarily defined boundaries.

Specific definitions of soil and nonsoil are given in Chapter 2.

Because soil occurs at the surface of the earth as a continuum with variable properties, it is necessary to decide on a basic unit of soil to be described, sampled, analyzed, and classified. Such a unit was defined by United States' pedologists (Soil Survey Staff 1960) and is accepted in Canada. It is called a pedon and is the smallest, three-dimensional body at the surface of the earth that is considered as a soil. Its lateral dimensions are 1–3.5 m and its depth is 1–2 m. Pedon is defined more specifically in Chapter 2.

Nature and purpose of soil classification

Soil classification systems are not truths that can be discovered but methods of organizing information and ideas in ways that seem logical and useful (Soil Survey Staff 1960). Thus no classification system is either true or false; some systems are more logical and useful for certain objectives than others. A classification system reflects the existing knowledge and concepts concerning the population of soils being classified (Cline 1949). It must be modified as knowledge grows and new concepts develop.

Both the theoretical and practical purposes of soil classification have been discussed in the literature (Cline 1949, 1963; De Bakker 1970). The general purpose of soil classification in Canada may be stated as follows:

To organize the knowledge of soils so that it can be recalled systematically and communicated and that relationships may be seen among kinds of soils, among soil properties and environmental factors, and among soil properties and suitabilities of soils for various uses.

The related purposes of soil classification are: to provide a framework for the formulation of hypotheses about soil genesis and the response of soil to management, to aid in extending knowledge of soils gained in one area to other areas having similar soils, and to provide a basis for indicating the kinds of soils within mapping units. Soil classification is essential to soil surveys, to the teaching of soils as a part of natural science, and to meeting the practical needs related to land use and management.

The overall philosophy of the Canadian system is pragmatic; the aim is to organize the knowledge of soils in a reasonable and usable way. The system is a natural or taxonomic one in which the classes (taxa) are based upon properties of the soils themselves and not upon interpretations of the soils for various uses. Interpretations involve a second step that is essential if the information is to be used effectively. If the taxa are defined on properties and the boundaries of these classes or of combinations of them are shown on a map, any interpretation based upon properties implied in the class definitions can be made.

Misconceptions about soil taxonomy

Misconceptions about the functions of a system of soil taxonomy are evident periodically. Some of these are listed to warn users of the Canadian system against unrealistic expectations.

1. It is a misconception that a good system results in the assignment of soils occurring close together to the same taxon at least at the higher categorical levels. This is neither possible nor desirable in some areas. Pedons a few metres apart may differ as greatly as pedons hundreds of kilometres apart within a climatic region.
2. Another common misconception is that a good national system provides the most suitable groupings of soils in all areas. This is not possible because criteria based upon properties of the whole population of soils in the country are bound to be different from those developed on the basis of properties of soils in any one region. Criteria developed for a national system will inevitably result in areas where most of the soils have properties that straddle the boundary line between two taxa.
3. The idea that if the system was soundly based there would be no need for changes every few years is erroneous. As new areas are surveyed, as more research is done, and as concepts of soil develop, changes in the system become inevitable to maintain a workable taxonomy.
4. Another unfortunate hope is that a good system will ensure that taxa at the order level at least can be assigned unambiguously and easily in the field. Actually in a hierarchical system the divisions between orders must be defined just as precisely as those between series. With pedons having properties close to class boundaries at any taxonomic level, classification is difficult and laboratory data may be necessary.
5. The assumption is made by some that a good system permits the classification of soils occurring within mapped areas as members of not more than three series. Clearly this is not reasonable because the number of taxonomic classes occurring within a mapping unit depends upon the complexity of the pattern of soils in the landscape, on the scale of the map, and on the narrowness of class limits.
6. The idea that a good system is simple enough to be clear to any layman is erroneous. Unfortunately, soil is complex and although the general ideas of the taxonomy should be explainable in simple terms the definitions of taxa must be complex in some instances.
7. Another misconception is that a good system makes soil mapping easy. Ease of mapping depends more upon the complexity of the landscape, the access, and the predictability of the pattern of soils within segments of the landscape than upon taxonomy.

Attributes of the Canadian system

The development of soil taxonomy in Canada has been toward a system with the following attributes:

1. It provides taxa for all known soils in Canada.
2. It involves a hierarchical organization of several categories to permit the consideration of soils at various levels of generality. Classes at high categorical levels reflect, to the extent possible, broad differences in soil environments that are related to differences in soil genesis.
3. The taxa are defined specifically so as to convey the same meaning to all users.
4. The taxa are concepts based upon generalizations of properties of real bodies of soils rather than idealized concepts of the kinds of soils that would result from the action of presumed genetic processes. The criteria chosen define taxa in accordance with the desired groupings of soils. The groupings are not decided upon initially on the basis of arbitrary criteria.
5. Differentiae among the taxa are based upon soil properties that can be observed and measured objectively in the field or, if necessary, in the laboratory.
6. It is possible to modify the system on the basis of new information and concepts without destroying the overall framework. Periodically, however, the entire framework of the system will be reevaluated.

Although taxa in the Canadian system are defined on the basis of soil properties, the system has a genetic bias in that properties or combinations of properties that reflect genesis are favored as differentiae in the higher categories. For example, the chernozemic A and the podzolic B imply genesis. The reason for the genetic bias is that it seems reasonable to combine at high categorical levels soils that developed their particular horizonation as a result of similar dominant processes resulting from broadly similar climatic conditions. Classification is not based directly on presumed genesis because soil genesis is incompletely understood, is subject to a wide variety of opinion, and cannot be measured simply.

Bases of criteria for defining taxa at various categorical levels

The bases of differentiation of taxa at the various categorical levels are not clear cut. In a hierarchical system of soil classification, logical groupings of soils that reflect environmental factors cannot be obtained by following any rigid systematic framework in which all taxa at the same categorical level are differentiated on the basis of a uniform specific criterion such as acidity or texture. The fact that criteria must be based on soil properties rather than directly on environmental factors or use evaluation was recognized by some pedologists half a century ago (Joel 1926). The general bases of the different categorical levels can be inferred from a study of the system and these are presented below. They apply better to some taxa than to others; for example, the statement for order applies more clearly to Chernozemic and Podzolic soils than to Regosolic and Brunisolic soils.

Order. Taxa at the order level are based on properties of the pedon that reflect the nature of the soil environment and the effects of the dominant, soil-forming processes.

Great group. Great groups are soil taxa formed by the subdivision of each order. Thus each great group carries with it the differentiating criteria of the order to which it belongs. In addition, taxa at the great group level are based on properties that reflect differences in the strengths of dominant processes or a major contribution of a process in addition to the dominant one. For example, in Luvic Gleysols the dominant process is considered to be gleying, but clay translocation is also a major process.

Subgroup. Subgroups are formed by subdivisions of each great group. Therefore they carry the differentiating criteria of the order and the great group to which they belong. Also, subgroups are differentiated on the basis of the kind and arrangement of horizons that indicate: conformity to the central concept of the great group, Orthic; intergrading toward soils of another order, e.g. Gleyed, Brunisolic; or additional special features within the control section, e.g. Ortstein.

Family. Taxa at the family level are formed by subdividing subgroups. Thus they carry the differentiating criteria of the order, great group, and subgroup to which they belong. Families within a subgroup are differentiated on the basis of parent material characteristics such as texture and mineralogy, and soil climatic factors and soil reaction.

Series. Series are formed by subdivisions of families. Therefore they carry all the differentiating criteria of the order, great group, subgroup, and family to which they belong. Series within a family are differentiated on the basis of detailed features of the pedon. Pedons belonging to a series have similar kinds and arrangements of horizons whose color, texture, structure, consistence, thickness, reaction, and composition fall within a narrow range. A series is a category in the taxonomic system; thus it is a conceptual class in the same sense as an order.

A pedon is a real unit of soil in the landscape; a series is a conceptual class with defined limits based on the generalization of properties of many pedons. A particular pedon may be classified as a series if its attributes fall within the limits of those of an established series. However, it is not, strictly speaking, a series because the attributes of any one pedon do not encompass the complete range of attributes allowable within a series. Thus, it is not correct to study part of a pedon and to declare, "this is X series." Rather it should be stated, "this pedon has properties that fall within the limits of the X series," or "this pedon is classified in the X series."

Type. Type is no longer a category in the Canadian system. Surface texture may be indicated as a phase.

Relationship of taxonomic classes to environments

A general relationship exists between kinds of environments and taxa at various levels in the system. This follows from the basis of selection of diagnostic criteria for the taxa; the primary basis at the higher levels is properties that reflect the environment and properties resulting from processes of soil genesis. Although the system may look like a key with classes defined precisely but arbitrarily on the basis of specific properties, it is one in which the taxa reflect, to as great an extent as possible, genetic or environmental factors.

The Podzolic order, for example, is defined on the basis of morphological and chemical properties of the B horizon. However, these properties are associated with humid conditions, sandy to loamy parent materials, and forest or heath vegetation. Although the great groups within the order are defined on the basis of the amounts of organic C and extractable Fe and Al in the B horizons, they have broad environmental significance. Humic Podzols are associated with very wet environments, high water tables, periodic or continuous reducing conditions, hydrophytic vegetation, and often a peaty surface. Ferro-Humic Podzols occur in areas of high effective precipitation, but they are not under reducing conditions for prolonged periods. Humo-Ferric Podzols occur generally in less humid environments than the other great groups in the order. There is an interrelation of climatic and vegetative factors and parent material and relief in determining the occurrence of these classes of Podzolic soils. Similarly, there are general relationships between other orders, great groups, and soil environmental factors. However, these relationships are much less clear for some Regosolic and Brunisolic soils than they are for most soils of other orders. At lower categorical levels in general, relationships between soil taxa and factors of the soil environment become increasingly close.

Relationship of the Canadian system to other systems of soil taxonomy

The numerous national systems of soil taxonomy might be looked upon as indications of the youthfulness of soil science. Knowledge of the properties of the soils of the world is far from complete, therefore it is not possible to develop an international system of classification for the whole population of known and unknown soils. Probably even after such a system has been developed, national systems will remain in use because they are familiar and are thought to be more useful for the restricted population of soils within the country. The soil units defined for the FAO-Unesco world soil map project are useful in international soil correlation, but they do not constitute a complete system of soil taxonomy (FAO 1974). The closest approach to a comprehensive system of soil taxonomy is that produced by the USDA Soil Survey Staff (1975), which has been under development since 1951. Like previous U.S. systems, it has had a major influence on soil taxonomy in Canada and elsewhere.

The Canadian system of soil taxonomy is more closely related to the U.S. system than to any other. Both are hierarchical and the taxa are defined on the basis of measurable soil properties. However, they differ in several respects. The Canadian system is designed to classify only soils that occur in Canada and is not a comprehensive system. The U.S. system has a suborder, which is a category that the Canadian system does not have. In the Canadian system Solonetzic, Gleysolic, and Cryosolic soils are differentiated at the highest categorical level as in the Russian and some other European systems. These soils are differentiated at the suborder or great group level in the U.S. system. Perhaps the main difference between the two systems is that all horizons to the surface may be diagnostic in the Canadian system, whereas horizons below the depth of plowing are emphasized in the U.S. system. This may be a consequence of the fact that 90% of the area of Canada is not likely to be cultivated.

SUMMARY

The Canadian system is a hierarchical one in which the classes are conceptual based upon the generalization of properties of real bodies of soil. Taxa are defined on the basis of observable and measurable soil properties that reflect processes of soil genesis and environmental factors. The development of the system has progressed with the increasing knowledge of the soils of Canada obtained through pedological surveys carried out over a 60-yr period. The system has been influenced strongly by concepts developed in other countries, but some aspects are uniquely Canadian. The system is imperfect because it is based on a limited knowledge of the vast population of soils in the country. However, the system does make possible the assignment of the soils throughout Canada to taxa at various levels of generalization and the organization of the knowledge about soils in such a way that relationships between factors of the environment and soil development can be seen. It is possible to define the kinds of soils that occur within units on soil maps, and to provide a basis for evaluating mapped areas of soil for a variety of potential uses.

18

THE MAIN TROPICAL SOILS OF BRAZIL

Prof. R. Costa de Lemos
Universidade de Santa Maria

This paper is based on the results of the 4th Technical Meeting of the Division of Pedology and Soil Fertility, and corresponds to the 2nd Draft of the Classification of Brazilian Soils (M. N. Camargo and J. Bennema).

Here, we consider mainly the well-drained soils actually found in Brazil.

By studies made in Brazil, it has been found that wide classes of soils may be characterized on the basis of the collection of morphological, physical, chemical and mineralogical properties which they possess.

However, it must be said that some of these classes are differentiated in relation to physical and chemical characteristics, which cannot necessarily be observed in the field.

At a high level of classification, the soils of Brazil can be subdivided into two wide classes: soils with a Latosolic B horizon (approximately equivalent to the "oxic horizon" of the 7th Approximation) and soils with a textural B horizon (approximately equivalent to the "argillic horizon" of the 7th Approximation). The soils with a cambic B horizon seem to be less important, occurring in limited mountain zones located at higher altitudes, where mesothermic climatic conditions predominate.

Soils with a Latosolic B horizon

At a lower level of classification the soils with a latosolic B horizon are divided into the following classes:

I.   Brown Latosols of Altitude.

II.   Other Latosols with a cation exchange capacity $>$ 6.5 mE/100 g of clay (after carbon correction) and a percent base saturation $<$ 50% in B2.

III. Other Latosols with a cation exchange capacity $>$ 6.5 mE/100 g of clay (after carbon correction) and base saturation in $B_2 > 50\%$.

IV.   Other Latosols with a weak A horizon and in addition high base saturation or the base saturation increases with depth in the lower part of B and below, up to values higher than 50%.

V.   Other Latosols with cation exchange capacity $<$ 6.5 mE/100 g of clay (after carbon correction).

I.   Brown Latosols of Altitude

The purpose is to separate the Latosols of mountain areas or cold and humid plateaux. A specific definition of these soils cannot at the moment be stated.   The yellow or yellowish colours they show, even when they have a high amount of iron oxides; the strong or very strong development of the A horizon;   the particular cracking of the soils in road cuts, and the high values of moisture equivalent, in most cases, for samples not previously dried for analysis, can be used as basic characteristics of these soils for identification purposes.

There are intergrades of these soils to Acid brown soils and Andosols.   The distinction of the former is made by the absence of primary minerals easily decomposed (including silicate clay of lattice 2:1) and with the latter by the absence of allophanes or when present in solum, the allophanes appear only in small percentages.

This class corresponds to the more cold and humid part of Udox in the 7th Approximation.

An example of this class is the profile no. 49 of Vacaria.

Data are actually insufficient to establish a subdivision of these Latosols and the tendency is to subdivide them in the future, according to their silica/aluminium ratio and exchangeable Alxxx and in addition the iron oxide content (perhaps also in relation to Al_2O_3/Fe_2O_3).

II.   Other Latosols with Cation Exchange Capacity $\geqslant$ 6.5 mE/100g of Clay (after carbon correction) and Base Saturation $<$ 50% in B2

The purpose is to set apart among the Latosols outside the cold and always humid zones, that is, from class 1, those with a low content of bases that are well developed as Latosols according to the clay activity.   This distinction substitutes the characteristic formerly used of SiO_2/Al_2O_3 above 1.6.   The clay activity seems better related with the climatic conditions than the silica/aluminium ratio and also is a better indication of the stage of development of a Latosol.   The silica/aluminium ratio will be used only in a lower level in the Latosols in class V.

The Latosols with a high cation exchange capacity content and low base saturation are mainly found in two cases: zonal soils outside the true tropical climate and as intergrades between the Latosols in class V and Acid brown soils. Outside Brazil they possibly include intergrades between Latosols and Andosols.

The soils in this class are characterized, mainly by their relatively high content of exchangeable Al, usually above 4mE/100 g of clay, (without carbon correction). Other characteristics which predominate but are not exclusive to these Latosols, include the dry consistence of the B horizon which is hard with a relatively strong compactness for Latosols, and the usual presence of clay coatings in clay textural profiles. The Erixim soils are characteristic of this class.

III. Other Latosols with a Cation Exchange capacity $\geq$ 6.5 mE/100 g of Clay (after carbon correction) and Base Saturation in $B_2 \geq 50\%$

The purpose is to set apart among the Latosols occurring in a subhumid zone, those with a medium to high base saturation, except those Latosols with a very low cation exchange capacity of the clay fraction.

Only a very small part of the Latosols studied belong to this class, occurring mostly within the transition to semi-arid zones, under "caatinga" vegetation and probably more frequently in the semi-deciduous forest and perhaps semi-evergreen forest.

In addition to these soils, others may or may not belong in this class III, depending on the criteria adopted for boundaries between the soils with a Latosolic B horizon and soils with a textural B, or between latosolic soils and podzolic soils. Many soils having boundaries near these limits, for example, Terra Roxa Estruturada, considered here as soils with a textural B but elsewhere considered as a latosolic soil, should be included in this class, according to the last criteria.

This class includes some of the soils called Ustox in the 7th Approximation.

Subdivisions of this class of Latosols can be made, based on the iron content, which seems to be well related to colour.

The following classes can be differentiated:

III. 1. Latosols with a high iron content (dusky red colours), mainly originating from basic rocks. Ex Latosol Roxo of Ituverava. SP-37.

III. 2. Latosols with a medium iron content (dark red colours) originating from other rocks.

The MGIV No. 17 profile is an example of this subclass.

IV.   Other Latosols with a weak A horizon and in addition high Base Saturation
or the Base Saturation increases with depth in the lower B and below up to
values above 50%

The purpose is to differentiate the Latosols in semi-arid regions from other
Latosols of more humid zones.   The principal characteristic for this differentiation
is the weak development of the A horizon.   The colours (moist) of the A_1 horizon
observed up to the present in these soils under natural vegetation in the semi-arid
zone are: 10 YR 5/4;   7.5 YR 5/5;   7.5 YR 4/4; 10 YR 5/3.   The carbon contents
are low, although the soils are studied under natural vegetation.

Base saturation is normally high.   In case of medium textures the base saturation is considerably lower in the upper portion of the B horizon, but increases with
depth.

This class is actually tentative since only a few profiles of the Latosols of the
semi-arid zone have been observed.

This class corresponds to the Idox of the 7th Approximation.

V.   Other Latosols with Cation Exchange Capacity $\leq$6.5 mE/100g of Clay (after
carbon correction)

These are, in Brazil, all Latosols not included in class I, II, III and IV.   They
are "normal" Latosols with low clay activity, occupying wide extension on old relief
surfaces in the humid and sub-humid tropical regions. They may, or may not, be
subject to a dry season but neither the base saturation nor the clay activity is intimate
ly related to the climatic conditions under which they occur.   In this case the development of relief and the kind of parent material are more important.   The development
of the A horizon is only partially related to prevailing climate, and presumably, is
due to other conditions not completely known.

Most of the Latosols, in tropical Brazil are in this class, in which are included:
a) soils with low silica/aluminium ratio;   ($\leq$ 7 containing high quantities of free
aluminium oxide (gibbsite) and b) soils with high silica/aluminium ratio containing
almost no free aluminium oxide and frequently containing only small quantities of iron
oxides.

These latter are tropical kaolinites that, however, probably due to well crystal-
lized forms and high content of Kaolin have a very low clay activity and are especially
found in the older marine, fluvial or lake beds, and are, for example, of wide extent
in the Amazon basin.

Below we find an explanation of the subdivision of this class which has been used to establish units in soil survey legends in Brazil.

V.1 Latosols with a thick strongly developed A horizon (if of clay texture with more than 1% of C, up to 1 m deep), which have low base saturation and C/N ratio generally above 12.

V.11 With colours in the B2 horizon redder than 5 YR and values of 4 or less. The soils in subclass V.11 are related with schist rocks such as phillites, slates and various mica schists.
Ex. Furnas Profile No. 10 of Campos Gerais.

V.12 With colours in the B2 horizon of 5 YR or yellower, or with values above 4, derived from acid igneous and alkaline rocks.
Ex. SP profile No. 67 of Brangaca Paulista.

V.13 With B2 colours of 5 YR or yellower related to sediments of plateau type.

Ex. Humic Red Yellow Latosols, "chapada" phase of Northeast Minas Gerais.

V.2. Other Latosols with a high iron content, mainly derived from basic rocks with $Fe_2O_3 > 18\%$ if of clay texture.

V.21 Ortho class, with low exchangeable Al content and colours 2.5 YR or redder.

V.211 Clay texture non-concretionary.
Ex. Profile 33 – SP.

V.212 Clay texture concretionary. No profile known up to present. Perhaps will be important in Minas Gerais.

V.213 Medium texture. If there are soils of this type, they have not yet been observed.

V.22 Class with relatively higher content of exchangeable Al that constitutes intergrades to Latosols of class II.

V.23 Other Latosols with colours yellower than 2.5 YR.
Ex. A profile from south of Bahia. These soils have some relation to Latosols of Class I, which in parts, are constituted of yellow Latosols that may have high iron content. These yellow colours may in both cases be due to the humid conditions in the regions where these soils occur.

251

V.3   Other Latosols with a medium iron content, with $Fe_2O_3 > 8\%$ and $< 18\%$ if of clay texture and if medium texture $\%Al_2O_3 /\% Fe_2O_3 > 2$ or molecular ratio of $Al_2O_3/Fe_2O_3 < 3.14$.

V.31   Ortho class with dark red colours, having in the B2 horizon values of 4 or 5 and high chromas.

V.311   Clay texture non-concretionary
Ex. profile SP 40.

V.312   Clay texture concretionary.   No profile observed yet. Possibility of some soils of Central Brazil belonging to this class.

V.313   Medium texture
Ex. Profile SP 46.

V.32   Other Latosols with colours ; yellower   than 5 YR.   This class remains undivided until further knowledge and analytical data are obtained.

V.4   Other Latosols with a low iron content with red to yellow colours, of value 4 or higher in the B2, with $Fe_2O_3 < 9\%$ if of clay texture and of medium texture, $\% Al_2O_3/Fe_2O_3 > 2$ or molecular ratio of $Al_2O_3/Fe_2O_3 > 3.14$.
It includes a great part of the undifferentiated Red Yellow Latosols recently identified.

V.41   Ortho class, with an A horizon moderately developed.

V.411   Related to sediments of coastal plains plateau type and derived from sediments of low fertility having predominantly yellow colours with silica/aluminium; > 1.5 and iron content comparatively lower than other soils within the Latosols of the V.4 class.

V.4111   Heavy clay texture.   Profile SBCS 3.

V.4112   Clay texture
Ex. Campos RJ.

V.4113   Medium texture.

V.412   From the Northeast "chapadas" and "Planalto de Conquista" of Bahia derived from transported material, low in fertility, having predominantly yellow colours, with silica/aluminium ≥ 1.5 and normally with an iron content equal or comparatively lower than the other Latosols of V.4.

252

V. 413 Other soils derived from pseudo-autochthonous materials generally more fertile, having yellow or reddish colours with silica/aluminium normally lower and usually iron contents comparatively higher than the other Latosols of V. 4.

V. 4131 Clay texture
Ex. Red Yellow Latosol of SP.

V. 4132 Medium texture
Ex. Red Yellow Latosol, sandy phase.

V. 42 Classes with an anthropogenic A horizon.
In this case the cation exchange capacity values and the base saturation values must be taken from the lower part of the B horizon, in the zone free from the influences of the anthropogenic part super-imposed. The soils included are the Terra Preta de Indio de Santarém. The soils known to belong to this class, actually are all related to the sediments of the coastal plains of the Amazon Region.

Soils with a textural B horizon with one or more of the following characteristics

1) Cation exchange capacity $\leq$ 24 mE/100 g of clay (after carbon correction);

2) Al^{+++} saturation $>$ 50% or base saturation $<$ 35% in B horizon;

3) Non-consolidated plinthite, existing as a continuous formation in or immediately below the textural B horizon.

1. Soils with $Al^{+++} >$ 50% or with base saturation $\leq$ 35% in B horizon

1.1 Soils with an A horizon very strongly developed.
Ex. Rubrozem of Curitiba.

1.2 Soils with an A horizon weakly, moderately or strongly developed but not very strongly developed.

1.21 Ortho class, with profiles well differentiated and cation exchange capacity value low to high, but medium in most cases, absence of non-consolidated plinthite existing as a continuous formation above 1.25 m.

1.211 Derived from basic rocks
Ex. Profile RS 2 Oasis.

1.212 Derived from granite and gneisses, that is, acid crystalline rocks.

253

          1.2121 With much gravel or stones in A horizon
                 Ex. SP 15.

          1.2122 Absence or with few gravel or stones in A horizon
                 Ex. RJ 13.

       1.213 Derived from argillitic-shales, in most cases with high Al
             content in B horizon and with cation exchange capacity value
             generally high to low.

             Ex. Red Yellow Podzolic, Piracicaba variation.

       1.214 Derived from sandstone
             Ex. SP 6, 7, and 8.

       1.215 Related to sediments of plateau type.   Profile of this soil has
             not been observed, but probably occurs in the coastal zone of
             the Northeast and Bahia.

1.22 Intergrades to Latosols of Class II with profiles not well differentiated,
      cation exchange capacity value somewhat low (after carbon correction),
      below 12 mE/100 g of clay and in addition absence of non-consolidated
      plinthite existing as a continuous formation above 1.25 m.

       1.221 Derived from or related to basic rocks.
             Ex. RS 5.

1.23 Intergrades to Latosols from Class V with profiles not well differentiated,
      with low or very low cation exchange capacity value (after carbon
      correction) and absence of non-consolidated plinthite existing as a con-
      tinuous formation above 1.25 m.

       1.231 Derived from acid crystalline rocks.
             Ex. SP 65.

       1.232 Derived from sandstone
             Ex. SP 9.

       1.233 Related to sediments of plateau type.
             Profiles have not been observed.

1.24 Intergrades to Groundwater laterite (from class of Hydromorphic soils),
      with non-consolidated plinthite existing as a continuous formation above
      1.25 m.
      Ex. RJ 16.

2. **Soils with medium to high Base Saturation in the B horizon   35%**

2.1 Soils with an A horizon very strongly developed.
Includes soils that have a very dark colour in A horizon or when
the colour is dark but not very dark, extends more deeply into
the profile.
Ex. RS 9.

2.2 Soils with an A horizon moderately to strongly developed, but not
weakly or very strongly developed.

2.21 Ortho class, with well differentiated profiles, not very deep
and absence of non-consolidated plinthite, existing as a continuous formation above 1.25 m.

2.211 Derived from basic rocks and gneisses of basic and
intermediary character with the respective limestones,
if any, interlayered.
Ex. Red Yellow Mediterranean RJ.

2.212 Derived from limestone.
Ex. Red Yellow Mediterranean SP.

2.213 Derived from acid crystalline rocks.

2.2131 With many gravel or stones in A horizon.
Ex. Podzolized soils with gravel SP.

2.2132 Absence or with small amount of gravel in
A horizon.

Ex. PVA SP. Profile No. 1

2.214 Derived from argillitic shales.

2.215 Derived from sandstone
Ex. Podzolized soils, Marilia variation.

2.216 Sediments related to plateaux and plains. Up to now,
only recognized soils included in this class occurring
in the transition area from sub-humid zone to semi-
arid, most profiles with plinthite, but below 1.25 m.
Ex. MGIV no 3 Pedra Azul.

2.22 Intergrades to Latosols of Class III and V, with profiles not
well differentiated and absence of non-consolidated plinthite,
existing as a continuous formation above 1.25 m.

255

2.221   Derived from basic rocks and gneisses of basic and
intermediary character with the respective limestone,
if any, interlayered.

Ex. Terra Roxa Estruturada. BP Profile 31.

2.222   Derived from limestone.   Up to now no profiles of
this class have been observed.

2.223   Derived from acid crystalline rocks.
Ex. PVA-LVA.

2.224   Derived from sandstone
Ex. Podzolized soils, lime variation SP.

2.225   Related to sediments of plateau type.   Up to now, no
profiles of this type have been observed, but the classification remains tentatively established for further confirmation.

2.23   Intergrades to Ground water laterite soils (Hydromorphic soil class)
with non-consolidated plinthite existing as a continuous formation
above 1.25 m.

2.3   Soil with weakly developed A horizon with high base saturation, when
clayey with base saturation $>$ 50% throughout the profile, always high in
C and D horizons (tentatively base saturation $>$80%).   The purpose is
to set apart in this class   soils that are related to conditions of semi-
arid environment.   The vesicular character of A must be studied in more
detail.

2.31   Ortho class, with profiles well differentiated, not very deep and
absence of non-consolidated plinthite, existing as a continuous
formation above 1.25 m.
Ex. Aracuai.

2.32   Intergrades to Latosols of Class IV with profiles not very well
differentiated and absence of non-consolidated plinthite, existing
as a continuous formation above 1.25 m.   Up to now the only soils
observed are related to medium textured sediments.

2.33   Intergrades to Groundwater Laterite(Class of Hydromorphic soils)
with non-consolidated plinthite existing as a continuous formation
above 1.25m.
Ex. Some soils of North eastern Brazil.

3. Soils with t extural B horizon with cation exchange capacity$>$24 mE/100 g of clay (after carbon correction) having base saturation $>$ 35% or Al +++ saturation $\leq$ 50% in B horizon and absence of non-consolidated plinthite existing as a continuous formation in or immediately under the textural B.

The classification of this group of soils is still incomplete because, up to now, only a small portion of them have been studied in Brazil.

1. Soils with Chernozemic horizon

    (A horizon well developed, with base saturation $>$ 50%, C/N ratio $<$13 etc., Mollic epipedon of 7th Approximation).

    1.1 Soils with B horizon Dark red or Dusky red.

        1.11 Ortho class, with textural B horizon well developed (not lithosolic).

            1.111 Derived from basic rocks and gneisses of basic and intermediary character, with respective limestones, if any, interlayered.
            Ex. Ciríaco.

        1.12 Intergrades to Lithosols

            1.121 Derived from basic rocks
            Ex. Charrua.

    1.2 Soils with colours in B horizon more yellow than 2.5 YR and chromas in general of 4 or less; frequently with slickensides in the lower part of profile.

        1.21 Ortho class, with textural B horizon well developed (non-Lithosolic).

            1.211 Derived from gneisses of basic and intermediary character with respective limestones, if any, interlayed.
            Ex. Jordânia.

        1.22 Intergrades to Lithosols

            1.221 Derived from gneisses of basic and intermediary character.
            Ex. Itapé, Bahia.

        1.23 Intergrades to Grumusols, with black colours in upper part of B horizon.

            Ex. RS 11.

2. <u>Soils with a prominent A horizon but not Chernozemic.</u>

   (A horizon with base saturation $<$ 50%).

   The elements available are still insufficient to establish subdivision of this class which will continue to be undivided at this moment.
   Ex. Santa Maria RSI.

3. <u>Soils with a moderately developed A horizon</u>

   (With base saturation $<$ 100% in C and D).

   3.1 Soil with colours of B horizon reddish and mottled.
        Ex. Reddish Brown Podzolic.

   3.2 Soil with colours of B horizon yellowish, near to 7.5 YR with chromas higher than 4.
        Ex. Cepec.

4. <u>Soils with weakly developed A horizon, massive or near massive and very hard when dry, with percent base saturation = 100% in C and D horizons</u> (they are soils of very dry regions)

   Present data available are not sufficient to make a subdivision, but probably can be subdivided into:

   4.1 Ortho class, with textural B horizon well developed.

   4.2 Intergrades to Lithosols.

   4.3 Intermediary to Solonetz.

   4.4 Intermediary to Grumusols.

19

Reprinted from pages 171–173, and 196–201 of *Systeem van bodemclassificatie voor Nederland: De hogere niveaus*, Center for Agricultural Publications and Documentation, Wageningen, The Netherlands, 1966, 217 p.

Summary of 'System of soil classification for the Netherlands. The higher levels', especially of the chapters 4 and 6

H. de Bakker and J. Schelling

Foreword

The system of soil classification for the Netherlands is the result of team-work by many members of the Netherlands Soil Survey Institute during the past decade.

1. The development of soil classification

Some parts of the history are sketched in a short survey: the first Dutch soil scientist (Dr. W. C. H. Staring, 1808–1877), the physiographical school of Edelman, the foreign development especially in the United States, and their influence on the establishment of the present Dutch system.

2. The establishment of the system

A number of the specific problems occurring in the establishment of a system of soil classification are elucidated and a general survey is presented.

3. Pedogenic background

The pedogenic background of the system is explained in a descriptive manner. This chapter, however, relates to Dutch conditions only and tries to explain Dutch phenomena. Of several divergences from generally accepted concepts we summarize two instances. 1. Although podzols and humus podzols in particular should have A2 horizons, visible A2 horizons are lacking in many Dutch hydropodzol soils. 2. Dark A1 horizons are much in evidence, although they should not occur under our climatic conditions (vide: mineral earthy layer and earth soils).

4. The differentiating criteria of the system

Kinds of soil

Classification according to texture
Size limits and names of separates:
below 0.002 mm lutum separate, abbreviated lutum; this term was proposed by Mohr in 1910 and introduced in the Netherlands by Zuur. The term lutum is now preferred to clay which is also a textural class.
0.002–0.05 mm silt separate; used untranslated in Dutch.
<0.05 mm loam separate; traditionally used for the fine fraction in lutum-poor sediments, e.g. loess.
0.05–2 mm sand separate
>2 mm gravel separate.
A classification still in use in the Netherlands (not in the Soil Survey Institute), has only two boundaries 0.016 and 2 mm, below 0.016 mm the separate is called 'slib', 0.016–2 mm: sand separate and over 2 mm: gravel separate.

The sand separate is divided in fine and coarse sand separate, and these two are subdivided as follows:

0.05 –0.105 mm extremely)
0.105–0.15 mm very   } fine sand separate
0.15 –0.21 mm moderately)
0.21 –0.42 mm moderately }
0.42 –2   mm very   } coarse sand separate.

In the Netherlands there are hardly any soils formed on consolidated rock. Roughly half of the mineral soils are formed in alluvial sediments, mostly marine clay and to a lesser extent river clay. The other half of the mineral soils are derived from aeolean sediments. The latter are mostly loam-poor and slightly loamy cover sands; a small part consists of the transitional sediments between cover sand and loess, viz. the loamy sands and sandy loams, whereas most of the loess comes under the silty loam class. Only boulder clay and other heavy textured pleistocene sediments, which form the subsoil in a minor part of the country, are scattered throughout the triangle.

Most soils clearly belong to two different universa, and as a consequence there are two classification triangles, one for aeolean and one for the other sediments, with an overlap in the sand corner (Fig. 1 and 2).

Approximate translation of the terms:
Nomenclature of the aeolean classes

leemarm zand   – loam-poor sand
zwak lemig zand   – slightly loamy sand
sterk lemig zand   – very loamy sand
zeer sterk lemig zand – extremely loamy sand
zandige leem   – sandy loam
siltige leem   – silty loam
kleiige leem   – clayey loam

Nomenclature of the nonaeolean classes

kleiarm zand   – clay-poor sand
kleiig zand   – clayey sand
silt   – silt
zeer lichte zavel   – very light 'zavel'
matig lichte zavel   – moderately light 'zavel'
zware zavel   – heavy 'zavel'
lichte klei   – light clay
matig zware klei   – moderately heavy clay
zeer zware klei   – very heavy clay

A few samples fall outside the field marked in grey. Samples left of this field may be termed 'sandy', and those to the right 'silty'.

Soil samples coming under the different subclasses of the sand class may be subdivided according to the coarseness of the sand separate. This is done with help of the median of this separate (M50), viz. the diameter below and above which half the weight of the sand separate falls. The names are the

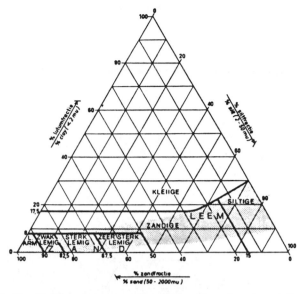

Fig. 1. Classification and nomenclature of aeolian deposits, sand as well as finer textured material ('loam-triangle'). The field marked in grey comprises the majority of samples from the cover sand and loess areas. For approximate translation of terms see page 4.

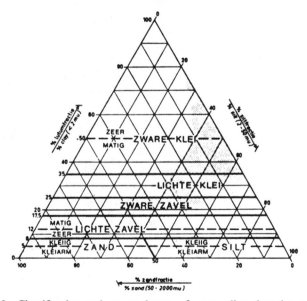

Fig. 2. Classification and nomenclature of nonaeolian deposits, sand as well as finer textured material ('clay-triangle'). The field marked in grey comprises the majority of samples from the river and marine clay areas. For approximate translation of terms see page 4.

261

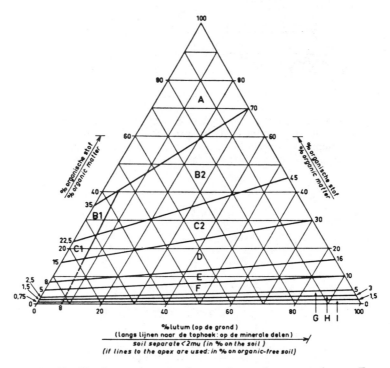

Fig. 3. Classification and nomenclature of organic matter classes. For terms see page 7.

same as those of the above mentioned subclasses of the sand separates, but of course without the addition 'separate'. For instance, many cover sands come under the class of slightly loamy moderately fine sand.

The coarseness of the sand fraction of heavier samples is defined in the same way, but of course used as an adjective. For instance, many young sea clay deposits come under the textural class: extremely fine sandy light 'zavel'.

Classification according to organic matter content

Agronomists always have given much attention to the organic matter content of soils, especially to that of the tilth. The soils of the Netherlands also differ much in organic matter content. For these two reasons this factor plays a fairly important part in this system of soil classification.

The boundaries between the organic matter content classes are not ordinary independent percentages, but are dependent on the lutum content. Experience in the field has shown that when samples have the same amount of organic matter and a different content of lutum, the heavier (more clayey) samples are less humic than the lighter (less clayey) samples.

This trend is not yet fully understood, but it is assumed that samples lying on the sloping lines of the diagram (Fig. 3) have the same or nearly the same organic matter content by volume.

Nomenclature of organic matter classes

I	extremely		
H	very	poor in humus	
G	moderately		1)
F	moderately	humose	
E	very		
D	rich in humus		

C1	peaty sand	
C2	peaty clay	
B1	sandy peat	2)
B2	clayey peat	
A	peat	

1) subdivided into textural classes according to Fig. 1 and 2.
2) not further subdivided into textural classes.

Horizon nomenclature

This is a slight modification of the USDA system of 1951. There is a deviation, for instance, for peat soils: a moulded top layer (in the nomenclature of this system a peaty earthy layer) is an A1 horizon; the oxidized subsurface soil a C horizon and the nonaerated subsoil a G horizon. The subscript *an* is especially used for gradually raised A1 horizons ('plaggent soils; in this system 'enk' earth soils); the subscript *v* is used to denominae' the illuvial horizon in peat soils (peaty B horizon).

[*Editor's Note:* Material has been omitted at this point.]

263

Outline of the soil classification system for the Netherlands. The higher levels

Order	*Suborder*
1 Peat soils	1.1 Earthy peat soils
	1.2 Raw peat soils
2 Podzol soils	2.1 Moder podzol soils
	2.2 Hydropodzol soils*

* Suborders 2.2 and 2.3 together are humuspodzol soils.

Group	Subgroup
1.1.1 Clayey earthy peat soils	1.1.1.1 'Aar' peat soils 1.1.1.2 'Koop' peat soils
1.1.2 Podzolic earthy peat soils	1.1.2.1 'Bouwte' peat soils
1.1.3 Clay-poor earthy peat soils	1.1.3.1 'Bo' peat soils 1.1.3.2 'Made' peat soils
1.2.1 Initial raw peat soils	1.2.1.1 'Vliet' peat soils
1.2.2 Podzolic raw peat soils	1.2.2.1 'Mond' peat soils
1.2.3 Ordinary raw peat soils	1.2.3.1 'Weide' peat soils 1.2.3.2 'Waard' peat soils 1.2.3.3 'Meer' peat soils 1.2.3.4 'Vlier' peat soils
2.1.1 Moder podzol soils	2.1.1.1 'Holt' podzol soils with a sand cover 2.1.1.2 'Loo' podzol soils 2.1.1.3 'Hoek' podzol soils 2.1.1.4 'Horst' podzol soils 2.1.1.5 'Holt' podzol soils
2.2.1 Peaty podzol soils	2.2.1.1 'Moer' podzol soils with a clay cover 2.2.1.2 'Moer' podzol soils with a sand cover 2.2.1.3 'Dam' podzol soils 2.2.1.4 'Moer' podzol soils
2.2.2 Ordinary hydropodzol soils	2.2.2.1 'Veld' podzol soils with a clay cover 2.2.2.2 'Veld' podzol soils with a sand cover 2.2.2.3 'Haar' podzol soils 2.2.2.4 'Veld' podzol soils

Order	Suborder
	2.3 Xeropodzol soils*
3 Brick soils	3.1 Hydrobrick soils
	3.2 Xerobrick soils
4 Earth soils	4.1 Thick earth soils
	4.2 Hydroearth soils

* Suborders 2.2 and 2.3 together are humuspodzol soils.

Group	Subgroup
2.3.1 Xeropodzol soils	2.3.1.1 'Haar' podzol soils with a sand cover 2.3.1.2 'Kamp' podzol soils 2.3.1.3 'Heuvel' podzol soils 2.3.1.4 'Haar' podzol soils
3.1.1 Hydrobrick soils	3.1.1.1 'Beemd' brick soils 3.1.1.2 'Kuil' brick soils
3.2.1 Xerobrick soils	3.2.1.1 'Berg' brick soils 3.2.1.2 'Del' brick soils 3.2.1.3 'Rooi' brick soils 3.2.1.4 'Daal' brick soils 3.2.1.5 'Rade' brick soils
4.1.1 'Enk' earth soils	4.1.1.1 Brown 'enk' earth soils 4.1.1.2 Black 'enk' earth soils
4.1.2 'Tuin' earth soils	4.1.2.1 'Tuin' earth soils
4.2.1 Peaty earth soils	4.2.1.1 'Plas' earth soils 4.2.1.2 'Broek' earth soils
4.2.2 Sandy hydroearth soils	4.2.2.1 Brown 'beek' earth soils 4.2.2.2 'Goor' earth soils 4.2.2.3 Black 'beek' earth soils
4.2.3 Clayey hydroearth soils	4.2.3.1 'Lied' earth soils 4.2.3.2 'Tocht' earth soils 4.2.3.3 'Woud' earth soils 4.2.3.4 'Leek' earth soils

Order	Suborder
	4.3 Xeroearth soils
5 Vague soils	5.1 Initial vague soils
	5.2 Hydrovague soils
	5.3 Xerovague soils

Group	Subgroup
4.3.1 'Krijt' earth soils	4.3.1.1 'Krijt' earth soils
4.3.2 Sandy xeroearth soils	4.3.2.1 'Akker' earth soils 4.3.2.2 'Kant' earth soils
4.3.3 Claycy xeroearth soils	4.3.3.1 'Hof' earth soils
5.1.1 Initial vague soils	5.1.1.1 'Gors' vague soils 5.1.1.2 'Slik' vague soils
5.2.1 Sandy hydrovague soils	5.2.1.1 'Vlak' vague soils
5.2.2 Clayey hydrovague soils	5.2.2.1 'Drecht' vague soils 5.2.2.2 'Nes' vague soils 5.2.2.3 'Polder' vague soils
5.3.1 Sandy xerovague soils	5.3.1.1 'Duin' vague soils 5.3.1.2 'Vorst' vague soils
5.3.2 Clayey xerovague soils	5.3.2.1 'Ooi' vague soils

MAJOR SOIL GROUPS IN JAPAN

Yasuaki MATSUZAKA

National Institute of Agricultural Sciences,

Tokyo, Japan

BASIC SOIL CLASSIFICATION

Japan is situated between latitude 24° to 46° N and longitude 123° to 146° E and has a total area of about 38 million ha.   Approximately 1/7 of the total area is utilized as agricultural land. A soil survey for the agricultural land was initiated in the 1950's using modern and scientific techniques. Up to 1975, the survey of about five million ha of agricultural land, which corresponds to approximately 85 percent of the whole agricultural area, was completed and soil maps scaled 1/50,000 have already been published.

In the survey, soil series were taken as the basic unit of classification and mapping, and 309 soil series have so far been established. The following criteria were set for the identification of a soil series: (A) Profile characteristics; i.e. humus content, soil color, depth and thickness of gravel layer and/or consolidated bedrock, depth of clay pan, presence of mottles and concretions, texture, presence of a well developed structure, presence and depth of peat or muck horizon and gleyed horizon; (B) Soil acidity; e.g. If the pH(KCl) of the subsurface soil is less than 4.2 or if Y_1 is more than 10, the soil is designated as "strongly acidic" ; (C) Mode of sedimentation; and (D) Parent materials of the soil.

Soil series having the same diagnostic horizon(s) and a similar genetic process were grouped into "Soil Group" The list of Soil Groups currently recognized and their correlations with Great

Groups of New Taxonomy of USDA and the Soil Units of the world soil map of FAO/UNESCO are shown in Table 1.

Table 1. Correlation of Soil Groups in Japan with the Great Groups of New Taxonomy and the Soil Unit of World Soil Map

Soil Group	No. of Soil Series	Great Groups of N.T.	Soil Units of W.S.M.
01 Lithosols	2	Udorthents	Dystric (Eutric) Lithosols
02 Sand-dune Regosols	1	Udipsamments	Fluvisols
03 Andosols	61	Dystrandepts (Eutrandepts, Durandepts, Vitrandepts)	Humic (Mollic, Ochric, Vitric) Andosols
04 Wet Andosols	48	Aquic Dystrandepts	Gleyic Andosols
05 Gleyed Andosols	15	Andaquepts	Gleyic Andosols
06 Brown Forest Soils	23	Dystrochrepts	Dystric (Eutric) Cambisols
07 Gray Upland Soils	13	Haplaquepts	Dystric (Eutric) Gleysols
08 Gley Upland Soils	11	Haplaquepts	Dystric (Eutric) Gleysols
09 Red Soils	7	Hapludults	Orthic Acrisols
10 Yellow Soils	23	Hapludults, (Aquic Hapludults)	Ortic (Gleyic) Acrisols
11 Dark Red Soils	2	Rhodudalfs	(Rhodic Luvisols)
12 Brown Lowland Soils	17	Udifluvents, Haplaquepts	Eutric (Dystric) Fluvisols, Eutric (Dystric) Gleysols
13 Gray Lowland Soils	37	Haplaquepts	Eutric (Dystric) Gleysols
14 Gley Soils	35	Haplaquents, Haplaquepts	Eutric (Dystric) Gleysols
15 Muck Soils	10	Saprists	Dystric (Eutric) Histosols
16 Peat Soils	5	Fibrists	Dystric (Eutric) Histosols

PRODUCTION CAPABILITY CLASSIFICATION

Production capability classification is one of the interpretative classifications based on basic soil classifications, and it can indicate the kinds and the extent of limitations that impede the crop production. In this classification, the soils are grouped into four classes on the basis of the presence of restrictive or impeding factors against normal crop production. Each class is defined as follows:

Class I: Soils that have no or only a few limiting factors or hazards for crop production and/or risks of soil damage, and are regarded as either being naturally fertile or having high potentiality for crop production without any ameliorative practices.

Class II: Soils that have some limiting factors or hazards and/or risks of soil damage, and require some ameliorative practices to achieve good production.

Class III: Soils that have many limitations or hazards and/or risks of soil damage, and require fairly intensive ameliorative practices to achieve good production.

Class IV: Soils that have more limitations than those in Class III, but can be utilized for production of some crops under very intensive ameliorative practices.

The inherent soil characters (independent factors) which are used for decision of production capability classes are as follows:

For Paddy Rice	*For Upland Crops*
Thickness of top soil (t)	Thickness of top soil (t)
Effective depth of soil (d)	Effective depth of soil (d)
Gravel content of top soil (g)	Gravel content of top soil (g)
Permeability (1)	Ease of plowing (p)
State of redox potential (r)	Wetness of land (w)
Inherent fertility (f)	Inherent fertility (f)
Content of available nutrient (n)	Content of available nutrients (n)
Presence of harmful substances (h)	Presence of harmful substances (h)
Frequency of disaster (a)	Frequency of disaster (a)

These independent factors are graded as I, II, III and IV by taking into consideration supplementary soil properties (dependent factors) each of which is also ranked into four grades. The production capability class of the soils is decided by the lowest class value of the enumerated independent factors. Table 2 shows area of each Soil Group and the percentage distribution among the production capability classes.

DETAILED EXPOSITION OF EACH SOIL GROUP

The Soil Groups found in Japan and their main utility and the methods of their amendment are briefly discussed here.

1. Lithosols

Lithosols are found distributed on the gentle to steep slopes in mountainous or hilly regions. The surface horizon is shallower than 30 cm. and covers a gravel layer and/or consolidated rock. These soils are partly used as common upland crop fields or as orchards, but usually productivity is low.

2. Sand-dune Regosols

Sand-dune Regosols are coarse textured and immatured soils mainly appearing on sand-dune and sand-bars along sea coasts. They are partly used for windbreak forests or fields for common upland crops or flowers. The productivity is generally low because of low inherent fertility. Protection from wind erosion and application of organic manures are required to increase productivity.

3. Andosols

The parent materials of Andosols are wholly or mostly derived from volcanic ejecta, which are widely distributed throughout Japan. The area covered by Andosols in Japan is estimated to be about 5.5 - 6.0 million ha which correspond approximately to 1/6 of total area of the country. They are characterized by the soft, black to blackish brown and humus-rich surface (and subsurface) horizon(s). Other physical and chemical properties of Andosols include: rich in hygroscopic water, low bulk density, high carbon-nitrogen ratio, high phosphorus absorption capacity, high cation exchange capacity and low silica-alumina ratio. These properties are mostly derived from their peculiar clay mineralogical composition in which allophane is dominant.

According to thickness of humic horizons and content of humus, Andosols are divided

272

Table 2. Area of Soil Groups and Percentage of Production Capability Class

Soil Group	Paddy field					Upland crop field & orchard				
	Class (%)				Total	Class (%)				Total
	I	II	III	IV	area (ha)	I	II	III	IV	area (ha)
01 Lithosols	-	-	-	-	0	0	2.7	56.7	40.6	14,935
02 Sand-dune Regosols	-	-	-	-	0	0	1.2	44.3	54.5	23,779
03 Andosols	0	23.0	75.7	1.3	17,314	0	35.3	63.0	1.7	986,456
04 Wet Andosols	0.1	59.6	39.0	1.3	278,584	0	14.1	84.0	1.9	98,733
05 Gleyed Andosols	0	51.8	48.2	0	43,408	-	-	-	-	1,602
06 Brown Forest Soils	-	-	-	-	5,353	0.2	33.2	54.1	12.5	406,256
07 Gray Upland Soils	0	53.5	44.5	2.0	79,183	0	17.3	80.8	1.8	47,051
08 Gley Upland Soils	0	33.2	66.6	0.3	39,561	-	-	-	-	3,213
09 Red Soils	-	-	-	-	434	0	14.4	79.2	6.4	35,171
10 Yellow Soils	0	70.2	29.3	0.5	148,125	0.1	27.1	62.5	10.3	179,307
11 Dark Red Soils	-	-	-	-	24	0	8.4	90.8	0.8	15,147
12 Brown Lowland Soils	0.4	72.8	26.5	0.3	145,135	0.1	56.1	40.5	3.3	228,246
13 Gray Lowland Soils	0.6	72.5	26.5	0.4	1,061,233	1.1	46.0	49.8	3.2	76,392
14 Gley Soils	0	47.4	52.5	0.1	882,376	0	18.2	63.5	18.3	21,069
15 Muck Soils	0	57.4	42.6	0	73,692	-	-	-	-	1,690
16 Peat Soils	0	47.3	52.7	0	113,144	0	1.5	33.8	64.6	27.859
Total	0.3	60.4	38.9	0.4	2,887,566	0.2	33.5	59.7	6.5	2,169,174

into the following 5 sub-groups: (1) Thick high humic Andosols: having a thick (more than 50 cm) surface horizon containing more than 10 percent humus, (2) Thick humic Andosols: having thick (more than 50 cm) surface horizon containing 5 to 10 percent humus, (3) High humic Andosols: having a surface horizon of less than 50 cm in thickness containing more than 10 percent humus, (4) Humic Andosols: having a surface horizon of less than 50 cm in thickness containing 5 to 10 percent humus, and, (5) Light colored Andosols: having a light colored surface horizon containing less than 5 percent humus or having a shallow humic surface horizon. Light colored Andosols are thought as immatured Andosols or to be truncated soils by surface erosion.

Andosols are found on the diluvial upland or at the foot of volcanos. These areas are used as forest lands, grass lands, common upland crop fields or orchards. Productivity of Andosols is generally low because of their particular properties such as a high phosphorus absorption capacity and low bulk density. However, Andosols' productivity can be increased by such techniques as a heavy dressing of phosphatic fertilizers, supply of bases including minor elements and application of organic manures.

4. Wet Andosols

Wet Andosols are the hydromorphic Andosols. Wet Andosols are poorly drained and have prominent iron mottles that have developed under the conditions of alternating reduction and oxidation. These soils are mainly distributed on alluvial lowlands, valley bottoms or depressions in mountainous or hilly areas. They are further divided into 5 sub-groups in a similar way as Andosols (3). They are mainly used as paddy field. The productivity is usually low. However, with certain soil amendment such as proper drainage and fertilization practices including the heavy application of phosphatic fertilizers, productivity may be raised to the same level as other lowland soils.

5. Gleyed Andosols

Gleyed Andosols are also hydromorphic Andosols which have been developed under perma-. nent or nearly permanent water-saturated conditions and are characterized by a gleyed horizon. $\alpha\alpha'$-dipyridyl is a very useful reagent for the detection of gleyzation. Gleyed Andosols are mainly distributed on valley bottoms and are used as paddy fields. Generally their productivity is low, but the practices such as artificial drainage and heavy application of phosphatic fertilizers may serve to increase productivity.

6. Brown Forest Soils

Brown Forest Soils have blackish brown to dark brown surface horizons and yellowish brown subsurface horizons. The humus content in the surface is generally less than 5 percent. In these soils there is little evidence of downward movement of the sesquioxides and clays. Therefore, their subsurface horizon can be recognized as color B horizon. The reaction of these soils is sometimes strongly acidic. They are widely distributed on the slopes of mountains or at the foot of hills, and sometimes on flat to undulated uplands. The greater part of them is used as forest lands while rather limited areas are used as commom upland crop fields or orchards. The productivity is generally low. The following practices are required to increase productivity: liming, supply of bases and minor elements, application of organic manures, and protection from erosion.

7. Gray Upland Soils

Gray Upland Soils have gray or grayish brown surface and/or subsurface horizons with iron mottles and occasionally manganese concretions. These morphologies show that these soils developed under a hydro-genetic process. This process will be discussed in more detail later in the section on Gray Lowland Soils (13). These soils lie scattered on flat or gently sloping uplands which have a relatively high ground water table or else stagnant water. They are used mostly as paddy fields. The productivity is generally medium to high except that the soil has a very heavy or very coarse texture.

8. Gley UPland Soils

Gley Upland Soils have gley horizon(s) which have developed under strongly reductive condtions. These soils are mainly distributed on nearly level diluvial uplands or on mountain slopes and are used as paddy fields. The productivity is generally low because of poor drainage.

9. Red Soils

Red Soils have the diagnostic red colored surface and/or subsurface horizons. The horizon should have the hue of 5YR or even redder, the chroma of above 6 and the value of above 3. Red Soils are mainly distributed on undulating uplands or terraces at elevations of less than 200 m along the seacoast. These soils are assumed to be a kind of paleosols that had been developed under high temperature during the Pleistocene period and have kept the original red color in spite of constant weathering. In these soils, some argillation may produce a textural B horizon, but usually the eluvial horizon can hardly be recognized because of the truncation by erosion or cultivation. Frequently, leaching of bases results in making these soils acidic. These soils are used as forest lands, grass lands, orchards and common upland crop fields. Their productivity is low because of their inherent low fertility. Application of organic manures, liming and supply of bases and minor elements are necessary to increase the productivity.

10. Yellow Soils

The genesis, location, distribution and physical and chemical properties of Yellow Soils are almost the same as Red Soils. Yellow Soils have yellow horizons that have the hue of 7.5YR or yellower, the chroma of over 6 and the value of above 3. Generally Yellow Soils are considered to be younger than Red Soils. Yellow Soils are distributed widely in the whole country and the total area occupies a larger area than Red Soils. They are used mostly as forest lands and partly as orchards and common upland crop fields. Where the irrigation facilities are available, Yellow Soils are also used as paddy fields. In the case of paddy fields, the chroma of soil falls below 6 and iron mottles develop gradually. The productivity of Yellow Soils is rather high in paddy fields, but they are usually low in common upland crop fields. For the upland fields, the practices of fertilization and soil amendment as in the case of Red Soils are necessary.

11. Dark Red Soils

The parent materils of Dark Red Soils are calcareous or ultra-basic rocks such as serpentine and gabbro. The hues of the subsurface horizons are mostly about 5YR; both the chroma and the value are 4 or less. The degree of base saturation is mostly high. Distribution of these soils is extremely scattered throughout Japan. They are used as forest lands, grass lands, orchards and common upland crop fields. The productivity is generally low.

12. Brown Lowland Soils

Brown Lowland Soils are derived from relatively recent alluvial deposits and have yellowish brown surface and/or subsurface horizons which have the hue of 7.5YR or yellower, and both the chroma and value of above 3. Generally, their colors indicate that these soils developed under little or no influence of water and so remain the original colors of alluvial deposits. They are common on riverine or marine alluvial plains, at valley bottoms or alluvial fans. Compared with Gray Lowland Soils or Gley Soils, however, Brown Lowland Soils occur at somewhat higher elevation, which is generally well-drained. The total area occupied by them is very large. It is extensively used as common upland crop fields, orchards and paddy fields. When used as a paddy field, a characteristic change may occur under the influence of irrigation water producing iron mottles and occasionally manganese concretions in the subsurface horizon. The productivity of Brown Lowland Soils is generally medium to high. But the coarse textured soils have the defects of nutrient deficiency caused by relatively rapid leaching.

13. Gray Lowland Soils

Gray Lowland Soils characterized by having gray or grayish brown surface and/or subsurface horizons are one of the representative paddy soils distributed widely on alluvial lowlands. The gray horizon has a neutral color or has the hues of 2.5Y, 5Y or 7.5Y, the chromas of below 3 and the values of above 3. The grayish brown horizon has the hues of 10YR, 7.5YR, 5YR or 2.5YR, the chromas of below 3 and the values of above 3. These horizons are degenerated ones that have developed through the hydrogenetic process, in which percolation of irrigation water or fluctuation of the groundwater tabel predominantly affected the degeneration. Temporary waterlogging during the rice growing season causes the reductive condition of soils and after harvest the soils recover the oxidative condition. Repetition of these reduction and oxidation stages bring on graying of the soil which develop gray or grayish brown horizons with iron mottles and occasionally manganese concretions. In short, it can be said that these horizons hold both characters of eluviation and illuviation.

These soils are distributed very widely and are mainly located on riverine and marine alluvial plains. They are utilized mostly as paddy fields and partly as common upland crop fields. The productivity is usually medium to high. But it should not be forgotten that the graying of the soils results in the loss of mineral nutrients from the plow soles. In fact, not a few "Akiochi (degraded)" rice fields are found in coarse textured Gray Lowland Soils. In the case of "Akiochi" rice fields, intensive soil amendment and fertilization practices such as dressing of young iron-rich soils, and supply of bases and organic manures are necessary for their improvement.

14. Gley Soils

Gley Soils are the typical lowland soils developed under strongly reductive conditions and are characterized by having gley horizons. In gley horizons, most irons are reduced to ferrous iron, the soils showing greenish or bluish colors. Generally, the hue of this horizon is as blue as 10Y or bluer. Gley Soils are divided into 3 sub-groups according to the horizon sequences with special reference to the depth of gley horizons. When there is no air penetration with downward movement of water, no iron rusty mottle develops. In this case the soils are designated as strongly-reduced Gley Soils. When there is some air penetration, tubular-shaped iron mottles develop. Such soils are designated as iron-mottled Gley Soils. Gley Soils are widely distributed on riverine or marine alluvial plains or valley bottom plains and are used mostly as paddy fields. Gley Soils, especially strongly reduced ones, are generally poorly productive soils. In these soils, rice roots are frequently damaged by inhibited respiration. Gley Soils can greatly be improved by artificial drainage. The air supply through drainage will meet the demand for oxygen by root respiration. Continuous and intensive drainage of the Gley Soils will result in converting them gradually into the soil group of the drier categories.

15. Muck Soils

Muck Soils are one type of the organic soils having surface and/or subsurface muck horizons that are composed of mixture of mineral and organic materials. The organic materials are derived from aquatic or semi-aquatic plant residues in which plant fibers are still visible to the naked eye. These horizons are black to blackish in color and contain more than 10 percent of humus. Muck Soils are located on alluvial plains, back marshes or depressions in mountainous and hilly regions which have a high ground water table. They are mostly used as paddy fields. Productivity is generally low. Drainage and application of organic manures are necessary to increase the productivity.

16. Peat Soils

Peat Soils are organic soils having surface and/or subsurface peat horizons. Peat horizon is mainly composed of low, intermediate or high peat that has been produced through partial decomposition of aquatic or semi-aquatic plants. They must contain more than 20 percent of humus. Peat Soils are widely scattered throughout the country and are found in the areas of old marshes and swamps. Low Peat Soils are partly used as paddy fields. The productivity is generally low because of the wetness and the deficiency of mineral nutrients. In Hokkaido, low, intermediate and high Peat Soils are commonly found. Most of them are not cultivated, but sometimes they are utilized as paddy field or grazing land.

21

THE SOIL CLASSIFICATION SYSTEM OF THE FEDERAL REPUBLIC OF GERMANY

E. MÜCKENHAUSEN

Director

Institut für Bodenkunde

Bonn — Germany

PREFACE

In 1952 the author presented a first draft of a classification system of soils of he Federal Republic of Germany during the congress of the German Soil Science Society. This draft was only tentative. The author realized that a soil classification system could only be developed by teamwork. Therefore a commission on soil classification was elected by the German Soil Science Society in 1952, including F. Heinrich, W. Laatsch, E. Mückenhausen and F. Vogel (chairman). The author was authorized continually to revise the draft first presented. The members of the German Soil Science Society were asked to cooperate on the development of the soil classification system. Since 1952 the Commission on Soil Classification met once or twice every year. At each meeting the suggestions sent in by colleagues were discussed. After each meeting a new draft or additions to each of the latest were made by the author. In March 1961 the Commission on Soil Classification decided to publish the latest draft, including for each soil type described its genesis and characteristics.

The soil classification system of the Federal Republic of Germany presented here follows closely the publication « Development, Properties and Classification System of Soils in the Federal Republic of Germany », published in 1962 (E. MÜCKENHAUSEN, 1962). Of course only the principles of this system can be pointed out. It is not possible to describe all the soils of Germany here.

GENERAL PRINCIPLES FOR A SOIL CLASSIFICATION SYSTEM

Today we *no longer* consider soils as natural objects defined by only *some* known static properties. Rather we consider soils to be natural bodies developed according to certain natural laws, subjected to alteration by certain still continuing processes, furthermore having specific characteristics derived from their parent material. The stages of soil evolution thus formed and fixed during a longer period, we would like to group systematically according to their pedogenetic and lithologic properties. This means that the soils are grouped together on the basis of common characteristics in higher categories and subdivided into lower categories according to significant differences. A soil classification system defined this way seems for us to reflect the present stage of knowledge.

It could be objected that all properties of the soils are not yet known and therefore a final classification system can not yet be achieved at present. It is true that we do not yet know all properties of soils, we think however that our knowledge of characteristics, dynamics and genesis of soils is *sufficient* so that a classification system can be worked out.

Obviously every classification system has its shortcomings and needs to be completed and revised. Every classification system reflects the ideas of the time when it was developed. It requires improvement when research — in this case research on soils — advances.

We deliberately want the classification system to be based on soil genesis. It has to be underlined that soil genesis is a process and not a property. Soil genesis produces characteristics which are partly visible (horizons) and partly not (reactions). Furthermore processes in the soil which we call dynamics are due to pedogenetic and lithologic factors. If we consider all these characteristics in the classification system we will have groupings which include soil genesis as well as properties inherited by the parent material. Not always do morphologic characteristics reflect pedogenetic processes. There are soils which have the same or similar morphology but a different genetic background. This case may be rare but it ought to be possible to separate soils with the same or similar morphology but with different genesis in anyone of the categories in the system. This separation seems to be necessary because a different set of genetic processes must have somehow caused different characteristics which we perhaps do not recognize at the moment. Pedogenetic considerations automatically lead to considerations on soil dynamics, which help us to recognize characteristics developed in the soil.

One could object that emphasizing soil genesis too much will lead to speculations and place soil science on an unsafe basis. But there is a great difference between speculation and a working-hypothesis. Field research on soils without a sound hypothesis does not work. What we mean is comparative scientific work that starts with a hypothesis. It permits a comparative way of looking at the soil. If the hypothesis does not prove right during the work, it at least has made the approach easier. We have to remember that strati-graphic geology working in this way for 200 years has been very successful. Without a genetic viewpoint stratigraphic geology would have been very unproductive. This example we ought to take into consideration if we are to deal with the question how far soil genesis should be emphasized in a soil classification system. It is obvious that a soil classification system in a certain area can be made in this way only if the area is known sufficiently and if we have a good general view about the distribution of the soils occuring in it, about their horizon sequence, their dynamics and their genesis. We have this general view in Central Europe. If there is an area in which the soils are not well enough known, it may be necessary to be satisfied with a *simpler* classification until sufficient knowledge is gained for a detailed classification system.

NUMBER, DEVELOPMENT AND DESIGNATION OF THE CATEGORIES IN THE SOIL CLASSIFICATION SYSTEM .

The more recent proposals for a classification system provide 5 to 7 categories (J. J. BASINSKI, 1959). Today there seem to be three possibilities for grouping soils.

1. The soils are put into major groups in the highest category. Starting with the highest category several lower categories are formed.

2. The grouping of soils begins with the lowest category proceed-ing from this basis to higher categories.

3. Starting points are the stages of transformation of the litho-sphere (W. L. KUBIËNA, 1953), i.e. the soil types (in the defini-tion of Russian and many European soil scientists). These are put into the center of the system and grouped into higher categories on the one hand and into lower categories on the other. The smaller the area, the soils of which have to be grouped, the less important are the higher categories. For that reason two categories placed above the soil type seem to be sufficient for the Federal Republic of Germany.

Following KUBIËNA (1953) the highest category is tentatively named « Abteilung » (section), the next lower category « Klasse » (class). Three pedogenetic categories subordinated to the « Typ » (soil type) seem to be required. They are called « Subtyp » (subtype), « Varietät » (variety) and « Subvarietät » (subvariety). The « Form » (form) supplements the pedogenetic categories through *textural* and *lithologic characteristics*.

There is one difficulty : one cannot avoid taking into account the *dominating influence* of texture and parent material, i.e. lithologic characteristics, in higher categories. This is the case e.g. with soil types like Rendzina, Pararendzina and Pelosol. On the other hand it is desirable to group the numerous locally occuring soils, characterized by texture and parent material, in a lower category like soil scientists in the USA and Great Britain do in their « series ». The problem is where to draw the boundary. In which case are lithologic characteristics significant enough to be considered on the level of soil types and where are they unimportant enough to be considered in a lower category — in our system in the « Form » ? This is a real problem, which evidently all soil scientists, concerned with soil classification, cannot solve perfectly. In Germany, especially considering the concept of W. LAATSCH, we decided to take into account lithologic characteristics in a higher category only if by their influence individual soil types like Rendzina, Pararendzina and Pelosol have been developed. In all other cases we consider *lithologic characteristics on the level of « Form »*. This is done the same way, if an abnormal horizon sequence is due to texture, as is the case with the Parabraunerde developed on sand, which has a *B-horizon in bands* (the same soil type with a sandy loam texture has a continuous B-horizon). Today it is still questionable if we will name this soil a « Sand-Parabraunerde » in future.

In his soil classification system W. L. KUBIËNA (1953) has provided the following categories : « Abteilung » (section), « Klasse » (class), « Unterklasse » (subclass), « Typ » (type), « Subtyp » (subtype), « Varietät » (variety), « Subvarietät » (subvariety).

We deviated from this grouping by dropping the « Unterklasse » and adding the « Form » (form). The form itself is not a category of the pedogenetic system. It completes the pedogenetic categories with lithologic characteristics. Following W. L. KUBIËNA (1953) the Commission on Soil Classification of the German Soil Science Society has established the following six pedogenetic categories and the lithologic form, designated by letters, respectively numerals.

Abteilungen (sections):	capital letters, e.g. A	
Klassen (classes):	small letters, e.g. a	
Typen (types):	roman numerals, e.g. I	
Subtypen (subtypes):	roman numerals in parenthesis, e.g. (I)	*Pedogenetic categories*
Varietäten (varieties):	arabic numerals, e.g. 1	
Subvarietäten (subvarieties):	arabic numerals in parenthesis, e.g. (1)	
Formen (forms) :	arabic numerals with asterisk, e.g. 1*	Lithologic completion of the pedogenetic categories

In agreement with many soil scientists in our country and in foreign countries the following characteristics of a soil are significant for its classification.

1. Direction and degree of percolation, that is, the migration of dissolved and colloidal substances as well as other materials capable of migration.
2. Differences in the individual soil profiles (including the organic horizons) as far as they are due to soil genesis.
3. Interstitial structure (percolation system) depending on parent material. It is very important with respect to soil development and moisture regime.
4. Specific soil dynamics as the result of the conditions mentioned under 1 to 3.

These *criteria, inherent to the soil itself,* automatically include the most important physical, chemical and biologic characteristics of the soil. According to these criteria the pedogenetic categories mentioned above are formed as follows.

1. Abteilungen

The « *Abteilungen* » (sections) include soils with the same main direction of percolation. Bog soils are separated as one section by their peculiarity, although this is not justified with respect to percolation.

2. Klassen

The « *Klassen* » (classes) include soils with the same or similar horizon sequence. They may also contain soils where the same specific dynamics are significant enough to justify the peculiar position in the second highest category.

3. Typen

In the « *Typen* » (types) soils having a characteristic horizon sequence and specific properties in single horizons are put together. Following W. LAATSCH and E. SCHLICHTING (1959) they represent the *transformation products of the lithosphere*, caused by specific pedogenetic processes or specific peculiarities of the parent material.

4. Subtypen

The « *Subtypen* » (subtypes) are *qualitative* modifications of the types. They are subdivisions of the type according to *deviations* from the central concept of the type, e.g. if a Braunerde contains calcium carbonate in the A- and B-horizon (=kalkhaltige Braunerde, $CaCO_3$-Braunerde).

In many cases subtypes include soils intergrading between two, or occasionally more, given types. Their names are formed by a combination of the two type-names typical for the soil. The dominating type is put at the end of the name, e.g. a Gley-Podzol is a Podzol with gleying in the subsoil, a Podzol-Gley is a Gley with podzolization in the upper horizons. Moreover there are many transitional formations in Central Europe. The most common are Braunerde intergrading to Parabraunerde and Podzol, moreover Parabraunerde intergrading to Pseudogley and Podzol.

5. Varietäten

The « *Varietäten* » (varieties) are the quantitative modifications of the subtype indicating differences in degree (weakly, moderately, strongly pronounced) of certain pedogenetic characteristics, e.g. « weakly pronounced chernozem », « strongly pronounced Gley-Podzol with weakly pronounced gleying » (in this case « strongly pronounced » refers to Podzol, i.e. the pedogenetic predominating type, « weakly pronounced gleying » refers to Gley, i.e. the pedogenetic subordinated type).

6. Subvarietäten

The « *Subvarietäten* » (subvarieties) include all pedogenetic peculiarities of the varieties in *quality and quantity*. They are the modifications of the varieties, e.g. « weakly pronounced chernozem, A-horizon 27 cm thick », or « strongly pronounced Gley-Podzol with weakly pronounced gleying, 25 cm weakly decomposed organic material ».

On the level of the subvariety, the factor water, among others, is differentiated in detail.

The « *Formen* » (forms; the name is derived from « local form ») can be established with each category, mostly of the categories 3 to 6 (type, subtype, variety, subvariety). They are defined by the lithologic characteristics (texture and parent material) in addition to the pedogenetic characteristics. An example for a « Form » as a subdivision of the variety is : « strongly pronounced Gley-Podzol with weakly pronounced gleying, developed from outwash gravelly sand (Riss) ». An example for a « Form » as a subdivision of the type is : « Braunerde developed from basalt, stony clay loam ».

Pedogenetic and lithologic characteristics cause the immense number of soil individuals (forms) of the pedosphere. The « Form » of the subvariety corresponds roughly to the soil series in the U.S.A. and other countries.

According to these general rules an exemple is given to show how a soil of the Braunerde-class is put into the system :

Abteilung (section) :	A	Terrestrial soils.
Klasse (class) :	a	Braunerden.
Typ (type) :	I	Parabraunerde.
Subtyp (subtype) :	(I)	Parabraunerde, low in base saturation.
Varietät (variety) :	1	Strongly eluviated Parabraunerde, low in base saturation.
Subvarietät (subvariety):	(1)	Strongly eluviated Parabraunerde, low in base saturation, 12 cm thick, weakly decomposed acid organic matter, weak gleying at 80 cm.
Form (form) (as a subdivision of the subvariety) :	1*	Strongly eluviated Parabraunerde, low in base saturation, 12 cm thick, weakly decomposed acid organic matter, weak gleying at 80 cm, fine sandy loam, derived from loess (Würm).

Designating the *form as a subdivision of the type* we would briefly say : Parabraunerde from loess (Würm).

In the publication « Development, Properties and Classification System of Soils in the Federal Republic of Germany » a certain *completeness* has been attempted *only for the higher categories, including the subtype*. Illustrating the categories *variety* and *subvariety* only some examples are given, because it is hardly possible to enumerate the immense number of soils in these categories. Furthermore this attempt would not make much sense, because it is impossible to reach even an approximate completeness. On 230 typewritten pages 47 soil types and 157 subtypes are described, including some examples of varieties, subvarieties and forms.

COMPILATION OF THE HIGHER SYSTEMATIC SOIL CATEGORIES OF THE FEDERAL REPUBLIC OF GERMANY

Following W. L. KUBIËNA (1953) we have grouped the soils of the Federal Republic of Germany into four « Abteilungen » (sections).

A Terrestrial soils

Klassen (classes)	*Typen (types)*
a Terrestrische Rohböden (terrestrial soils with beginning of soil formation)	I Alpiner Rohboden (alpine Rohboden)
	II Arktischer Strukturboden (arctic structure soil — tundra soil)
	III Syrosem (extremely shallow lithosol)
b A-C-Böden (soils with A-C-profile unweathered in the subsoil)	I Ranker
	II Rendzina
	III Pararendzina
c Steppenböden (dry prairie soils)	I Tschernosem (chernozem)
	II Brauner Steppenboden (brown prairie soil)
d Pelosole (highly clayey soils derived from clay-sediments)	I Pelosol
e Braunerden	I (Typische) Braunerde
	II Parabraunerde
	III Fahlerde (pale Parabraunerde with very strong eluviation)
f Podsole (podzols)	I Podsol (podzol)
	II Bändchen-Podsol (podzol with thin iron pan)
g Terrae calcis	I Terra fusca
	II Terra rossa
h Plastosole (palaeosols highly clayey and high in kaolinite)	I Braunlehm
	II Graulehm
	III Rotlehm
i Latosole (latosols)	I Roterde (red latosol)
	II Gelberde (yellow latosol)
k Staunässeböden (soils with occasional wetness due to low permeability)	I Pseudogley
	II Stagnogley (Pseudogley wet over a long period)
l Terrestrische anthropogene Böden (terrestrial man made soils)	I Plaggenesch (plaggen soil)
	II Hortisol (old garden soil)
	III Rigosol (very deep mixed soil, e.g. in vineyards)

B Semiterrestrial soils

Klassen (classes)

a Auenböden (soils in the valleys with a strongly fluctuating water table)

b Gleye

c Marschen (marshes)

C Subhydric soils

D Bog soils ·

Typen (types)

I Rambla (Auenboden with beginning soil formation)

II Paternia (young Auenboden)

III Borowina (rendzinalike Auenboden)

IV Tschernosemartiger Auenboden (chernozemlike Auenboden)

V Vega (brown Auenboden)

I Gley (low humic gley)

II Nassgley (wet gley)

III Anmoorgley (humic gley)

IV Moorgley (gley intergrading to bog soil)

V Tundragley

I Seemarsch (marsh influenced by seawater)

II Brackmarsch (marsh influenced by brackish water)

III Flussmarsch (marsh influenced by river water)

IV Moormarsch (marsh with peat in the subsoil)

Typen (types)

I Protopedon

II Gyttja

III Sapropel

IV Dy

Typen (types)

I Niedermoor (bog soils with vegetation depending on ground water)

II Übergangsmoor (bog soils intergrading between Niedermoor and Hochmoor)

III Hochmoor (bog soils with vegetation depending on rainfall, highly acid)

Altogether the system contains 14 classes and 47 types as shown. Subordinated to the types 157 *subtypes* have been described in the classification system.

285

PROBLEMS OF A COMPREHENSIVE SOIL CLASSIFICATION SYSTEM

Many a soil scientist may wonder how a worldwide valid soil classification system can be achieved if a soil classification system is worked out in several countries at the same time more or less independently. This is a problem indeed. First however a classification system has to be attempted by intensive research on soils, in order to get a frame for grouping the soils. This is primarily important for a soil survey. On the other hand we will never get a worldwide soil classification system if the soils of many small areas have not been classified before. In doing this our knowledge will grow. Only if we have detailed knowledge of all regions of the world, a worldwide comprehensive soil classification system can be developed. It is a fact that we do not have this knowledge at present. At the moment we are in the stage of investigating and systematical grouping of individual soil provinces. The soil classification system of the Federal Republic of Germany presented now is nothing more than a part of this investigation and an attempt at grouping. In his book W. L. KUBIËNA (1953) gave a picture of the soils of Europe and we have set ourselves the task to investigate systematically a part of this larger area. Perhaps the new classification systems, worked out by the Dokutchaiev Institute, Moscow (1957, 1958), and the Soil Survey Staff of the USA (1960), are a beginning for a worldwide soil classification system.

We ought to try to co-ordinate partial investigations as soon as possible in order to approach a soil classification system valid for a larger area. For this purpose soil scientists of all countries have to contact each other. This is a task, that has to be tackled by the soil societies of the individual countries and the International Society of Soil Science. As a good start we may consider the initiative of New Zealand, which has made soil classification a central theme for the sessions of the IVth and Vth Commissions of the International Society of Soil Science in 1962. Neighbouring countries, especially, are supposed to co-operate on this task. A hopeful beginning has been made already between Belgium, the Netherlands, Austria, Switzerland and our country. The co-operation between the Austrian and the German Soil Science Society could not have been demonstrated better than on the meeting in Vienna in 1961, where we also reported on the soil classification system of the Federal Republic of Germany.

On this meeting in Ghent several countries co-operate again. Under the approved chairmanship of Prof. Tavernier a full succes will be sure.

LITERATURE

BASINSKI, J. J. — The Russian Approach of Soil Classification and its Development.
J. Soil Sci., 10, 14-26. Oxford, 1959. (This work contains an extensive list of literature).

KUBIËNA, W. L. — Bestimmungsbuch und Systematik der Böden Europas. Stuttgart, 1953.

LAATSCH, W. & SCHLICHTING, E. — Bodentypus und Bodensystematik.
Zeitschrift f. Pflanzenernährung, Düngung, Bodenkunde, 87, (132), H. 1, 97-108. Weinheim, 1959.

MÜCKENHAUSEN, E. (in Zusammenarbeit mit F. Heinrich, W. Laatsch & F. Vogel). — Entstehung, Eigenschaften und Systematik der Böden der Bundesrepublik Deutschland.
DLG-Verlags-GmbH. Frankfurt/M., 1962. (This book contains an extensive list of literature).

ROSSOW, N. N., KARAWAJEWA, N. A. & RODE, A. A. — Erste Plenarsitzung der Kommission für Bodennomenklatur, -systematik und -klassifikation in der Akademie der Wissenschaften der UdSSR (russ.).
Podschwowedenie, 8, 60-65. Moskau, 1957.

— — Zweite Plenarsitzung der Kommission für Bodennomenklatur, -systematik und -klassifikation in der Akademie der Wissenschaften der UdSSR (russ.).
Podschwowedenie, 9, 109-115. Moskau, 1958.

SOIL SURVEY STAFF (C. E. Kellogg, G. D. Smith). — Soil Classification — A comprehensive system. 7th approximation.
Soil Survey Staff, Soil Conservation Service, United States Dept. Agric. Washington, D. C., 1960.

THE SOIL CLASSIFICATION SYSTEM OF
THE FEDERAL REPUBLIC OF GERMANY

Summary

Pursuing the course that has been set by W. L. Kubiëna (1953), the soils of the Federal Republic of Germany are classified in a classification system with higher and lower categories. Following categories are distinguished: sections, classes, types, subtypes, varieties and subvarieties all as pedogenetic categories; the form (series) completes those 6 pedogenetic categories on the base of lithologic features (texture and parent material). The type occupies a central position in the classification system. The types are grouped in higher categories and split up in lower ones. The features that are decisive for the system are here discussed. The single categories are characterized by certain features inherent to the soil itself. According to these principles the soils of the Federal Republic of Germany are listed in a system of 4 sections, 14 classes, 47 types and 157 subtypes in the book of E. Mückenhausen (1962). For varieties, subvarieties and forms however only examples are given in this book.

LE SYSTEME DE CLASSIFICATION DES SOLS
DE LA REPUBLIQUE FEDERALE ALLEMANDE

Résumé

En suivant W. L. Kubiëna (1953), les sols de la République Fédérale Allemande ont été classés dans un système à catégories supérieures et inférieures. Les catégories suivantes sont distinguées: divisions, classes, types, sous-types, variétés et sous-variétés, en tant que catégories pédogénétiques, complétées par la forme, basée sur des caractères lithologiques (texture et roche-mère). Le type occupe une position centrale dans le système de classification. Les types sont groupés dans des catégories supérieures et subdivisés dans des catégories inférieures. Les caractéristiques qui sont essentielles pour la systématique sont décrites. Les diverses catégories sont caractérisées par certains caractères inhérents au sol. D'après ces règles fondamentales les sols de la République Fédérale Allemande sont divisés, dans le livre de E. Mückenhausen (1962), en 4 divisions, 14 classes, 47 types et 157 sous-types; en ce qui concerne les variétés, sous-variétés et formes, le livre ne donne que des exemples.

HET BODEMKLASSIFIKATIESYSTEEM
VAN DE DUITSE BONDSREPUBLIEK

Samenvatting

In navolging van W. L. Kubiëna (1953) worden de bodems van de Duitse Bondsrepubliek geklasseerd in een systeem met hogere en lagere kategorieën. Volgende kategorieën worden onderscheiden: afdelingen, klassen, typen, subtypen, variëteiten en subvariëteiten als pedogenetische kategorieën, aangevuld door de ‹ vorm›, die gebaseerd is op lithologische kenmerken (textuur en moedergesteente). Het type neemt in het klassifikatiesysteem een centrale positie in. De typen worden in hogere kategorieën samengebundeld en in lagere kategorieën gesplitst. De eigenschappen die voor de systematiek doorslaggevend zijn worden besproken De afzonderlijke kategorieën worden door bijzondere, aan de bodem inherente eigenschappen gekenmerkt. Volgens deze basisregels worden de bodems van de Duitse Bondsrepubliek in het boek van E. Mückenhausen (1962) systematisch ingedeeld in 4 afdelingen, 14 klassen, 47 typen en 157 subtypen; van de variëteiten, subvariëteiten en vormen worden in het boek slechts voorbeelden gegeven.

DAS BODENKLASSIFIKATIONSSYSTEM DER
BUNDESREPUBLIK DEUTSCHLAND

Zusammenfassung

In Anlehnung an W. L. Kubiëna (1953) werden die Böden der Bundesrepublik Deutschland in ein Klassifikationssystem mit höheren und niederen Kategorien eingeordnet. Folgende Kategorien werden aufgestellt: Abteilungen, Klassen, Typen, Subtypen, Varietäten und Subvarietäten als pedogenetische Kategorien; die Form ergänzt diese 6 pedogenetischen Kategorien durch lithologische Merkmale (Textur und Ausgangsgestein). Der Typ steht zentral im Klassifikationssystem. Die Typen werden in höhere Kategorien zusammengefaßt und in niedere Kategorien aufgegliedert. Es werden die Merkmale aufgeführt, die für die Systematik entscheidend sind. Die einzelnen Kategorien sind durch bestimmte, bodeneigene Merkmale charakterisiert. Nach diesen Grundregeln sind in dem Buch von E. Mückenhausen (1962) die Böden der Bundesrepublik Deutschland in 4 Abteilungen, 14 Klassen, 47 Typen und 157 Subtypen systematisiert. Für Varietäten, Subvarietäten und Formen sind in diesem Buch nur Beispiele gegeben.

DISCUSSION(*)

G. Aubert

Le Prof. Mückenhausen peut-il nous préciser sa position quant aux trois classes de Terrae calcis, Plastosols et Latosols ? Ne peut-on les considérer seulement comme des matériaux originels et classer les sols qui se sont développés dessus ? Ne sont-ils pas assez vieux et donc assez développés pour avoir les caractères reconnaissables d'autres classes de sols ?

Les Terrae calcis, les Plastosols et les Latosols en tant que sols fossiles n'occupent que de petites superficies en Europe Centrale. Ces sols se sont formés au Pré-Pléistocène (et aussi partiellement pendant le Pléistocène) par suite de processus d'altération très intenses; ils résistent fort bien aux altérations faibles de l'Holocène (éventuellement du Pléistocène) et ont de ce fait peu changé. On peut donc les considérer encore comme des sols des climats pré-pléistocènes (en partie interglaciaires). Dans les cas où un sol à dynamique du climat actuel se soit formé, on peut distinguer le type de sol actuel; p.ex. un Pseudogley s'est souvent formé dans le Graulehm. Dans ce cas on peut dire : Graulehm - Pseudogley ou Pseudogley sur Graulehm. Dans le premier cas le Graulehm est le type pédogénétique, dans le second cas on considère le Graulehm comme matériau de départ. C'est une question de point de vue laquelle des deux définitions on considère comme étant le plus utile; toutes les deux sont cependant exactes.

H. Lobova

1. Est-ce que la Terra fusca est un sol d'altitude ?

   La Terra fusca se trouve sur les calcaires et dolomies de toute l'Europe Centrale sur de petites superficies, et d'après Kubiëna également dans les Alpes calcaires.

2. Quelle est la différence entre la Terra fusca et la Braunerde ?

   La Terra fusca se développe sur des calcaires; elle a un « Lehmgefüge » (dans le sens de Kubiëna) et est de ce fait très plastique.

   Par contre, la Braunerde se forme sur des roches silicatées calcarifères et non calcarifères; elle montre surtout un « erdiges Gefüge » (dans le sens de Kubiëna) et est moins plastique.

3. Est-ce que la Terra fusca est actuelle en Espagne (cf. Kubiëna) ?

   Ces dernières années, Klinge a étudié en détail la Terra fusca et la Terra rossa en Espagne; il est d'avis que ces Terrae calcis sont essentiellement fossiles, c.-à-d. préholocènes.

4. Est-ce que la Terra fusca a un horizon B textural ?

   La Terra fusca typique a un profil A-(B)-C, donc pas d'horizon B textural. Par contre, la Terra fusca lessivée a un profil A_1 - A_3 - Bt - C par suite d'un lessivage de l'argile; elle a donc un B textural.

5. Est-ce que la Terra fusca a encore des carbonates dans le profil ?

   Normalement la Terra fusca est dépourvue de carbonates dans le solum. Si cependant elle en contient encore, ceci est dû à un remaniement du solum ou à un apport secondaire de carbonates par l'eau de pente.

6. A quelle époque s'est formé le Latosol ?

   Le Latosol du Vogelsberg (Hessen) s'est formé au Tertiaire sur un basalte et un tuff basaltique tertiaires.

(*) Prof. Dr. Dr. E. Mückenhausen's replies to questions and comments are printed in *italics*.

I. V. TIURIN

1. How large are the areas of Terrae calcis, Plastosol and Latosol ?

The Terrae calcis chiefly occur only in cracks and dolines on limestones and dolomites. Through solifluction, material of Terrae calcis has been mixed with Braunerde and Parabraunerde material on loess and Rendzina material over areas of several square kilometers, e.g. on the Alb in Southern Germany and on the Obere Muschelkalk in the Triasic basin of Trier. A rather large area of Terra rossa (several ha) occurs south of Aachen on lower-carboniferous limestone.

The Plastosols are found as Graulehm in the Rheinische Schiefergebirge, and less in the other German middle mountains; they occupy areas of 1-100 ha. Typical Braunlehm and Rotlehm occupy only small areas of less than 1 ha. On the other hand intergrades between Pseudogley and Braunlehm have an extension of several square kilometers.

The Latosols are restricted to the Vogelsberg area and occupy a few square kilometers.

2. What is the distribution of the Pelosols ?

The Pelosols originate from clays and claymarls of the Mittlere Muschelkalk, the Keuper, the Jura, the Cretaceous and the Tertiary. They occur chiefly in South-west Germany, in Eastern Westphalia and in the Southern Eifel. The areas are numerous but not large; however on the whole they occupy many square kilometers.

J. DUPUIS

1. En France, jusqu'à maintenant, nous avons considéré les « Pelosols » comme une variété de « sols bruns » (Braunerden). Pourquoi en faire la distinction à un niveau aussi élevé que celui des classes ? Ne pourrait on le faire seulement au niveau des types, soit dans la classe b (sols AC) en un type IV, soit dans la classe e également en un type IV.

Comme le Pelosol ne s'adaptait dans aucune autre classe par suite de sa séquence d'horizons et de ses caractères spécifiques, nous l'avons mis dans une classe spéciale. Le Pelosol typique a un profil A-P-C. L'horizon P est un horizon d'altération du matériau originel argileux; il a la couleur de ce matériau originel et peut être gris, rougeâtre, violet ou brun. En plus, il y a des profils à séquence A-C, A_1-A_3-P-C et A-gP-C. Ces séquences d'horizons seules exigent l'introduction d'une classe spéciale. A ce moment nous ne savons pas encore si nous devons distinguer plusieurs types dans la classe des Pelosols, ou si des sous-types suffisent. Des recherches de terrain nous l'apprendront.

2. Les Pelosols existent-ils sur des argiles calcarifères ? Dans ce cas, sont-ils calcaires en surface et quelle est la structure de l'horizon A ?

Les Pelosols se forment également sur des sédiments argileux calcarifères, p. ex. des marnes calcaires. Les formations de sols jeunes sur ce matériau de départ sont calcarifères jusqu'en surface; il s'agit du Pelosol calcaire à profil A-C. La structure de l'horizon A est dans ce cas polyédrique ou granulaire, après l'hiver souvent très friable par suite de l'action du gel.

R. MAIGNIEN

Comment peut-on distinguer *morphologiquement* un Gley oxydé d'un Pseudogley ?

Dans beaucoup de cas on ne peut pas ou difficilement faire la distinction entre l'horizon Go d'un Gley et l'horizon à taches de rouille g_2 d'un Pseudogley, quand on les

291

considère en eux-mêmes, c.-à-d. sans avoir connaissance des autres horizons de ces deux types de sol.

De ce fait la question du Dr. Maignien est très pertinente. Quand on étudie cependant le profil en entier, on peut les distinguer aisément. Dans le gley typique l'horizon Go est suivi d'un horizon Gr; en plus il y a une nappe phréatique permanente dans la partie inférieure du profil. Le Pseudogley ne manque pas seulement l'horizon réduit, mais en plus l'horizon à taches de rouille g_x s'étend souvent à grande profondeur; dans la plupart des cas il s'arrête à un matériau originel (roche) peu altéré qui a sa propre couleur et qui ne contient pas de nappe phréatique. La stagnation de l'eau y est superficielle et temporaire.

R. STEFFENS

1. Gehört der Pelosol-Pseudogley, dessen Abbildung Sie heute Morgen zeigten, zu der Pelosol-Klasse oder zur Klasse der Staunässeböden ?
   Wo liegt in diesem Falle der Übergang ?

   Der Pelosol-Pseudogley gehört zur Klasse der Staunässeböden. Er ist ein Subtyp des Pseudogleyes und besitzt ein Wasserregime, das vorwiegend dem des Pseudogleyes ähnlich ist (die obersten 20 cm noch durchlässig). Daneben besitzt dieser Subtyp auch Eigenschaften des Pelosols, welche durch die tonige Textur bedingt sind. Ist jedoch der tonige Boden im gequollenen Zustand dicht bis zur Oberfläche, so entspricht das Wasserregime mehr dem des Pelosols. In diesem Falle sprechen wir von Pelosol.

2. Haben Sie schon einen « Textur B » bemerkt in den schweren Tonböden auf Keuper ?

   T. Diez hat in den Pelosolen aus dem Feuerletten des Keupers in Mittelfranken eine schwache Tonverlagerung festgestellt, und damit ist ein schwach ausgeprägter Textur-B-Horizont verbunden. Da die Textur des ganzen Solums sehr tonreich ist, kann der Textur-B-Horizont nicht gut erkannt werden. Eine Tonverlagerung in den tonreichen Böden des Keupers wurde auch in England festgestellt.

J. SCHELLING

You have stated that your system of soil classification is a genetic one. It seems that the distinction within the Bog soils of High Moor and Low Moor soils is essentially based on characteristics of the parent material and not on pedogenetic criteria.

When we consider the process that takes place when an A_{oo} is transformed into an A_1 in mineral soils, the loss of the original plant structure is one of the most striking features.

In peat soils we can find, under biologic influence, the same process of decomposition of the original plant material that loses its original plant structure. This may be considered as the formation of an A_1 layer.

Furthermore, we know that a vertical transport of amorphous organic matter through a peat soil may take place and give way to a deposition of this material in deeper layers.

This has been confirmed with C^{14} datings. These accumulations may be considered as B horizons.

The peats that show such a humus infiltration and accumulation may be considered as subtypes of the types of Low Moor and High Moor.

F. Mancini

Which are the main differences between the « Terra fusca » and the « Braunlehm »?

Are they so big to allow to put these soils into two classes of the German system?

The Terra fusca originates from limestones and with the Terra rossa belongs to the class of the Terrae calcis. The « Braunlehm » however is formed on silicate rocks and belongs with the « Graulehm » and the « Rotlehm » to the class of the Plastosols. The dynamics of the Terrae calcis are always influenced by the calcareous subsoil; they seem to have a wider genetic range than the Plastosols, according to the influence of the limestone. The dynamics of the Plastosols (and of the Braunlehm also) are less influenced by the parent rock; they seem to be genetically resticted to a humid warm climate; their range of distribution is smaller than that of the Terrae calcis. A common characteristic of Terrae calcis and Plastosols is the so-called « Lehmgefüge » (sensu Kubiëna's), i.e. they both have iron and clay substances that are easily dispersed, producing a compact fabric and a high plasticity.

R. Marechal

I should like to ask Prof. Mückenhausen to give us some details about a group which can be very important for Western European countries, namely the Fahlerde. Is it always possible to make a clear distinction between Fahlerden and Podzol-Parabraunerde intergrades ? Are the Fahlerden more or less corresponding to the « sols podzoliques » of the classification of Prof. Aubert.

The Fahlerde is a soil type of the « Braunerdeklasse ». It has a thick A_3 horizon and a somewhat compact B horizon rich in clay with a well developed polyedric structure and clay skins on the peds' surfaces. The clay movement is more important than in the Parabraunerde. Beside the clay movement there is also a clay destruction due to the low pH; this pedodynamic pecularity distinguishes the Fahlerde from the Parabraunerde. It is difficult to separate both soil types on their morphological features; for this we need more experience.

It is almost certain that the « sols podzoliques » are for the greatest part identical to the Fahlerde. If the « sol podzolique » shows an ashy A_2 horizon (similar to that of the podzol), we would call it a « podzolic Fahlerde ». If the « sol podzolique » also has a pseudogleyification in the B horizon and in the lower part of the A_3, we would call it a « Pseudogley-Fahlerde ».

P. Ryan

1. How do you distinguish between *Pararendzina* and *Rendzina-Braunerde*. Is it on the basis of purity of parent material, or otherwise ?

   A Pararendzina has an A-C profile and originates from calcareous silicate rocks (hard and loosened), especially on loess, boulder clay and calcareous sandstone. A Rendzina-Braunerde is a more evoluated form of the Rendzina (originates from limestone) in a temperate warm and humid climate and has an A-(B)-C profile.

2. Is the *Podsol-Parabraunerde* a simple transitional sequence or is it a bisequal effect due to vegetation change (human interference) ?

   The Podsol-Parabraunerde (mostly a Podsol-Fahlerde) has an A_0-A_1-A_2-B-A_3-B-C profile. The first four horizons belong to the podzol formation. In Central Europe *the Podsol-Parabraunerde (the podzolisation) occurs chiefly under the influence of* Caluna vulgaris *and* Vaccinium myrtillus, *and also under coniferous forest. Man often but not always has caused this vegetation.*

3. Would it not be more logical to express transition profiles as :

   a) *Parabraunerde - Podsol* rather than *Podsol-Parabraunerde*

   b) *Ranker - Podsol* rather than *Podsol-Ranker.*

   We have Parabraunerde-Podsol and Podsol-Parabraunerde. The former is closer related to the Podsol (with a podzol B), the latter to the Parabraunerde (here the podzol B is lacking). Further on we distinguish a Ranker-Podsol and a Podsol-Ranker. The former is closer related to the Podsol (with a podzol B), the latter to the Ranker (here the podzol B is lacking).

4. Has Prof. Mückenhausen experienced Gleys with Go horizon underlying a Gr horizon in the subsurface depths ?

   A G_0 horizon beneath a Gr horizon may occur when the deeper G_0 horizon has a permeable texture through which runs ground water rich in oxygen; above the latter the texture is more or less impermeable, with only little water movement and probably also with organic matter, so that reduction (Gr horizon) can take place.

22

Reprinted from *7th Intern. Congr. Soil Sci. Trans.*, New Zealand, 1962, pp. 291–297

THE FACTUAL CLASSIFICATION OF SOILS AND ITS USE IN SOIL RESEARCH

K. H. NORTHCOTE,

Division of Soils, C.S.I.R.O., Adelaide, Australia

In 1956 I was given the task of preparing an Atlas of Australian Soils. For this work a suitable classification of soils had to be selected. All existing classifications were rejected because they failed to recognise SOIL AS A SEPARATE THING capable of classification as other things are. Subsequent research, starting from first principles, has resulted in "a factual key for the recognition of Australian soils" (Northcote, 1960a). This key is being used to classify soils for the Soils Atlas, two sheets of which have been published (Northcote, 1960b, 1962) and two more are in preparation.

The philosophy and principles behind the factual approach to classification may be summarised in the words of Hegel, as quoted by Tomkeieff (1954), when he said: "Things ARE, but he who can THINK what they are is their master." Reasoning from this standpoint resulted in the following definition of soil: SOIL IS THAT NATURAL DYNAMIC SYSTEM WITHIN THE SURFACE OF THE LITHOSPHERE COMPOSED OF MINERAL AND ORGANIC MATERIALS DEVELOPED *in situ* BY PHYSICAL, CHEMICAL, AND BIOLOGICAL PROCESSES INTO ORGANISED PROFILES OF LAYERS MORE OR LESS PARALLEL TO THE EARTH'S SURFACE.

Various factual classifications of soil are possible. The above definition of soil suggests that factual classifications could be based on (i) the common soil processes or (ii) the resultant organised profiles. Understanding and agreement about soil processes are strictly limited, so that ORGANISED PROFILES select themselves as the concept around which a factual classification can be made. Organised profiles, termed PROFILE FORM herein, are placed in logical sequences of mutually exclusive groupings by employing the methods of definition, logical alphabet, and bifurcation, as illustrated in Figs. 1 and 2. Although this work is not finished, it is of a sufficiently definite character to use. This is a quality inherent in factual classifications, because groupings are recognised by specific properties. Practical use will therefore provide data to develop the scheme further.

The Usefulness of the Factual Key in Soil Research

The factual key has provided a classification of soils based solely on soil properties. Experience has shown that the key will operate

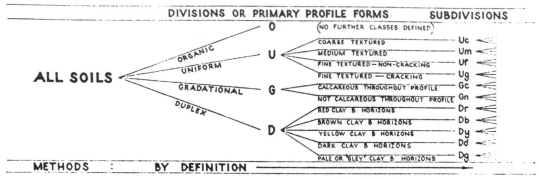

Fig. 1—Diagrammatic presentation of the divisions and subdivisions of the factual
key.

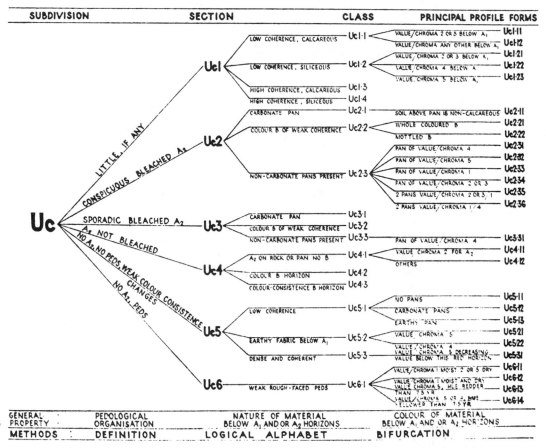

Fig. 2—Diagrammatic presentation of the sections, classes, and principal profile
forms of subdivision Uc.

simply and smoothly. In the booklets which have accompanied each map sheet, Table 2 (Northcote, 1960b, 1962) is eloquent testimony of the confusion avoided by the use of the key. In this table the principal profile forms (that is, the key symbols) are compared with the various classification names previously used by other soil workers. Nor is anything lost by using symbols instead of words; indeed, there is gain, for actual properties are distinguished by the symbols, whereas the older classification names such as podzol, krasnozem, and so forth have stood for more or less intangible concepts and ideas (see also Leeper, 1952). The key has shown, therefore, considerable powers of resolution and reconciliation, as implied by Table 2. An example will serve to illustrate. In Tasmania and southern Victoria there are deep, friable, red acidic soils termed krasnozems or red loams (Stephens, 1956). They are classified by the factual key as Gn 4.11. At Dookie in northern Victoria, Downes (1949) reported a red loam soil, the Dookie clay loam, profiles of which on examination in 1960 were classified as Gn 3.12 and Gn 3.13. That is, the Dookie clay loam, although similar to the Gn 4.11 red loam of southern Victoria in that it is a friable red soil, has: (i) shiny, smooth-faced peds in the B-horizon typical of the presence of clay skins, whereas the Gn 4 soils do not; (ii) neutral or alkaline soil reaction trends, whereas the Gn 4 red loams are acid. More recently, in 1961, it was found that red friable soils at Inverell in northern New South Wales, termed euchrozem by Hallsworth, Colwell, and Gibbons (1953), were classified by the factual key as Gn 3.12 and Gn 3.13. These authors had pointed out that while the euchrozem bears a superficial resemblance to krasnozem, their structure and analytical features are quite different. This is borne out by the factual key. Although not absolutely essential to the present discussion, it may be of interest to point out that both the Inverell and Dookie Gn 3 soils occur in sub-humid, wet-dry climates with average annual rainfalls of 29 in. (740 mm) and 21 in. (530 mm) respectively, and are used for wheat production; whereas the Gn 4 soils enjoy a humid climate with an average annual rainfall over 35 in. (900 mm) and are used for pasture production for dairy-farming and speciality crops, such as potatoes. As data are collected it will be possible to make more precise inductive generalisations about the principal profile forms than those made using the older great soil groups.

New light is brought to bear on the distribution of soils in relation to environmental factors, and this in turn should inspire fresh enquiry into the effect of environmental factors on the development of soil profiles. For example, on Sheet 2 of the Soils Atlas (Northcote *et al.*, 1962) the following succession of areally dominant soils exists along the transect from Mt Baw Baw in the Victorian Alps to the town of Shepparton on the northern Victorian plain: (i) Um 7.11 soils: high alpine plains and sub-alpine forests over 3,000 to 3,500 ft (900-1,000 m); average annual rainfall over 55 in. (1,400 mm); mean temperature about 40–46°F (4–8°C); (ii) GN 4.31 soils: slopes of mountain mass; rain forest; average annual rainfall over 40 in. (1,000 mm); mean temperature about 50°F (10°C); (iii) DY 3.41 and DR 2.21 soils: lower range country about 1,000–2,000 ft (300–600 m); dry sclerophyll forests: average annual rainfall about 30 in. (750 mm); mean temperature less than 60°F (16°C); (iv) DY 3.42 and DR 2.32 soils: proluvial aprons and hills fringing ranges; dry sclerophyll forests: average annual rainfall about 25 in. (625 mm); mean temperature less than 60°F (16°C); (v) DR 2.33 soils: northern Victorian plain about 400 ft (120 m); savannah woodland; average annual rainfall 20 in. (500 mm) or less; mean temperature 60°F (16°C) or higher. Great Soil Group maps such as those of Prescott (1944) or Stephens (1961) have shown a succession of three great soil groups over the same transect. They were: (i) High Moor (Peats) and/or Alpine Humus soils; (ii) Podzols (Prescott) or various Podzolic soils (Stephens); (iii) red-brown earths. Certainly these maps are of a broader scale than the Atlas, but even allowing for this, they miss the significant area of gradational (GN) soils. Thus the occurrence of areally dominant soils with U profile form at the highest elevations, passing to GN soils, then to acid DY and DR soils, to neutral DY and DR soils, and finally to alkaline DR soils at the lower elevations, becomes a new and significant fact in our thinking about pedogenesis in relation to environmental factors in Victoria. The occurrence of the D soils seems to coincide with the occurrence of seasonally wet-dry climatic conditions. This fact should raise questions regarding weathering processes.

Butler (1957) has drawn attention to three distinct attitudes that soils workers have towards soils and has termed these the pedologic, geographic, and edaphic concepts. The question of soil classification in regard to each is dealt with below.

1. By the pedologic concept Butler (loc. cit.) means "the impartial study of all the attributes of soils", and for this purpose the soil profile is considered the "essential soil system" and therefore the basis of study. The relevance and importance of the present factual classification, based as it is on profile form, should be clear at once. Indeed, it would be true to say that soil studies within the pedologic concept have been hindered in the past by the lack of a general-purpose classification uncomplicated by geographic and/or edaphic considerations. The factual key fulfils this need. Of course, there is also need for particular-purpose classifications relating to parent materials, soil processes, and so forth. As these become available and are integrated with the factual key, so inductive generalisations on a firm basis will be possible.

The podzol-podzolic confusion will serve to illustrate some further points. Stace (1961), in his supplement to Stephens's (1961) Manual of Australian Soils, has recorded three podzol profiles. On the factual key they have been classified as Uc 2.33; that is, they are uniformly coarse-textured soils with a conspicuous bleach and a dark hardpan below this. In the same publication, Stace has recorded three yellow podzolic profiles which have been classified by the factual key as Dy 2.21, Dy 3.41, and Gn 3.74. There is some diversity here. The first two profiles are duplex and the third is gradational. Without discussing details, it is obvious from the key symbols that there is a great difference between the nature of the profile form of the podzols and of the yellow podzolics. Since profile form is the final expression of the interaction of all soil processes, can the same process be considered as dominant in both these groups? Great doubt must be expressed at the use of the word "podzolic" for the latter group. Again, Downes (1949) included the Warrenbayne sandy loam in his group of podzolised soils. On the factual key, this soil has been classified as Dr 2.21. This is the most common kind of so-called "red podzolic" soil in southern Australia. But on the basis of profile form its affinities are seen to be with the red-brown earths, Dr 2.23, of South Australia rather than with podzols, Uc 2.33. Clearly then the direction of research into soil processes will receive new inspiration from the profile form affinities expressed through the factual key. Many more examples of a similar character could be given.

2. The geographic concept, according to Butler (*loc. cit.*), "is concerned with integrating all the data that bear upon the comprehension and use of the countryside. These data include soil, climate, topography, lithology, and vegetation". The results are usually expressed by areas placed on maps which represent a generalised uniformity. Such a simplification is very valuable and should assist in the understanding and use of land. But unfortunately the geographer, be he soil surveyor, ecologist, or otherwise, has had to use classifications of a compromising nature. In the realm of soil classification, for example, he has had to use some classification based on the great soil group concept, which, with its environmental implications, compromises the results from the outset. For studies within the geographic concept to be valid and useful, each geographic component such as soil, climate, and vegetation needs to have its own independent classification before satisfactory integration can be achieved. In this regard, the factual key has provided an independent classification of the soil component, and it will be useful therefore in studies within the geographic concept. Incidentally, it may be worth stressing the fact that the Soils Atlas has relied on the factual key only to classify the soils. Their geographic distribution is portrayed through the delineation of landscape units based on generalised topographies. Therefore actually existing soils in terms of profile features, and not theoretically probable soils, are shown on the Atlas Maps.

3. The edaphic concept is concerned with soil as a medium for the growth of plants. This is not an easy field of study, since "its point of departure—a plant's requirement—is not known with sufficient precision for relevance to be given to specific values of soil properties"

(Butler, *loc. cit.*). It seems likely that the plant-growth potential of a site, which depends on other environmental factors besides soil and may even depend on politics and economics, will never be known perfectly. Whatever the soil-plant worker has achieved is only an approximation to this potential, and it seems likely to remain so. But even to get this, he must have unbiased and uncompromised data to work with. It seems that there have never been classifications of soils, for example, helpful for studies within the edaphic concept. The soil-plant worker has been placed at a disadvantage. Of course,. he will need particular-purpose classifications, and therein he has not been as creative as he might have been. Nevertheless, his principal lack has been a general-purpose classification which should assist him to gain better perspective. Undoubtedly, the factual key can help. Contact between the factual key and studies within the edaphic concept are through physical properties which influence, for example, aeration, available moisture, and the tilth of the soil. That some intermediate step is necessary, involving determination of what these physical properties are, is obvious. Once this is achieved, a general-purpose classification such as the factual key, based as it is on physical properties, can be employed to make inductive generalisations. It seems necessary to point out that all too often soil-plant workers have tried to achieve these desirable ends without having the involvement of this intermediate step. Once this need is faced, and there is no way around it short of magic, then the usefulness of a general reference classification of soils will become clearer in the realm of edaphic studies. Many edaphic studies are made as nutritional plot trials, such as that of Anderson and McLachlan (1951) on the residual effect of phosphorus on pasture on the Southern Tablelands of New South Wales. In such work soils have been classed as great soil groups, in this case yellow and red podzolics. Further data regarding the nature of the soils may or may not be given; in this case the colour, texture, and pH of the immediate surface soil and the presumed parent materials of each of four sites are stated. Results have usually shown differences due to the soils; in this case, yields over a four-year period "were very different on the four soils". In Australia there is a great deal of such work which, because the soils are inadequately described, or classified, or both, cannot be used to its full extent. Obviously, the factual key can help to rationalise such studies so that results can be applied more widely.

Conclusion

The claim that this factual key can be used as a general-reference classification for many purposes is based not simply on its successful functioning in the few cases that have been cited, but also on the philosophy and principles behind it and the method of its construction. No preconceived ideas have gone into its formation. The factual key is therefore an uncommitted classification; that is, it does not pretend to have the answers before research is carried out. Because of this, research into the nature of soils, using the factual key for purposes of orderly classing and arrangement, will allow knowledge acquired from one soil to be applied to like soils.

ACKNOWLEDGMENT

The writings of G. W. Leeper on classification were a source of inspiration during the formative phase of this work.

References

ANDERSON, A. J.; McLACHLAN, K. D. 1951: The Residual Effect of Phosphorus on Soil Fertility and Pasture Development on Acid Soils. *Aust. J. agric. Res. 2*: 377–400.

BUTLER, B. E. 1957: A Diversity of Concepts about Soils. *J. Aust. Inst. agric. Sci. 24*:14–20.

DOWNES, R. C. 1949: A Soil, Land-Use, and Erosion Survey of Parts of the Counties of Moira and Delatite, Victoria. *C.S.I.R.O., Aust. Bull. 243*.

HALLSWORTH, E. G.; COLWELL, J. D.; GIBBONS, F. B. 1953: Studies in Pedogenesis in New South Wales. V. The Euchrozems. *Aust. J. agric. Res. 4*: 305–325.

LEEPER, G. W. 1952: On Classifying Soils. *J. Aust. Inst. agric. Sci. 18*:77–80.

NORTHCOTE, K. H. 1960a: Factual Key for the Recognition of Australian Soils. *C.S.I.R.O., Aust. Soils Div. div. Rep. 4/60*.

—— 1960b: Atlas of Australian Soils. Sheet 1. Port Augusta-Adelaide-Hamilton Area. With explanatory data. 50 pp. C.S.I.R.O., Aust.

NORTHCOTE, K. H.; NICHOLLS, K. D.; DIMMOCK, G. M. 1962: Atlas of Australian Soils. Sheet 2. Melbourne-Tasmania Area. With explanatory data. C.S.I.R.O., Aust.

PRESCOTT, J. A. 1944: A Soil Map of Australia. *C.S.I.R.O., Aust. Bull. 177*.

STACE, H. C. T. 1961: Morphological and Chemical Characteristics of Representative Profiles of the Great Soil Groups of Australia. C.S.I.R.O., Aust. Soils Div. 230 pp.

STEPHENS, C. G. 1956: A Manual of Australian Soils, 2nd ed. C.S.I.R.O., Aust., Melbourne. 54 pp.

—— 1961: The Soil Landscapes of Australia. *C.S.I.R.O., Aust. Soil Pub. 18*: 43 pp.

TOMKEIEFF, S. I. 1954: "A New Periodic Table of the Elements Based on the Structure of the Atom." Chapman and Hall, London. 30 pp.

23

Systems of Polar Soil Classification

J. C. F. Tedrow

[*Editor's Note:* Tables 15–1 and 15–2a, b, c, and d are included as an addendum in order to reflect concepts of polar soil classification as proposed in the works of Dr. J. C. F. Tedrow. These tables summarize Tedrow's approach based on his lifelong study of soils in cold regions.]

This chapter reviews some developments in polar soil science over the past forty-odd years. The period is significant for several reasons. During these years Soviet investigators made a number of detailed studies of frozen soil characteristics. Most of their investigations were conducted in the East European tundras (some reports dealt with northern forested sectors rather than with arctic or polar areas, properly speaking). Furthermore, proposals for all-encompassing classification schemes for polar soils were presented in this period, notably by the Soviet investigators and by the United States and Canadian governments.

Grigor'ev (1930) was one of the first to point out that the high arctic has a strong continental climate. The landscape is mostly well drained, in contrast to the wetter regions to the south, and is colonized by "a small number of xerophytic flowering plants very closely related to those in the steppe of the warm temperate belt." Discussing probable changes in plant distribution during the glacial epochs, Grigor'ev stated that the meadow steppe (a subdivision of the arctic steppe) must have occupied larger areas during maximum glaciation than at present, especially to the north.

Liverovskij (1934) grouped polar soils as follows, based on work in the Bolshezemelskaya Tundra (the stretch of land between the northern Ural Mountains and the Barents Sea) and the Malozemelskaya Tundra (the coast of Siberia directly south of Kolguev Island, *ca.* 68°N, 50°E):

    I. Tundra soils of the Lake-bog-solonchak (saline) series.
        A. Soils of the subarctic tundra. Tundra gley-bog soil formation, which includes: (1) concealed gley soil, (2) gley-ocher-spotty soil, (3) gley soil with a continuous gley horizon, and (4) peaty gley soil.

> B. Soils of the arctic tundra. The above four soils and Solonchak (saline) soil.
> II. Tundra soils of the eluvial series.
>   A. Relic forest podzolic tundra. Relic podzolic soils from the postglacial climatic optimum.
>   B. Recent podzolic soils.
>   C. Dark-colored tundra soils.

Sochava (1933), who studied the effects of relief and parent material on tundra soil properties in the Anabar basin, observed weakly podzolic soils on summits and well-drained slopes where the parent material consisted of subsands; lower positions, on the other hand, were covered with Tundra soils which had a 4-inch organic surface and a gley condition extending to the permafrost table. A low degree of leaching was indicated by the fact that the upper portion of the Tundra soil was leached of carbonates, whereas carbonates persisted in the lower horizons. Carbonates accumulated on the crusts of the frost boils.

Gorodkov's report (1939) on the broad aspects of genetic soils, soil processes, and soil taxonomy of the northern sectors of Eurasia shattered the old myths about a special tundra process in the northern areas. It may be noted, however, that a number of investigators (including some from the Soviet Union) did not subscribe to all of Gorodkov's views and that this is still the case in some quarters. Gorodkov elected to connect the tundra landscape to the podzolic zone only because the automorphic podzolic type of soil formation is zonal, whereas the gley (tundra) soils are intrazonal. He advocated distinguishing patterned ground formation from soil formation: "In speaking of the arctic soil cover, we include only those sectors with gleyey and Podzol soils, and [not] the so-called structural, polygonal, and other arctic 'soils.'. . . In our opinion, the latter are only products of the mechanical influences of natural arctic conditions and, therefore, should be included in the forms of microrelief, not [among the] soils."

Gorodkov recognized the error of the concept of a universal type of soil formation in the polar regions when he stated: "only Liverovskij tries to find a special tundra soil-forming process, and he sometimes classifies the soils as a special variant of the gley-marsh process. In confirming this [opinion], he draws on the data from Zaitsev and Afanasiev that are analogous to his own, forgetting that the work of both men relates to soils of the forest zone and not to the arctic . . . it must be partly for this reason that in pedological literature one still finds [mention of] a special category of tundra soils resulting from a special type of soil formation. . . . There was some excuse for a special classification of tundra soils [in the time] of Dokuchaev and Sibirtsev, when information on natural conditions of the tundra zone was very scarce and completely incorrect; at present we have no basis for considering the arctic as something special by way of soils."

Gorodkov's great work is further enhanced by his understanding of the polar desert conditions of the high arctic. He believed that the nature of the soil processes in the tundra and polar desert zones is gleyic and podzolic and that they weaken towards the pole, with the podzolic process weakening first. Traces of podzolic soils are found under lichen tundra in the Anabar tundra as far north as the 71st parallel, especially on sandy ground.

Several investigators, including Grigor'ev, Frishenfeld (as quoted by Sochava, 1933), and Sheludiakova (1938), spoke of "arctic steppe" conditions but were criticized by colleagues. Yet the term arctic steppe appears to be very appropriate and correct as applied to the polar desert.

With the publication of Gorodkov's work, the main principles about soil processes in northern latitudes were well stated on a qualitative basis. There remained, however, the tasks of providing more details of tundra soil morphology and relating tundra soil to the terrestrial soils in the contiguous areas to the south and to patterned ground formation.

Systems of the Soviet Union

During the post-World War II years, there was a renewed effort by Soviet investigators to set forth concrete diagnostic features in soils as a basis for systematic subdivision of soils. "Unity of origin" and "unity of processes of transformation and migration of substances" were emphasized (Gerasimov, 1952). Gerasimov stated: (1) Every genetic soil type should describe a definite phase in the development of the soil-forming processes, including its transitional stages; (2) every genetic soil type is characterized by specific peculiarities in the biological cycle of substances as it occurs against the background of the greater geological cycle; and (3) every genetic soil type corresponds to a certain level of natural fertility which is determined by the degree of development of the soil's air-moisture regime and the concentration of plant nutrients.

This is a fundamental approach. The question is, do we at present have sufficient information on the processes of the polar regions to apply these ideas fully, or must we await further investigations? There have been several attempts by Soviet investigators to establish a general classification that would provide for arctic soil conditions. Proceeding from the earlier work of Prasolov, Zakharov, and, more recently, Gerasimov, there has been a concentrated effort to support a polygenic approach to determining the origin of the soil "type." The validity of such an approach rests with the assumptions that synthesis and decomposition of organic matter constitute the essential process of soil formation and that the universality of the series of stages through which the soil passes in its developmental cycles can be established.

Rozov (1956), expanding on the views of Sibirtsev, defined the soil as the product of interactions between living organisms and the parent rock under various climatic and topographical conditions and listed the essential processes of the soil system as (1) synthesis of mineral compounds, including synthesis of a biological nature and the breakdown of minerals; (2) synthesis of organic matter, including its breakdown; (3) accumulation of organic and mineral substances and the elimination of such substances from the soil; (4) entrance and exit of water; and (5) exchange of heat energy. Rozov grouped all the soils of the globe into three general groups: (1) boreal and subboreal, (2) subtropical, and (3) tropical. The boreal and subboreal group he divided into the following classes (subgroups): arctic tundra, frozen taiga, forested taiga, moist maritime temperate zone forest, steppe, and desert.

Fig. 7-1 shows the general classification scheme Rozov proposed for the three classes generally present in polar lands. Within each class of soils are biogenic, halobiogenic, and lithobiogenic subclasses, and each subclass is further divided into series according to moisture conditions. Within the biogenic subclass of arctic tundra soils (Class 1) are listed arctic, tundra, and sod turf soils; this grouping, in effect, provides for the automorphic soils within various zones. Rozov also shows possible evolutionary pathways in soils by a series of flow diagrams (Fig. 7-1).

Since Rozov's classification lacks detail, it is difficult to make a full evaluation of its adaptability and utility. There is no provision for frost processes, and there is no indication as to how the author proposes to handle the distribution of patterned ground in field operations. Even so, the report merits serious consideration as an approach to polar soil classification.

Ivanova (1956b) considered the broad aspects of soil classification for all soils of the globe, and that portion of her scheme covering the northern regions is given in Table 7-1. A strong proponent of genetic classification, Ivanova stated that soils should be grouped according to the similarity of their soil-forming processes. Soils that develop under ecologically similar vegetation are listed as "soil formation groups" (subclasses).

CLASS \ SERIES	BIOGENIC SOILS			HALOBIOGENIC SOILS			LITHOBIOGENIC SOILS		
	WITH ATMOSPHERIC MOISTURE	WITH SPORADIC GROUNDWATER	WITH CONSTANT GROUNDWATER	WITH ATMOSPHERIC MOISTURE	WITH SPORADIC GROUNDWATER	WITH CONSTANT GROUNDWATER	WITH ATMOSPHERIC MOISTURE	WITH SPORADIC GROUNDWATER	WITH CONSTANT GROUNDWATER
CLASS 1 Arctic Tundra	A Arctic Tundra T_u Tundra S Sod turf		B^{tu} Boggy tundra	—	—	Sk^{tu} Tundra solonchak	—	—	—
CLASS 2 Frozen Taiga	T Taiga (ferruginous) T^p Pale yellow taiga	T^g Gleyic pale yellow taiga	B^f Frozen bog	Sd^f Frozen solod (degraded alkali)	Sd^{fg} Frozen gleyic solod			—	
CLASS 3 Forested Taiga	P Podzolic F Forest gray soil	P^b Podzolic boggy F^g Forest gray gleyic	B^h High bog	—	—	Sc Soddy carbonate soil	Sc^g Soddy gleyic carbonate soil	B^l Low boggy soil	

GROUP ASSN. ← —————— BOREAL -------- →

306

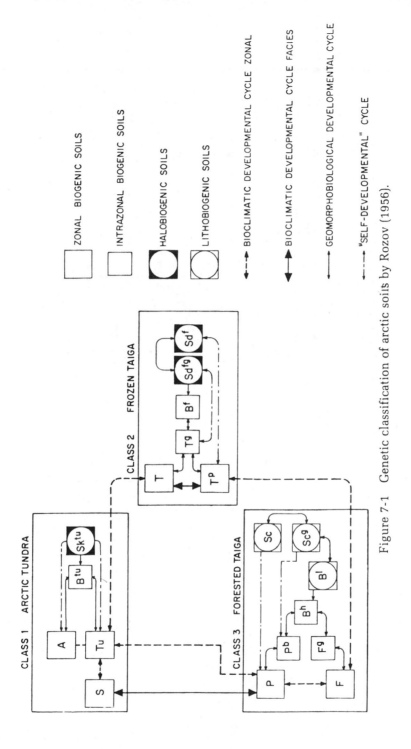

Figure 7-1   Genetic classification of arctic soils by Rozov (1956).

307

TABLE 7-1 Excerpts adapted from the soil classification scheme of Ivanova (1956b).*

Class of Soil	Soil Formation Group (subclass of soil)	Type of Soil		
		Automorphic	Automorpho-Hydromorphic	Hydromorphic
GLOBAL GROUP OF CLASSES OF BOREAL SOIL FORMATION				
Class I Tundra-arctic	1. Tundra-arctic	Arctic Tundra	— —	— —
	2. Subarctic turf	Turfy meadow	—	—
	3. Tundra-marshy	—	—	Marshy-tundra
	4. Arctic-saline (solon-chak)	—	Arctic-saline (solonchak)	—
Class II Boreal-permafrost taiga	1. Permafrost-taiga	Taiga-ferruginous Pale yellow taiga	— Pale yellow gleyey	— —
	2. Permafrost-marshy	—	—	Marshy-permafrost
	3. Permafrost-solonetz	Permafrost solods	Permafrost-gleyey solods	—
Class III Boreal-taiga and forest	1. Podzolic forest	Podzolic Gray forest	Podzolic-marshy Gray forest-gleyey	— —
	2. Turfy taiga	Turfy taiga (including the turfy-carbonated)	Turfy-gleyey	—
	3. Marshy	—	—	Marshy

*Compiled by Ivanova in collaboration with N. N. Rozov.

Paralleling the work shown in Table 7-1, the same author (Ivanova, 1956a) proposed the following soils specifically for the northern part of the U.S.S.R.:

Tundra-arctic zone
{
1) Arctic soils
2) Tundra soils
3) Turfy-meadow soils
}

Taiga forested zone
{
4) Podzolic soils
5) Podzolic-marshy soils
6) Marshy soils
7) Turfy-forest soils (high in bases)
8) Turfy-gleyey soils
}

Northern forest-steppe zone
{
9) Gray forest soils
10) Gray forest soils with gley
}

In all three zones
{
11) Floodplain alluvial soils
12) Floodplain turfy soils
13) Floodplain-alluvial marshy soils
}

Following the above general scheme, Ivanova (1956a) expanded her listing for the tundra-arctic zone as follows:

Arctic* Soils
  Arctic gley-turf soils
    Arctic gley-turfy solonchak soils
    Arctic gley-turfy soils
    Arctic gley-humus-turfy soils
  Arctic crypto-gley polygonal soils
    Arctic crypto-gleyey polygonal soils
    Arctic crypto-gleyey soils of stone polygons
Tundra+ Soils
  Tundra gleyey soils
    Tundra surface-gleyey soils
    Tundra humus-peaty-gleyey soils
    Tundra residual-gleyey (often carbonate-bearing) soils
  Tundra gleyey podzolized soils
    Tundra surface-gleyey podzolized soils
    Tundra humus-slightly peaty-gleyey podzolized soils
    Tundra humus-peaty-gleyey podzolized soils

---

*Approximates polar desert and subpolar desert soil zones as used in this volume.
+Approximates tundra soil zone as used in this volume.

Tundra humus-gleyey podzolized soils
Tundra residual-gleyey podzolized soils
Tundra illuvial-humus soils
Tundra illuvial-humus soils
Tundra humus-peaty-illuvial-humus soils
Tundra humus-illuvial-humus soils
Tundra surface-illuvial-humus soils
Turfy-Meadow Soils
Developed at the forest edge and most commonly found (on a percentage-of-the-landscape basis) in the oceanic regions of the boreal belt of Kamchatka and the northern Kuriles.
Type I. Turfy-meadow soils
Subtype 1. Gleyey turfy-meadow soils
Subtype 2. Turfy-meadow soils

Ivanova's classification scheme for all soils of the globe comprised a total of 40 soil formation groups and included the various intrazonal soils (Ivanova, 1956b). These groups were subdivided according to two principles: first, according to the transformation of substances and the general ecological conditions of soil formation (class of soil), and second, according to elements of soil formation (type of soil) (Table 7-1). The concept of the group was built on the first principle, and the concept of the type on the second, that is, on the genetic relations among the types. Type of soil expressed specific properties somewhat along the lines of a catenary arrangement: automorphic, automorphic-hydromorphic, and hydromorphic. Although there is some overlap between the two principles, the emphasis is placed on the soil formation groups, the types representing divisions of the groups.

Ivanova *et al.* (1961), describing the general geography and classification of polar soils, distinguished three facies within the Siberian tundra zone:

Western
a) Tundra-arctic soils of the moss-grass subzone. Mainly humic-tundra arctic gley soils.
b) Tundra gley soils of the moss-lichen subzone. Gleyed soils with permafrost at the 20-inch depth.
c) Tundra gley podzolic soils of the moss-shrub subzone. Gleyed soils with some oxidation in the gleyed horizons.
Central
Tundra soils without thixotropic properties. High oxidation in upper part of profile. pH values of 5 to 6, with 80 to 85 percent base saturation.

Eastern (oceanic climate)
   Slightly gleyed tundra-arctic soils. Weakly gleyed soils displaying the sod process.

The above generalizations parallel somewhat the earlier work of Gorodkov (1939), who distinguished western and eastern areas as well as north-south pedogenic gradients in Siberia.

The System of the United States Department of Agriculture

Since the early 1950's, investigators from the United States Department of Agriculture have developed a new system of soil classification that departs considerably from most recognized systems. The system accounts for the various qualitative processes — singularly or by group affinities. The terminology is loosely based on Greek and Latin roots. The first or highest unit of classification is the Order, of which there are 10 (Table 7-2). The second order, designated as the Suborder, is established by one of a number of factors, such as climate, time, material, or landscape elements. In the third order of classification, the Great Group, the Greek *cryo* (*kryo*-) is used as a prefix to designate a soil of the cold zone (usually where permafrost exists within 30 inches of the surface).

This system has not been tested extensively in polar regions. Some investigators, however, have used it in Alaska, particularly in the southern and western parts of the state (Rieger, 1966; Holowaychuk et al., 1966; Allan, 1969).

Rieger (1966), seeking to establish some international uniformity in soil terminology, suggested that the well-drained soils of the tundra and alpine regions be combined into one system built on soil morphology. Accordingly, he suggested that the system used by the United States Department of Agriculture (*Soil Classification, a Comprehensive System*, Seventh Approximation, 1960) be adopted. By this he meant that the variously designated Arctic brown soils (Tedrow & Hill, 1955), Alpine turf soils (Retzer, 1956), Alpine meadow soils (Retzer, 1956), and Sod-mountain soils (Rode & Sokolov, 1960) should be grouped together under the name Cryumbrepts, but that where such soils have a high base saturation they should be designated as Cryoborolls (Table 7-2).

Arctic soil research clearly shows that there are many soils with similar properties that can be found on a circumpolar basis, but, as Rieger stated, they are confused as a result of being identified by various names.

TABLE 7-2   **Excerpts from United States Department of Agriculture, "Soil Classification, a Comprehensive System," Seventh Approximation (1960), and as amended by the 1967 and 1968 supplements.**

Order	Suborder	Great Group*	Approximate Equivalents
Entisols†	Aquents‡ Orthents Psamments	Cryaquents§ Cryorthents Cryopsamments	Azonal soils and some Low Humic Gley soils
Vertisols			Grumusols
Inceptisols	Aquepts Andepts Umbrepts Ochrepts	Cryaquepts Cryandepts Cryumbrepts Cryochrepts	Ando, Sol Brun Acide, some Brown Forest, Low Humic Gley, and Humic Gley soils
Aridisols			Desert, Reddish Desert, Sierozem, Solonchak, some Brown and Reddish-Brown soils, and associated Solonetz
Mollisols	Aquolls Borolls	Cryaquolls Cryoborolls	Chestnut, Chernozem, Brunizem (Prairie), Rendzinas, some Brown, Brown Forest, and associated Solonetz and Humic Gley soils
Spodosol	Aquods Orthods Humods	Cryaquods Cryorthods Cryohumods	Podzols, Brown Podzolic soils, and Ground-Water Podzols
Alfisols			Gray-Brown Podzolic, Gray Wooded soils, Noncalcic Brown soils, Degraded Chernozem and associated Planosols, and some Half-Bog soils
Ultisols			Red-Yellow Podzolic soils, Reddish-Brown Lateritic soils of the United States and associated Planosols, and Half-Bog soils
Oxisols			Laterite soils, Latosols
Histosols	Fibrists Folists Hemists Saprists	Cryofibrists Cryofolists Cryohemists Cryosaprists	Bog soils

* With a cryic soil temperature regime or with permafrost, the prefix cry is used.
† Entisols: Recent soil.
‡ Aquents: The prefix of Entisol (Ent) becomes the suffix of Aquent. Aqu combines with ent to form Aquent.
§ Cryaquents: Cold, wet, recent soil.

The Canadian System

From time to time the National Soil Survey Committee of Canada meets to consider the subject of soil classification for Canada. Basically, the system followed does not depart very much from the ideas set forth by Dokuchaev and by Marbut, but it is more complete and has been refined (National Soil Survey Committee of Canada, *Proceedings*, 1968). The highest taxonomic units are the orders, of which there are 8 (Table 7-3). The Chernozemic and Solonetzic soils are not extended into the polar regions. The Podzolic and Brunisolic orders provide for the *normal* soils; the Regosolic order provides for various soils with little or no horizon development. Gleysolic soils include the gamut of poorly drained mineral soils with an organic layer up to 1 foot thick. The organic soils provide for peat and muck deposits. Within the Canadian system, *cry* is attached to the names of 10 subgroups. The names Cryic regosol and Cryic gleysol indicate similarity to Regosol and Gleysol but also that those soils are underlain by permafrost within a depth of 40 inches of the surface.

The Canadian system will probably be most useful in the subarctic regions and the main tundra belt. Day (1964), however, has used it in the high arctic. A major point in favor of the Canadian system is that it dispels the belief that a qualitative change takes place at the tree line, a notion that has plagued the literature for many years.

Other Systems of Polar Soil Classification

Fedoroff (1966a, b), working in the islands of the Svalbard group, considered not only local soil characteristics but also the overall problem of classifying polar soils. His work is of great importance because he deals with climatic elements, together with soil properties, frost action, and patterned ground, in determining regional patterns of soils. He divided the polar lands into zones based on climate (Table 7-4) and listed the major automorphic elements in each zone. Although his work lacks detail, it gets to the core of classification problems and criteria for the development of a unified soil classification system for polar soils.

In a general report on cryosols, Fedoroff (1966b) provided a critical review of the state of knowledge on frozen soils and then elaborated his own views. He stated that cryosols, even though greatly altered or disturbed by the action of the cold, always display some genetic characteristics of their own. For this reason, he divided cryosols into the following classes: Raw soils, "Little-evolved" soils, "Browned" soils, Podzol and podzolic soils, Halomorphic soils, and Hydromorphic soils.

TABLE 7-3   Excerpts from Canada Department of Agriculture, "The System of Soil Classification for Canada," 1970. (Day & McKeague, 1972.)

Order	Great Group	Subgroup
1. Chernozemic		
2. Solonetzic		
3. Luvisolic		
4. Podzolic		
	4·32	
		4·32-/7 Cryic Humo-Ferric Podzol. Podzol soils underlain by permafrost within 1 meter of the surface. Lower part of solum may be gleyed.
5. Brunisolic		
	5·2	
		5·21-/7 Cryic Orthic Eutric Brunisol. Soils with permafrost within 1 meter of the surface.
	5·4	
		5·41-/7 Cryic Orthic Dystric Brunisol. Soils with permafrost within 1 meter of the surface.
	5·43 Alpine Brown	
		5·43-/7 Cryic Alpine Brown. Soils with permafrost within 1 meter of the surface.
6. Regosolic		
	6·1 Regosol	
		6·1-/7 Cryic Regosol. Soils with permafrost within 1 meter of the surface. Lower part of profile may be gleyed.
7. Gleysolic		
	7·1 Humic Gleysol	
		7·1-/7 Cryic Humic Gleysol. Any Humic Gleysol with permafrost with 40 inches of mineral soil surface. May have up to 12 inches of consolidated peat on the surface.
	7·2 Gleysol	
		7·2-/7 Cryic Gleysol. Any gleysol with permafrost within 40 inches of the mineral surface. May have up to 12 inches of peat on the surface. Resembles the better-drained Cryic Humic Gleysol.
8. Organic		
	8·1 Fibrisol	
		8·1-9 Cryic Fibrisol. Fibric peat with permafrost in the control section.
	8·2 Mesisol	
		8·2-9 Cryic Mesisol. Dominantly mesic peat with permafrost in the control section.
	8·3 Humisol	
		8·3-9 Cryic Humisol. Dominantly humic peat with permafrost in the control section.

314

TABLE 7-4   Classification of cryosols according to climate, vegetation, and automorphic elements by Fedoroff (1966a).

Cryosol	Climatic Conditions	Nature of Pedogenic Development	Nature of Plant Cover	Classification — Group	Classification — Subgroup
Polar soils	Mean temperatures of warmest month of the year < +5°C.	Intense disintegration; considerable cryoturbation.	None	Arctic polygonal soils	Typical arctic polygonal soils
					Arctic polygonal soils intergrading with poorly differentiated soils of tundra
Northern tundra soils in humid climate (Spitsbergen)	Mean temperatures never > 10°C; sum of temperatures > +5°C never > 100°C. Precipitation > 250 mm.	Intense disintegration and cryoturbation; weak chemical and biochemical reactions.	Discontinuous in well-drained sites; continuous in imperfectly drained soils.	Northern tundra soils with humid climate	Well-drained soils
					Moderately drained soils on undifferentiated materials
					Moderately or poorly drained soils on differentiated materials
					Poorly or very poorly drained soils
Northern tundra soils in dry climate (eastern Siberia)	Precipitation near 100 mm; summer drought.	Disintegration and cryoturbation not very active; cation enrichment in horizons above permafrost.			
Southern tundra soils in humid climate (Kola Peninsula)	Sum of temperatures > +5°C more than 100°C. Precipitation greater than evaporation.	Some disintegration and cryoturbation; gley development.			
Southern tundra soils in dry climate (eastern Siberia)	Low precipitation; summer drought.	Weak disintegration and cryoturbation; cation enrichment in horizons above permafrost.			

Within the Raw soil class are four groups: (1) Lithosols with no organized features; (2) Regosols without development; (3) Regosols with a developed morphology, of which there are two subgroups, those with a closed geometric pattern and those with an open geometric pattern; and (4) Regosols on terminal moraines.

In the case of "Little-evolved" soils, Fedoroff suggests that a subclass of cryosols be distinguished, and that it be divided into four groups based on degree or nature of profile disturbance:

1) Cryosols with prominent unordered ice segregation in which the ice formation influences the vegetation. Within this group, hydromorphic features may be in evidence, and, accordingly, two subgroups can be recognized: (a) Cryosols with ice lenses, for example, palsas; and (b) Cryosols on thufur ground (small vegetated hummocks).

2) Cryosols with prominent ice segregation ordered in a network (Fig. 7-2 and Table 7-4). Within this second group are two subgroups: (a) Cryosols with a polygonal network of large size and a depressed center (Types C to E);* and (b) Cryosols with mounds whose ice cores are ordered in a network (Type F).

3) Cryosols ordered in a network without prominent ice segregation. These soils become partially dry during the course of the summer; and when they freeze, their water content is not sufficient to cause much ice segregation. The permagel (permafrost) is surficial, but locally can be as much as 4 feet deep. Hydromorphy may be manifested by rust spots. This third group has four subgroups: (a) Cryosols with small mounds separated by shallow cracks; (b) Cryosols with mud boils; (c) Cryosols with polygonal and circular networks of rocks; and (d) Cryosols with parallel forms of patterned ground (striated soils).

4) Cryosols without either ice segregation or well-developed patterned ground forms. Permafrost usually begins 4 feet below the surface. This fourth group is divided into three subgroups: (a) skeletal Arctic brown soil; (b) modal Arctic brown soil; and (c) Arctic brown soil with hydromorphic features at depth.

In the "Browned" soil class, Fedoroff (1966b) suggests the introduction of a new group, Brown boreal permagel soils. The group would be divided into several subgroups: Eutrophic brown permagel soils and Pale yellow permagel soils, as well as a Gray forest permagel soil, which differs from the modal Gray forest soil in that it has certain frost features.

---

*The ice-wedge polygon types to which Fedoroff refers are the types established by Drew & Tedrow (1962) according to degree of development. See Chapter 14, Fig. 14-1.

Main Forms of Cryoturbation		Forms of Cryoturbation According to Type of Drainage			
Type	Subtype	Well-drained or moderately well-drained medium (region)	Moderately or poorly drained medium (region)	Poorly drained medium (region)	Very poorly drained medium (region)
Soils without geometric network					Soils with ice lentils
Segregations of ice connected to plant cover				Thufur soils	Thufur soils
I. Soils with geometric network					
Type 1. Polygonal networks of ice wedges	Large-sized (from 15 to 30 meters in diameter) with ice wedges	Soils with ice wedges of Types A and B (Drew & Tedrow)	Soils with ice wedges of Types B through D	Soils with ice wedges of Types D and E	Soils with ice wedges of Type E
	Small-sized, which can be circular (from 0.20 to 1 meter in diameter) with little or no ice	Soils with ice wedges demarcating small mounds ──────→ (Thufur soils)			
Type 2. Polygonal networks with a mesh 1 meter in diameter. Differential segregation of fine materials from the exterior toward the interior of the mesh.	Polygonal network with stone rings	Soils with slightly developed stone rings	Soils with developed stone rings	(Thufur soils)	
	Circular networks with bands of stones	Soils with slightly developed stone rings	Soils with rock circles	(Thufur soils)	
	Circular networks with vegetated margins	Soils with continuous plant cover without microrelief	Soils with mud boils	(Thufur soils)	(Thufur soils) (Soils with ice wedges of Type F)
	Networks with parallel stripes	Soil stripes	Soil stripes		

----------- Frequently observed evolution: change of form without important drainage change.

————— Evolution linked to a change in drainage.

Figure 7-2 Forms of cryoturbation as proposed by Fedoroff (1966b).

Classes of Genetic Soils		
Zonal Soils	**Intrazonal Soils**	**Azonal Soils**
Soils of arctic tundras Tundra gley soils	Arctic solonchak Soils of arctic marshes	
Soils of bare peaks Straw-yellow taiga soils	Dark-colored soils of thermokarst depressions	Soils of frozen hummocky swamps

Figure 7-3   Genetic grouping of arctic soils in the U.S.S.R. as outlined by Gerasimov and Glazovskaia (1960).

As to the Podzol and Podzolic soils class, Fedoroff (1966b) states that it is necessary to include a group of Permagel podzols in the cold climate nonhydromorphic podzolic subclass. These are related to Non-permagel podzols by the discontinuity of the A_2 horizon and a tonguing effect. Stagno-gley soils would be included in the subclass of hydromorphic podzols.

Fedoroff proposes that three subgroups be recognized in the Halomorphic soil class: (1) soils with small mounds separated by wide, shallow cracks; (2) soils with mud boils; and (3) soils with polygonal and circular networks of small size. Others within the group of lixiviated ("leached") alkali soils are Permagel "solods" and Pale yellow "solodized" soils.

Hydromorphic soils are divided into two groups: those with unordered ice segregation and those with prominent ice segregation ordered in a network of Types D, E, and F (Fig. 7-2) and characterized by accumulation of organic matter as the summer temperatures rise.

Other factors considered in Fedoroff's classification proposal are the depth to the gley horizon, frost action, and patterned ground. Fedoroff's reports (1966a, b) are valuable because his classification schemes distinguish the soil processes from processes resulting from frost action, yet, in the final analysis, represent the soil as the product of both processes.

Gerasimov and Glazovskaia (1960) list two varieties of zonal soils for the arctic: the well-drained automorphic soils of the uplands and the hydromorphic soils of the poorly drained expanses of the tundra (Fig. 7-3).

Detailed treatment of soil characteristics in conjunction with soil classification in northern areas has been attempted by very few investigators. Targulian (1971) published the results of an exhaustive study of soil development in the arctic areas, stressing his own research in the East European tundras. He shows that soil formation in the tundra, forest-tundra, and northern taiga is subject to a slow rate of weathering of the minerals and a slow rate of organic matter decomposition. These processes are slowed down by the low temperatures. Once elements are released, however, they are mobile because of the high humidity. The nature of the soil-forming processes in these northern regions is such that the soil contains an acid medium, a leached and unsaturated soil, an insignificant chemical-mineralogical transformation product of the parent rock, as well as an accumulation of peat or mor humus with mobile organic compounds. Such a combination of properties does not occur in any other climatic zone of the globe.

Rather than the traditional zonal-facies soil classification system (Tundra, Cryogenic-taiga, etc.) generally set forth, particularly by Soviet investigators, Targulian suggests two general pathways of weathering and therefore recognizes two sets of soil evolution pathways: (1) nongleyic, coarse-textured ferrosialitic, illuvial Al-Fe-humic forms, which develop under free internal drainage conditions; and (2) gleyic-sialitic allochthonous-argillic forms, which occur under conditions of imperfect drainage. As indicated by Targulian, these two sets of weathering processes and corresponding soil formation are equally characteristic of and should be considered zonal in both the tundra and taiga regions. Their genesis and geographic distribution are determined not so much by the broad climatic zones as by local geologic-geomorphic conditions.

In developing his thesis for the recognition of two sets of qualitative processes, Targulian (1971) proposes that the two main varieties of his nongleyic soils formed on the sandy deposits and crystalline rocks be called (1) podburs; and (2) podzolic Al-Fe-humus soils. Podburs are the soils of the tundra and northern taiga. They lack a gleyed horizon and consist of a brown, nonpodzolized profile. The soils in this group include those described in the literature as Cryptopodzolic, Tundra illuvial-humus, Cryogenic-taiga ferruginous, Arctic brown, Subarctic brown forest, Brown wooded, Sod bare-rock, Brown melanized, and other names. It would include the northern Brown forest soils of Scandinavia and the various brown subarctic soils of Canada. According to Targulian, the following processes are characteristic of podburs: leaching, desilification, relative accumulation of aluminum and iron, transformation of layered silicates, illuvial Al-Fe-humus podzolization—generally not expressed morphologically—translocation of clay, and silt suspension. The podburs lack a podzolic horizon, and the A_0 horizon is immediately underlain by mineral, a ferrosialitized illuvial Al-Fe-humus B horizon. The absence of a podzolic horizon is not connected with the substitution of a new process for the podzolic one and can be explained by the extreme weakening, local or temporary slowing down, or lithogenic inexpressiveness of podzolization.

Podzolic Al-Fe-humus soils are formed by the same combination of processes as podburs, but they are characterized by a greater intensity and duration of the Al-Fe-humus podzolization. This accounts for the formation not only of an illuvial brown B horizon but of a bleached podzolic horizon as well. Examples of this group of soils are found in the valleys of the Brooks Range of Alaska (Brown & Tedrow, 1964).

The gleyic soils Targulian describes develop on slightly dissected plains on heavy-textured material or on stratified deposits. He proposes two varieties within the group: (1) Homogeneous gleyic, and (2) Differentiated gleyic. The Homogeneous gleyic soils are characterized by a mineral gleyic profile without any redistribution of clay, SiO_2, Fe_2O_3, or Al_2O_3, and are present in the tundra zone as well as the taiga zone. Profiles are formed by the combination of gleyzation and seasonal oxidation, cryoturbation, transformation of silicates and amorphous compounds in situ, and the accumulation of mobile compounds within the active layer. The Differentiated gleyic soils form within the tundra and northern taiga zones and have a well-expressed eluvial, often bleached, mineral horizon low in clay, Fe_2O_3, and Al_2O_3. Besides the effects of the gley process, these soils are characterized by an eluvial-gleyic transfer of iron and possibly aluminum, as well as transformation of mineral suspension downward.

Targulian's work (1971) agrees remarkably well with the ideas set forth on North America by this author and clarifies many of the ambiguities in the older literature relating to the imagined influence of the tree line on soil properties (Tedrow, 1970a). Of particular significance is the quantitative separation of the gley process, which in effect is a process-intensity separation.

Problems in Formulating an Ideal System

Although a number of investigators from various nations and schools are now focusing on the study of polar soils, there is no well-crystallized concept of a unified classification system or one that has proved acceptable to all. There appear to be some "centers of understanding" as to the genetic processes in soils generally, but there is little agreement on a system of taxonomy. The general rules of soil systematics set forth by Soviet investigators can serve as a basis for formulating a realistic system. I cite once again the review by Gerasimov (1952).

In depicting large-scale patterns of soil development, Glazovskaia (1967) discusses the parameters for consideration and states that the soil-forming processes, so defined, should meet two conditions: (1) they should reflect in the most universal form the end result of soil-forming processes and the result of interaction between organic life and the mineral basis of the soil; and (2) at the same time, these properties should serve as an indicator of present soil-forming tendencies and should relate to other soil properties.

It is not the intention here to debate the merits and shortcomings of various proposals on soil classification (as carried out in the publications of Smith, 1965, and Gerasimov, 1969), but it appears that if the polar soil classification problem is to be solved, it will have to be strongly based on principles of soil genetics.

The properties of the soils must be described before projections can be made about current processes, and if a discussion of systematics is to have value, it should show spatial relationships. Soil systematics, when applied to the polar sectors, becomes very complicated because the arctic environment is a dynamic one with changing climatic patterns and biological zones. There are certain spatial factors that can be followed, such as the embryonic soils on the recently drained lake basins, recently formed glacial moraines, landslides, and recent alluvium, but the main focus should be on current properties and processes within any given landscape. Accordingly, many of the following discussions in this volume are based on my own impressions from the field as well as on the literature.

Toward the end of the nineteenth century, as V. V. Dokuchaev's writings on soil classification began penetrating the scientific communities of Russia and somewhat beyond, he established 5 soil zones whose boundaries roughly approximated broad climatic-vegetative zones—among which was a tundra soil zone. As more data accumulated, the uniformity of soil characteristics he projected within the various zones came in question, and because of this, most of the zones have been divided and subdivided. Glazovskaia (1967), in a review of the history of the development of soil zonality from the time the 5 zones were proposed by Dokuchaev to the present, pointed that the *Fiziko-Geograficheskii Atlas Mira* (Moscow, 1964) lists some 93 subdivisions. Thus, she said (1966), scientists are recognizing more and more that soil belts and soil zones that they once thought were monolithic in character, that is, had homogeneous soil conditions, really encompass a mosaic of soils. In her view, only the polar belt and part of the boreal belt display any uniformity in terms of soils. Since Glazovskaia made these remarks, international investigations in the field have shown that the polar lands do indeed require many soil subdivisions. The more one examines soils of the polar landscapes in detail, the more apparent the great multiplicity of soil conditions.

TABLE 15-1 Comparison of categories used in soil classification in the Soviet Union and in the United States as of 1938 and changes proposed in this volume.

First Order	Second Order	Third Order	Fourth Order	Fifth Order
		AS PROPOSED IN THIS VOLUME		
Soil Zone (tundra, polar desert, etc.)	Great Soil Group	Genetic Soil Type (qualitative and quantitative variations)*	Genetic Soil Type + Wetness Factor (wetness)†	Genetic Soil Type + Wetness Factor + Patterned Ground (patterned ground)
	UNITED STATES DEPARTMENT OF AGRICULTURE AS OF 1938†			
Bioclimatic Zonation	Great Soil Group (genetic)	Soil Family (soils with similar profiles) / Soil Series	Soil Type	
	VARIOUS SOVIET INVESTIGATORS‡			
Global Group/ Class	Soil Type (genetic)	Soil Sub-type / Soil Genus	Soil Species	

*This category to be divided according to need, e.g., Family, Series, and Type, or Subtype, Genus, and Species, as listed below.
†Baldwin, Kellogg & Thorp (1938).
‡Gerasimov (1952); Ivanova (1956b); and others.

TABLE 15-2a   Polar soil classification, as proposed in this volume: Tundra soil zone.

First Order	Second Order	Third Order	Fourth Order	Fifth Order
TUNDRA SOIL ZONE	Well-Drained Soils 　Arctic brown soil* 　Podzol-like soil† Mineral Gley Soils' 　Upland Tundra soil‡ 　Meadow tundra soil‡ Organic Soils 　Bog soils§ Other Soils 　Ranker soil 　Rendzina soil 　Shungite soil 　Grumusols 　Lithosols 　Regosols 　Soils of the solifluction slopes‖	Separations based on textural and mineral properties of the parent material, etc.	Soil type + wetness factor (applies mainly to Tundra and Bog soils)	Soil type + wetness factor + patterned ground

*Equivalent to the Podburs of Targulian (1971) or the Cryumbrepts or Cryborolls of Rieger (1966).

†Equivalent to the Podzolic Al-Fe-humus soils of Targulian (1971).

‡This classification is both morphologic and genetic. Targulian (1971) recommends a separation of Tundra soils on the basis of genetic horizonation. Accordingly, Homogeneous gleyic and Differentiated gleyic soils as proposed by Targulian can be provided for in the second order of classification.

§Half-bog soils may also be recognized in this category (Tedrow et al., 1958).

‖Genetic identity is greatly obscured in these soils, and in classifying them one probably must revert to an edaphic-cryogenic nomenclature designating wetness and type of plant communities (and vegetated or nonvegetated), as well as the relative stability of the soil material. Edaphically the soil approximates Tundra, except in the comparatively dry, shallow conditions of the mountains, where it approximates Ranker.

TABLE 15-2b Polar soil classification, as proposed in this volume: Polar desert soil zone.

First Order	Second Order	Third Order	Fourth Order	Fifth Order
	Well-Drained Soils Polar desert soil Arctic brown soil *Mineral Gley Soils* Upland tundra Meadow tundra Soils of the hummocky ground	Separations based on textural and mineral properties of the parent material, etc.	Soil type + wetness factor (applies mainly to Tundra and Bog soils)	Soil type + wetness factor + patterned ground

POLAR DESERT SOIL ZONE	Soils of the polar desert–
	    tundra interjacence
	Organic Soils
	Bog soils
	Other Soils
	Regosols
	Lithosols
	Soils of the solifluction
	    slopes (may be a form
	    of gley soil but usually
	    well drained)			

324

TABLE 15-2c   Polar soil classification, as proposed in this volume: Subpolar desert soil zone.

First Order	Second Order	Third Order	Fourth Order	Fifth Order
	Well-Drained Soils	Separations based on tex-	Soil type + wetness factor	Soil type + wetness factor
	Polar desert soil*	tural and mineral prop-	(applies mainly to	+ patterned ground
	Arctic brown soil†	erties of the parent ma-	Tundra and Bog soils)	
	Mineral Gley Soils‡	terial, etc.		
	Upland tundra			
	Meadow tundra			
	Soils of the hummocky			
	grounds§			
SUBPOLAR	Soils of the polar desert–			
DESERT SOIL	tundra interjacence			
ZONE	*Organic Soils*			
	Bog soils			
	Other Soils			
	Regosols			
	Lithosols			
	Soils of the solifluction			
	slopes			

*Polar desert soil is not as arid in the subpolar desert zone as it is in the polar desert zone. There is generally more vegetation, and the B and C horizons, particularly the C, have more moisture.
† Has virtually the same morphology as the Arctic brown soil in the polar desert zone but with a thicker solum.
‡ This entire group would comprise the Arctic gley-turf and the Arctic cryptogley polygonal soils of Ivanova (1956a).
§ Has browner colors in the "solum" than is the case in the polar desert soil zone.

325

TABLE 15-2d Polar soil classification, as proposed in this volume: Cold desert soil zone (Antarctica).*

First Order	Second Order	Third Order	Fourth Order	Fifth Order
	Ahumic (Frigic)[†] Soils	Separations based on tex-tural and mineral prop-erties of the parent ma-terial, etc.	Not applicable for the mainland of Antarctica	Soil type + patterned ground
	Ultraxerous[†]			
	Xerous[†]			
	Subxerous[†]			
	Ahumisol[‡]			
	Evaporite Soils			
COLD	*Ornithogenic (Avian)*[†]			
DESERT	Soils[§]			
SOIL	*Other Soils*			
ZONE	Protoranker			
	Algae peats[†]			
	Hydrothermal soils[‖]			
	Regosols (recent soils)			
	Lithosols			

*The northern Antarctic Peninsula and the Antarctic Islands are likely to be allied to the polar desert, subpolar desert, and tundra zones for purposes of classification.
[†]Campbell & Claridge (1969).
[‡]MacNamara (1969a).
[§]Syroechkovsky (1959).
[‖]Ugolini & Starkey (1966).

REFERENCES

[*Editor's Note:* Only those references cited in the preceding excerpts are reproduced here.]

Allan, R. J., 1969. Clay mineralogy and geochemistry of soils and sediments with permafrost in interior Alaska. Report to Cold Reg. Res. and Eng. lab. (Hanover, N.H.). 289 pp. (mimeo).

Baldwin, M., Kellogg, C. E., and Thorp, J., 1938. Soil Classification. *Soils and Men*, Yearbook of Agriculture, U. S. Dept. of Agr. (Washington, D. C.), pp. 979–1001.

Campbell, I. B., and Claridge, G. G. C., 1969. A classification of frigic soils—the zonal soils of the antarctic continent. *Soil Sci.* 107:75–85.

Canada National Soil Survey Committee, 1968. *Proceedings, 7th Meeting*, Univ. of Alberta (Edmonton). 216 pp. (mimeo).

Day, J. H., and McKeague, J. A., 1972. *The System of Soil Classification for Canada*. Presented at the 22d Int. Geogr. Congr., Montreal. Canada Dept. of Agr., Soil Res. Inst. (Ottawa). 21 pp. (mimeo).

Drew, J. V., and Tedrow, J. C. F., 1962. Arctic soil classification and patterned ground. *Arctic*, **15**:109–116.

Fedoroff, N., 1966a. *Les sols de Spitsberg occidental*. Spitsberg, 1964, Cent. Nat. de la Rech. Sci. Rech. Coop. Programme 42 (Lyon), Chap. 10, pp. 111–228.

Fedoroff, N., 1966b. Les cryosols. *Sci. du Sol*, **2**:77–110.

Gerasimov, I. P., 1952. Nauchnye osnovy sistematiki pochv (Scientific fundamentals of soil systematics). *Pochvovedenie*, **11**:1019–1026 (U. S. Dept. of Com. trans., TT65–50060).

Gerasimov, I. P., 1969. Reviziya geneticheskikh osnov Dokuchayevskogo pochvovedeniya v nevoi Amerikanskoi klassifikatsii pochv i rabotakh po sostavleniyu mirovoi pochvennoi karty (Revision of the genetic principles of Dokuchaev's soil science in the new American classification of soils and in works on compilation of the world soil map). *Pochvovedenie*, **9**:3–17 (Scripta Technica, Soviet Soil Sci., pp. 511–524).

Gerasimov, I. P., and Glazovskaia, M. A., 1960. *Osnovy pochvovedeniia i geografiia pochv (Fundamentals of Soil Science and Soil Geography)*. Gosudarstvennoe Izatelstvo Geograficheskoi Literatury (Moscow). 489 pp. (Israel Program of Sci. Trans., TT65–50061, 382 pp.).

Glazovskaia, M. A., 1967. (General patterns in the world geography of soils). *Vest. Moskovskogo Univ., Ser. Geogr.*, **4**:11–27 (trans. *Soviet Geogr.*, **8(4)**:208–227). (Russian original pub. in 1966).

Gorodkov, B. N., 1939. Ob osobennostiakh prochvennogo pokrova arktiki (Peculiarities of the arctic topsoil). *Izv. Gosud. Geogr. Obshch.*, **71**:1516–1532.

Grigor'ev, A. A., 1930. Vechnaia merzlota i drevnee oledenenie (Permafrost and ancient glaciation). *Mater. Komis. po Izucheniiu Estestvennykh Proizvoditelnykh Sil*, Soviet Acad. Sci. **80**:43–104.

Holowaychuk, N., et al., 1966. Soils of Ogotoruk Creek watershed. *Environment of the Cape Thompson Region, Alaska* (N. J. Wilimivsky, ed.), U. S. Dept. of Com. (Springfield, Va.), Chap. 13, pp. 221–276 (LC 66–60018).

Ivanova, E. N., 1956a. Sistematika pochv severnoi chasti Evropeiskoi territorii SSSR (Classification of soils of the northern parts of the European U.S.S.R.). *Pochvovedenie*, **1**:70–88 (U. S. Dept. of Com. trans., OTS61–11495).

Ivanova, E. N., 1956b. Opyt obshchei klassifikatsii pochv (An attempt at a general soil classification). *Pochvovedenie,* **6**:82–102 (U. S. Dept. of Com. trans., OTS61–11495).

Ivanova, E. N., et al., 1961. Novye materialy po obshei geografii i klassifikazii pochv poliarnogo i borealnogo poiissa Sibiri (New materials on the general geography and classification of soils in the polar and boreal belts of Siberia). *Pochvovedenie,* **11**:7–23 (AIBS trans., pp. 1171–1181).

Liverovskij. J. A., 1934. Pochvy tundr Severnogo kraia (Soils of the tundras of the northern regions). *Tr. Poliarnaia Komissi,* Soviet Acad. Sic. (Leningrad), 19. 112 pp. (In Russian with English summary).

MacNamara, E. E., 1969a. Active layer development and soil moisture dynamics in Enderby Land, East Antarctica. *Soil Sci.,* 108:345–349.

Retzer, J. L., 1956. Alpine soils of the Rocky Mountains. *J. Soil Sci.,* 7:1–32.

Rieger, S., 1966. Dark, well-drained soils of the tundra regions in western Alaska. *J. Soil Sci.,* 17:264–273.

Rode, T. A., and Sokolov, I. A., 1960. K kharakteristike Gorno-Tundrovyke landshaftov Zabaikalia (The Transbaikal mountain-tundra landscapes). *Pochvovedenie,* **4**:47–56 (AIBS trans., pp. 384–391).

Rozov, N. N., 1956. K voprosu o printsipakh postroeniye geneticheskoi klassifikatsii pochv (Principles of elaborating a genetic classification of soils). *Pochvovedenie,* **6**:76–83 (U. S. Dept. of Com. trans., OTS61–31001).

Sheludiakova, V. A., 1938. Rastitelnost basseina reki Indogirki (The vegetation of the Indigirka River basin). *Sovetskaia Bot. (Leningrad),* **4–5**:43–79.

Smith, G. D., 1965. Lectures on soil classification. *Pédologie* (Ghent) (Spec. No.) **4**:121–134.

Sochava, V. B., 1933. Tundry basseina reki Anabary (Tundras of the Anabar River basin). *Izv. Vsesoiuzn Geogr. Obshch.,* **65(4)**:340–364.

Syroechkovsky, E. E., 1959. Rol zhivotnykh v obrazovanii pervichnykh pripoliarnoi oblasti zemnogo shara, na primere Antarktiki (The role of animals in primary soil formation under conditions of prepolar region of the globe, as exemplified by the Antarctic). *Zool. Zhur.* (Moscow), **38**:1770–1775.

Targulian, V. O., 1971. *Pochvoobrazovanie i vyvetrivanie v kholodnykh gumidnykh oblastiakh* (Soil Formation and Weathering in Cold Humid Regions). Soviet Acad. Sci. (Moscow). 266 pp.

Tedrow, J. C. F., Bruggemann, P. F., and Walton, G. F., 1968. Soils of Prince Patrick Island. *Arctic Inst. of No. Amer.* (Washington, D. C.) *Res. Paper,* 44. 82 pp.

Tedrow, J. C. F., and Hill, D. E., 1955. Arctic brown soil, *Soil Sci.,* **80**: 265–275.

Ugolini, F. C., and Starkey, R. L., 1966. Soils and micro-organisms from Mt. Erebus, Antarctica. *Nature,* **211**:440–441.

United Stated Department of Agriculture, Soil Conservation Service— Soil Survey Staff, 1960. *Soil Classification, a Comprehensive System, Seventh Approximation,* U.S. Govt. Ptg. Off. Washington, D. C.). 265 pp.

Part V

SPECIAL APPROACHES

Editor's Comments
on Papers 24, 25, and 26

24   ARKLEY
Factor Analysis and Numerical Taxonomy of Soils

25   PORTLAND CEMENT ASSOCIATION STAFF
Soil Classification Systems

26   RADFORTH
Suggested Classification of Muskeg for the Engineer

Special approaches to soil classification are directed toward
the heirarchial ordering of classes by non-Linnaean or coordinate
means, as in the case of numerical taxonomy (Paper 24), or they
reflect interest in particular soil properties of engineering significance. The approaches outlined in Papers 25 and 26, for example,
consider soil merely as a construction material.

Numerical taxonomy, as defined by Sokal and Sneath (1963),
involves "the numerical evolution of the affinity or similarity
between taxonomic units and the ordering of these units into
taxa on the basis of their affinities." The axioms of numerical taxonomy, derived from ideas of phenetic empirical taxonomy put
forward by botanists in the eighteenth century, are that: (1) taxa
are to be based on as many characters as possible; (2) every character carries equal weight in the analysis; (3) affinity equals overall
similarity; (4) different character correlations may be used to
separate distinct taxa; and (5) taxonomy is strictly empirical.
Clustering procedures, as employed in Paper 24, eliminate all so-
called redundant groupings so that the resulting dendogram
contains only monothetic, nonredundant groups. Arkley's approach thus marks an important departure from most previous
efforts and signals a new line of attack on the problem of soil
classification. Advantages of factor or cluster analysis, according
to Arkley (Paper 24), include the selection of a minimal set of soil
properties, those that are most appropriate for classification, and
that can then serve as the basis for a coordinate system of soil
classification. In addition to facilitating the problem of soil corre-

lation (see Editor's Comments on Paper 27 in Part VI), this approach seems to overcome the distinct disadvantage of dichotomous separations in categorical systems. Although numerical taxonomy has obvious merits, its general adoption would remove the process of soil classification one step further away from many users of the final results and possibly intimidate nonspecialists by this pedological "black box approach."

The civil engineer views soil as any unconsolidated material found between the ground surface and bedrock. It is seen as material that can be excavated and placed in an earthwork. In short, it is a material amenable to engineering purposes. This concept of soil, one quite different from that of pedologists, is reflected in the engineering classifications of the American Association of State Highway and Transportation Officials (AASHTO), American Society for Testing and Materials (ASTM), and the Federal Aviation Agency (FAA). The AASHTO system is commonly used by highway engineers, the ASTM system by foundation engineers, and the FAA system by engineers charged with the design of dams and airfields. The *PCA Soil Primer* (Paper 25) briefly summarizes particulars of the textural method of classification and other engineering tests.

In the AASHTO system, inorganic soils are classified into seven groups, designated A-1 through A-7, that are in turn divided into twelve subgroups. Organic soils are classified as A-8. A group index is appended to the group and subgroup classification of fine-grained soils. According to the ASTM or Unified System, coarse-grained soils are divided into gravel and gravelly soils (symbol G). The sands and sandy soils (M), inorganic clays (C), and organic silts and clays (0) make up the three groups of fine-grained soils that are further divided according to liquid and plastic limits. Highly organic soils are placed in one group (Pt).

The engineering classifications should be recognized for what they are, namely, ratings of potential behavior characteristics. Such systems only serve as starting points for description of soils under field conditions, because they do not consider properties of intact material as found in nature (Burmister, 1951). Even with these limitations much information concerning general soil properties can be inferred from classifications based on engineering tests.

Radforth's classification (Paper 26) represents an initial attempt to classify organic soils for engineering purposes. This effort is an important milestone, because such materials were long regarded by engineers as "unclassifiable." In the ASTM system, for example, all highly organic soils, such as those occurring as peats and mucks in swamps, are lumped together in one group (Pt) on the basis of visual identification. The needs of the engineer interested in trafficability, construction, and foundation engineering are obviously quite different from those geared not to the classification of organic soils *sensu stricto* but to the identification of organic terrain features using methods of aerial photo interpretation as well as field inspection and laboratory analyses. The engineering classification of muskeg, a Canadian term for organic terrain, is based on the physical condition (structure) of peat and its related mineral sublayer, topographic features, and surface vegetation. For those concerned with the complexity of terminology in *Soil Taxonomy*, there is hope for an equally contentious nomenclature in engineering fields due to the rise of what Radforth calls "palaeovegetography," the specialized field of applied botany on which muskeg interpretation depends.

REFERENCES

Burmister, D. M., 1951, Identification and Classification of Soils, *ASTM Spec. Tech. Publ.* **113**:3–24.
Sokal, R. R., and P. H. A. Sneath, 1963, *Principles of Numerical Taxonomy,* W. H. Freeman, San Francisco, 359 p.

24

Factor Analysis and Numerical Taxonomy of Soils[1]

Rodney J. Arkley[2]

ABSTRACT

A large computer and multivariate statistics were used to select a minimal number of soil properties with the greatest prediction value for use in soil classification. Key communality cluster analysis and principal axis factor analysis with Varimax rotation were applied to six sets of soil data containing from 21 to 66 profile characteristics. The analyses selected from four to seven factors or dimensions which were highly independent of each other and contained from two to four highly correlated soil properties. These factors accounted for 100% of the communality and 90% of the raw variance in all properties used in each set. In spite of the diversity of the soils (220 and 620 California, 59 and 86 World, and 148 Ohio soils) the same factors appeared repeatedly; i.e., properties related to reaction, hue and chroma, texture, color value and mottling, and profile differentiation in five analyses and solum thickness in four. Composite factor scores computed for each soil were also used for Numerical Taxonomy. The groupings formed were sufficiently similar to those of the New Classification System of the USDA to suggest that these methods may be very useful in soil classification. The use of these methods for the development of a coordinate system of soil classification is discussed.

Additional Key Words for Indexing: soil classification, cluster analysis, coordinate classification, non-Linnean classification.

C LUSTER ANALYSIS of objects (a procedure in Numerical Taxonomy) is the use of one of a number of statistical methods for determining the natural groups of objects which make up a population. A number of methods using various measures of similarity or difference between objects have been proposed (6, 7, 10).[3] Most of the methods used heretofore in Numerical Taxonomy avoid the question of the selection and weighting of properties used, by simply using as many properties as possible. This rather "shotgun" approach to the problem is intellectually unsatisfactory and the properties used may be heavily weighted by the selection of unduly large numbers of related properties either due to ease of measurement or the bias of the investigator.

Sarkar, Bidwell, and Marcus (1966) reduced the number of soil properties from 61 to 51, 40, 33, and 22 by successively eliminating soil properties according to decreasing levels of correlation coefficients in a correlation matrix of the original 61 properties. However, a more appropriate method for the analysis of such a correlation matrix is multivariate analysis, which includes factor analysis and a newer method, cluster analysis of variables. It is the purpose of this paper to show that multivariate

analysis of soil properties can be used to select a minimal subset of properties (probably less than 20) which can be grouped into a still smaller number of factors or independent dimensions, which then can be used as the basis for forming groups of soils of any desired degree of homogeneity or for a coordinate system of soil classification.

METHODS AND MATERIALS

Analysis of Variables (Soil Properties)

Data were processed for analysis by the same method used by Cipra et al. (1970). Two statistical procedures were then applied to one set of data (59 WORLD) in order to compare the results. One was Principal Axis factor analysis with Varimax rotation (Harmann, 1960); the other was Cumulative Communality cluster analysis (CC5) developed by Tryon and Bailey (1966) and outlined by Crovello (1968). Comparison of the results (Table 1, columns 5 and 6) indicates that they are so similar that either can be used with equal effectiveness. Since the CC5 method is less expensive of computer time, only CC5 was applied to the other sets of data.

Cluster Analysis of Objects (Soil Profiles)

The method used for finding natural groups of soils in this study was a new one called Condensation O-type Analysis also developed by Tryon and described briefly by Arkley (1968). It has the advantage over most previous methods used in Numerical Taxonomy in that it first isolates groups of related objects with a predetermined degree of homogeneity and then finds the relationships between groups and of unique soils which are not members of any group. The relationships are expressed as data for the preparation of a phenogram (dendrogram) using a centroid-sorting procedure in which each group is treated as if it were located at a centroid point represented by the average coordinates of all members of the group. Groups are merged progressively, on the basis of the nearest pair of groups in terms of Euclidean distance in multidimensional space. The centroid concept used here is akin to the "modal soil" concept used in soil correlation. The method used for preparing phenograms such as Fig. 1 is very similar to the Unweighted Pair Group Method using arithmetic averages described by Sokal and Sneath (1963), except that a few unique soils are attached to the most similar group after the basic phenogram is complete.

Data

Six different sets of soil data were subjected to analysis of variables. These data included laboratory measurements and/or morphological properties of soil profiles. The number of soils and of soil properties included in each set is shown in Table 1.

Two sets of data (220 CAL and 620 CAL) were taken from descriptions of California soils; the 620 CAL set used only "official" series descriptions. The other sets contained both descriptive and laboratory information. The OHIO set was furnished on punched cards by L. P. Wilding of Ohio State Univ. and consisted of data from about 30 profiles for each of five soil series. The 59 WORLD data were furnished by Cipra et al. (1970) and were taken originally from the *Soil classification. A comprehensive system* (Soil Survey Staff, 1960). The WORLD II and WORLD III data sets were prepared by adding data from the same source to the 59 WORLD set.

Descriptive properties, such as soil structure, were coded subjectively on numerical scales following essentially the same

[1] Contribution from Dept. of Soils & Plant Nutr., Univ. of California, Berkeley, Cailf. 94720. Received July 24, 1970. Approved Dec. 29, 1970.
[2] Soil Morphologist and Lecturer.
[3] Gower, J. C. 1969. The basis of numerical methods of classification. *In* J. G. Sheals (ed.) The soil ecosystem: Systematic aspects of the environment, organisms, and communities. p. 19–30. E. W. Classey, Ltd., 353 Hanworth Road, Hampton, MIDDX, Great Britain.

procedure described by Cipra et al. (1970). These included such properties as structure of the A1 and B2 horizons, prominence of clay skins in the B2 horizon, degree of mottling (contrast, abundance, size), Fe-Mn concretions, consistence of B2, Ap, and A1 horizons, and abruptness of horizon boundaries. Scatter diagrams of these variables plotted against related metric variables showed in a few cases that the subjective coding used was incorrect, and that a change in the numerical scale used would produce a higher linear correlation between a coded variable and several other metric variables. The result was that simple numerical scale values (such as 0, 1, 2, 3, 4, 5) were actually the most effective.

RESULTS AND DISCUSSION

The results of the analyses of variables (soil properties) are shown in Table 1. The most striking observation is that similar sets of dimensions and definitive soil properties were selected in spite of the fact that six different sets of data were used, the number of soils varied from 59 to 620, and the number of properties from 21 to 54. Some of the differences in the kinds of soil properties included in a dimension were the result of differences in the soil properties included in the data and due to the proportions of kinds of soils included in the analysis. This kind of variation would be expected to disappear if all kinds of soils and soil properties were adequately represented in the data.

The dimensions produced by cluster analysis are essentially independent; that is, the properties within a dimension are highly correlated, but poorly correlated between

dimensions. Properties not included in any dimension may be correlated with properties of more than one dimension and thus can be predicted by multiple correlation. However, there are generally a few properties which are not significantly correlated with any dimension or other property. For the purposes of classification, these may be ignored, or if considered to be important in themselves, they may be set up as independent dimensions, each defined by the single property. For example, of the 53 soil properties used in the analysis of WORLD III data (column 8) which are the kinds of properties used to separate orders in the 7th Approximation, nine properties were poorly correlated (communalities below 0.20). Two of these were poorly represented in the data; the remaining seven were: (i) percent organic carbon in the epipedon, (ii) maximum silt content of any B horizon, (iii) maximum pH of B minus minimum pH of A, (iv) change in color value from A to B, (v) change in hue from A to B, (vi) change in lime content from A to Ca horizon, and (vii) most distinct boundary between any two B horizons.

The variation in numbers of soils included in each set seems to have little influence upon the outcome, except it is generally true that the calculation of correlation coefficients requires at least 40 if they are to have high validity. The number of soil properties does affect the result, however; as the number of properties increases, it is clear that the likelihood of more kinds of dimensions being included

Table 1—Dimensions and key definer soil variables resulting from factor analysis of soil properties

		200CAL	620CAL	OHIO	59WORLD	59WFACTOR	WORLD II	WORLD III
Key definer soil variables								
Soil data set:		200CAL	620CAL	OHIO	59WORLD	59WFACTOR	WORLD II	WORLD III
No. of soils:		220	620	148	59	59	87	86
No. of properties:		34	23	54	21	21*	34	53
DIMENSION								
Reaction	A.	A pH B pH Ca thick Ca depth (.30)†	B pH Ca thick Ca depth (B-A) pH (.16)	B pH B base sat. B exch. H (.18)	B pH (.24)	B pH B exch. Na (.19)	B pH‡ B exch. Na‡	B pH B exch. Na B Na sat. % Months dry (.16)
	B.			A2 pH A2 base sat. (.15)				
Hue-chroma		B hue A hue A chroma (.21)	B hue A hue A chroma B chroma (.29)		B hue A chroma B chroma (.31)	B hue A chroma B chroma (.23)	B hue A chroma B chroma (.24)	B hue A1 chroma A2 chroma B chroma (.19)
Texture		B text. A text. A consist. B value (.18)	B text. B structure A consist. (.25)	A silt B silt (.13)	B clay B consist. B structure (.28)	B clay B consist. B structure B clayskins (.22)	B clay‡ B consist.‡	B clay Min. clay Epi. consist. (.17)
Color Value - mottling		Gley degree‡ Gley depth‡	Gley depth‡ Dark thick.‡		B org. C B value 0 thick. Gley degree (.24)	B org. C B value 0 thick. R depth (.19)	Gley depth B value Ep value Dark thick. (.30)	Gley depth Gley degree B value (.15)
Profile	A.	B/A text. A, B boundary B structure (.20)	B/A text. A, B boundary B clayskins (.29)	Clay gain B22 cl. gain B21 cl. gain (.18)	B/A clay‡ B silt‡	B/A clay B silt A thick. A org. C (.17)	B - A clay B/A clay B structure (.28)	B - A clay B/A clay B structure (.13)
	B.			Fine cl. gain B22 f. cl. gain B/C f. cl. (.20)				
Thickness	A.	R depth B3 thick C thick. (.17)	Solum R depth (.23)	Solum B2 thick. Ca depth (.13)	No data	No data	Solum B2 thick. (.24)	Solum B2 thick. (.21)
	B.			A thick. (.10)				
Cation exchange capacity/clay								B CEC/clay B org. C (.15)

* Analysis for principal axis factoring with Varimax rotation. All others by Key Communality Cluster Analysis.
† Numbers in parenthesis indicate the fraction of initial communality exhausted by the dimension. The fractions may add up to greater than 1.0 because there is some intercorrelation between dimensions.
‡ Dimensions added for cluster analysis of objects (soils) from visual inspection of residual correlation matrix.

in the data is increased. Also, it is clear that dimensions can be "forced" by including a number of highly correlated variables; this was the case in the OHIO data set, and the result was additional dimensions related to Reaction, Profile, and Thickness, of which only the two Thickness dimensions were independent (r < 0.20).

The main conclusion which is drawn from these analyses is that a large number of soil properties can be represented by a relatively few dimensions defined by some 20 or fewer properties. Also, the fact that the same kinds of dimensions appear in the various analyses suggests that they are likely to appear in any analysis which includes a wide range of kinds of soils.

Each of the analyses of variables indicated in Table 1 was followed by a cluster analysis of objects (soil profiles). The results of the analysis of the 220 CAL set have already been reported by Arkley (1968). The results of the 620 CAL analysis are too voluminous to describe here, but careful examination revealed that each "official" soil series could be distinguished from every other in one or more dimensions, with the exception of a very few which have since been reclassified.

The 148 OHIO profiles represented only five soil series: Miami, Celina, Crosby, Morley, and Blount, members of two closely related toposequences, with a good many properties in common. The dimensions found by analysis of variables are generally similar to those found in the other sets of data. The cluster analysis of the soils did not give a satisfactory grouping of these soils, apparently because differentia used in the classification into soil series, such as parent material and related clay minerology, depth and degree of mottling, and color were not included in the data or did not appear as dimensions in the analysis of variables.

Analyses of the other data sets taken from the 7th Approximation produced results with varying degrees of correspondence with the 7th Approximation classification. The phenogram resulting from cluster analysis of variables (CC5) of the 59 WORLD data was similar to that obtained by Cipra, Bidwell, and Rohlf (1970) using the same data and centroid-component factor analysis. However, Cipra agrees that the phenogram following (CC5) gave groupings which were more in accord with the 7th Approximation (personal communication). The results from

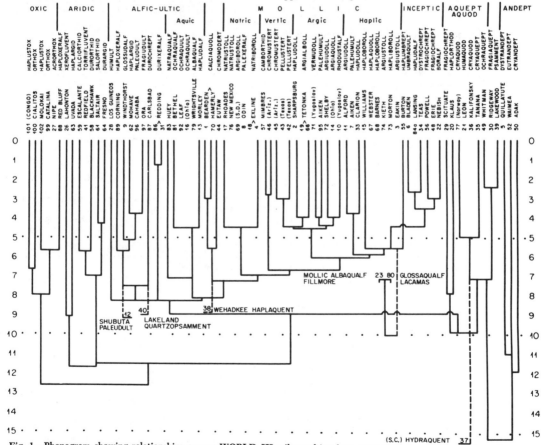

Fig. 1—Phenogram showing relationships among WORLD III soils resulting from composite scores on seven dimensions.

the WORLD III data set are shown in a two dimensional phenogram (Fig. 1). Such a diagram is an inadequate representation of a six dimensional hyperspace, but it does show a number of relationships. The most distinct groupings are the Andepts, Oxisols (OX), and Aridisols. The Mollisols and Vertisols form one broad group which is subdivided into reasonably homogeneous groups which might be called haplic, argic, natric, calcic, and aquic Mollisols and Vertisols. There is an aquic group which includes both Aquepts and Aquods; Alfisols and Ultisols are not separated but tend to fall together into two groups, one of which is clearly aquic. Also there are some distinctive soils which are indicated in the lower part of the chart. The pattern of groupings formed by this method is somewhat biased by the limited number of soils included in the analysis. The data included an unduly large sample of Mollisols, so that a number of subgroups could be formed. The remaining orders of soils were represented by smaller samples so that fewer distinct subgroups could be formed, and a number of unique soils were left without related soils with which groups could be formed. Also it is important to note that this set of soil groups was formed from only seven factor (or dimension) scores composited from the soil properties indicated in Table 1, column 8. In view of this fact, it is remarkable that this purely statistical procedure could produce a classification so similar to the 7th Approximation.

There are several conclusions from this study of importance to soil classification and correlation. First, factor or cluster analysis of variables can be used to select, by statistical criteria, a minimal set of soil properties most appropriate for the classification of any groups of soils. Second, these properties can often be combined into a relatively few factors or dimensions which can serve as the basis for a coordinate system of soil classification. For example, if a set of centroid (or modal) soils were established by defining a set of coordinates for each, then any soil could be classified by calculating its coordinates from soil data and assigning it to the centroid with the most similar set of coordinates. This could be done manually for a single soil or by computer for a large set of soils. Such a system would involve no dichotomous separations (i.e., greater or less than 35% base saturation), which is one distinct disadvantage of a categorical system for soil classification. Avery (1968) argued that the nature of soils makes them inherently unsuited for a categorical system of classification, and that a coordinate system would be more appropriate. Third a coordinate system of classification of the type described above would greatly facilitate soil correlation. Closely similar soil series could be readily identified, and unknown soils assigned to the most similar established soil series with ease. Fourth, the number of classes in a coordinate system can be predetermined at a level appropriate to the purpose. The number of classes (or centroids) is s^d, where s is the number of segments per dimension and d is the number of dimensions. For example, if each dimension is divided into just three segments, the number of centroids increases from 81 for a four-dimensional system to 65,536 for an eight-dimensional system. Clearly, an eight-dimensional system would provide quite enough classes for classifying the soils

of the world, especially if those soils with rather unique properties such as duripan, bisequal profiles, etc., are treated as separate groups.

The advantages of such a non-Linnean taxonomic system have been discussed by DuPraw (1964). He stated that "The primary goal of non-Linnean taxonomy is the recognition of individual organisms; its justification is the unrestricted transfer of . . . information from the known specimens which yield the information, to the populations of unknown specimens which they represent." In an analysis of honey bees using measurements of veins in the fore-wings only, DuPraw was able to identify 18 of 20 unknown specimens correctly as to specific area of origin. He used a coordinate system of classification, based upon discriminate function analysis of variables, rather than factor or cluster analysis of variables. DuPraw also demonstrated that dividing the population into broad groups on the basis of one or two dimensions followed by analysis of each group separately yields finer "resolution" than analysis of the total population. His results strongly support the idea that a coordinate system of classification can be extremely useful even for organisms controlled by genetic heredity. Such a system should be even more useful for soils, which have no such control.

LITERATURE CITED

1. Arkley, R. J. 1968. Statistical methods in soil classification. Int. Congr. Soil Sci., Trans. 9th (Adelaide, Aust.) 4:187–192.
2. Avery, B. W. 1968. General soil classification: Hierarchical and co-ordinate systems. Int. Congr. Soil Sci. Trans. 9th (Adelaide, Aust.) 4:169–176.
3. Cipra, J. E., O. W. Bidwell, and F. J. Rohlf. 1970. Numerical taxonomy of soils from nine orders by cluster and centroid-component analysis. Soil Sci. Soc. Amer. Proc. 34:281–287.
4. Crovello, T. J. 1968. Key communality cluster analysis as a taxonomic tool. Taxon. 17:241–258.
5. DuPraw, E. J. 1964. Non-Linnean taxonomy. Nature 202 (4935):849–852.
6. Harmann, H. H. 1960. Modern factor analysis. University of Chicago Press, Chicago. 471 p.
7. Lance, G. N., and W. T. Williams. 1967. A general theory of classificatory sorting strategies. I: Hierarchical systems. Computer J. 9:373–380.
8. Lance, G. N., and W. T. Williams. 1967. A general theory of classificatory sorting strategies. II: Clustering systems. Computer J. 10:271–277.
9. Sarkar, P. K., O. W. Bidwell, and L. F. Marcus. 1966. Selection of characteristics for numerical classification of soils. Soil Sci. Soc. Amer. Proc. 30:269–272.
10. Soil Survey Staff, USDA. 1960. Soil classification. A comprehensive system. 7th Approximation. US Government Printing Office, Washington. 265 p.
11. Sokal, R. R., and P. H. A. Sneath. 1963. Principles of numerical taxonomy. W. H. Freeman and Co., San Francisco. 359 p.
12. Tryon, R. C., and D. E. Bailey. 1966. The BCTRY computer system of cluster and factor analysis. Multivariate Behav. Res. 1:95–111.
13. Wilding, L. P., and E. M. Rutledge. 1966. Cation-exchange capacity as a function of organic matter, total clay, and various clay fractions in a soil toposequence. Soil Sci. Soc. Amer. Proc. 30:782–785.
14. Wilding, L. P., R. B. Jones, and G. M. Schafer. 1965. Variation of soil morphological properties within Miami, Celina, and Crosby mapping units in west-central Ohio. Soil Sci. Soc. Amer. Proc. 29:711–717.

Soil Classification Systems

Portland Cement Association Staff

In order that soils may be evaluated, it is necessary to devise systems or methods for identifying soils with similar properties, and then to follow this identification with a grouping or classifying of soils that will perform in a similar manner when their densities, moisture contents, and relations to water tables, climate, etc., are similar. Such procedures are common practice where a variety of soil types exists. A clear understanding of the relation of soil identification to soil classification is necessary to prevent confusion about many factors involved in soil work.

In general, certain soil tests such as gradation and liquid limit are used to assist in the identification of a soil. Then these same tests are used to assist in classification. Several systems are in use for both processes.

The primary purpose of soil identification is to describe a soil in sufficient detail to permit engineers to recognize it and, if need be, to obtain samples in the field. The soil identification system of the Department of Agriculture (described under "U.S. Department of Agriculture Classification System") is used widely.

The most widely used system of soil classification was devised a number of years ago by the U.S. Bureau of Public Roads (now Federal Highway Administration) for subgrade soils. In this system, AASHO M145, soils are classified in one of seven groups, A-1 through A-7. The U.S. Army Corps of Engineers adopted a classification system that uses texture as the descriptive term such as "GW–gravel, well graded"; "GC–clayey gravel"; and "GP–gravel, poorly graded." This classification was expanded in cooperation with the Bureau of Reclamation and was referred to as the unified soil classification system. It is now identified as ASTM D 2487. The U.S. Federal Aviation Administration has also adopted a system of classifications—E-1 through E-13. In these systems certain definite tests and test-result limits are used to classify a soil and illustrate the interplay between soil identification and soil classification.

U.S. DEPARTMENT OF AGRICULTURE CLASSIFICATION SYSTEM

A system of soil classification was devised by Russian agricultural engineers about 1870 to permit close study of soils with the same agricultural characteristics. Around 1900 this system was adopted by the U.S. Department of Agriculture Division of Soil Survey,* which has since classified and mapped the soils in most of the agricultural areas in the United States. Many agricultural and geological departments of state universities and colleges use a similar system.

The highway engineer found that this system and the resulting valuable soil information could be used in identifying soils, after which he could classify them for engineering purposes in his own work. Therefore, while the U.S. Department of Agriculture system is called a soil classification system for purposes of nomenclature and use by the agricultural engineer, it is used as a soil identification system by the highway engineer. This system of soil classification and identification is based on the fact that soils with the same weather (rainfall and temperature ranges), the same topography (hillside, hilltop, valley, etc.), and the same drainage characteristics (water-table height, speed of drainage, etc.) will grow the same type of vegetation and be the same kind of soil. This is illustrated by the fact that the black wheat-belt soils of the West are the same as the black wheat-belt soils of Russia, Argentina, and other countries.

---

*The Division of Soil Survey is now part of the Soil Conservation Service. Surveys dated after November 1952 were published by the Soil Conservation Service. Reports and maps dated 1932 to November 1952 were published by the Bureau of Plant Industry; those dated 1923 to 1931 were published by the Bureau of Chemistry and Soils; and those of earlier date were published by the Bureau of Soils.

The system is important basically because a subgrade of a particular soil series, horizon, and grain size will perform the same wherever it occurs since such important factors as rainfall, freezing, groundwater table, and capillarity of the soil are part of the identification system. In no other system in use are these important factors employed directly. The system's value and use can be extended widely as soon as the engineering properties, such as load-carrying capacity, mud-pumping characteristics, and cement requirements for soil-cement, are determined for a particular soil. This is because soils of the same grain size, horizon, and series are the same and will function the same wherever they occur. Hence, a North Carolina engineer and a Texas engineer, after each has identified a soil in his own area by this system, could exchange accurate pavement design and performance data.

This system can be used only as an initial step in soil classification since the engineering properties of a soil must be determined after it is identified.

For identification, this system first divides the soils of the United States into three main orders—zonal, intrazonal, and azonal—depending on the amount of profile development.*

The zonal soils are mature soils characterized by well-differentiated horizons and profiles that differ noticeably according to the climatic zone in which they occur. They are found in great areas where the land is well drained but not too steep.

Intrazonal soils are those with well-developed characteristics resulting from some influential local factor of relief or parent rock. They are usually local in occurrence. Bog soils, peats, and salt soils are typical examples.

Azonal soils are relatively young and reflect to a minimum degree the effects of environment. They do not have profile development and structure developed from the soil-forming processes. Alluvial soils of flood plains and dry sands along large lakes are examples.

Great Soil Groups

The three major divisions are subdivided into suborders and then further subdivided into great soil groups on the basis of the combined effect of climate, vegetation, and topography. For example, the great chernozem soil group is developed under grass vegetation in temperate subhumid areas, while the laterite group is formed in areas of abundant rainfall and high temperature. The great soil groups falling in the zonal, intrazonal, and azonal orders are given in Table. 1.

Soil Series

Soils within each great soil group are divided into soil series, and the soil series are further broken down into soil types.

*These three divisions of the top order replace the two categories (pedalfers and pedocals) previously used by the Department of Agriculture. See James Thorp and Guy D. Smith, "Higher Categories of Soil Classification: Order, Suborder, and Great Soil Groups," *Soil Science*, Vol. 67, January to June 1949, pages 117-126.

Similar soils within a great soil group that have uniform development (the same age, climate, vegetation, and relief) and similar parent material are given a soil series designation. All soil profiles of a certain soil series, therefore, are similar in all respects with the exception of a variation in the texture of the topsoil, or A horizon. The soil series were originally named after a town, county, stream, or similar geographical source, such as "Norfolk" or "Hagerstown," where first identified. This method of naming series is not necessarily used now since it may in some cases interfere with the Department of Agriculture's present system of correlating series over wide areas.

Soil Type

As already mentioned, the texture of the surface soil, or A horizon, may vary slightly within the same soil series. The soil series is, therefore, subdivided into the final classification unit, called the soil type. The soil type recognizes the texture of the surface soil and is made up of the soil series name plus the textural classification of the A horizon. For example, if the textures of the A horizon of a soil series named Norfolk are classified texturally as sand and sandy loam, the soil type in each case would be Norfolk sand and Norfolk sandy loam. Both of these soil types would have the same B and C horizons (parent material) and would have been found under the same conditions of climate, vegetation, and topography.

The basic textural groups based on particles smaller than 2 mm. in diameter as defined by the Department of Agriculture are given in Fig. 3. Three of the basic textural groups—sand, loamy sand, and sandy loam—are further subdivided as shown in Table 2. The terminology and size limits of the soil separates are given in Fig. 1. The textural soil group has a "gravelly" prefix if it contains 20 percent or more gravel. The basic textural class name, however, is based on the size distribution of the material smaller than 2 mm. in diameter. The sum of the percentages of each of the soil separates, therefore, equals 100 after the gravel material has been excluded.

The following are excellent references on the U.S. Department of Agriculture soil classification system:

"Soils of the United States," *Atlas of American Agriculture*, Part III, U.S. Department of Agriculture, 1935.

Charles E. Kellogg, *Development and Significance of the Great Soil Groups of the United States*, Miscellaneous Publication No. 229, U.S. Department of Agriculture, 1936.

Soils and Men, Yearbook of Agriculture 1938, U.S. Department of Agriculture.

F. F. Riecken and Guy D. Smith, "Lower Categories of Soil Classification: Family, Series, Type, and Phase," pages 107-115; James Thorp and Guy D. Smith, "Higher Categories of Soil Classification: Order, Suborder, and Great Soil Groups," pages 117-126; Olaf Stockstad and R. P. Humbert, "Interpretive Soil Classification Engineering Properties," pages 159-161, *Soil Science*, Vol. 67, January to June 1949.

Table 1. Soil Classification in the Higher Categories

Order	Suborder	Great soil groups
Zonal soils	1. Soils of the cold zone 2. Light-colored soils of arid regions	Tundra soils Desert soils Red desert soils Sierozem Brown soils Reddish-brown soils
	3. Dark-colored soils of semiarid, subhumid, and humid grasslands	Chestnut soils Reddish chestnut soils Chernozem soils Prairie soils Reddish prairie soils
	4. Soils of the forest-grassland transition	Degraded chernozem Noncalcic brown or Shantung brown soils
	5. Light-colored podzolized soils of the timbered regions	Podzol soils Gray wooded or Gray podzolic soils* Brown podzolic soils Gray-brown podzolic soils Red-yellow podzolic soils*
	6. Lateritic soils of forested warm-temperature and tropical regions	Reddish-brown lateritic soils* Yellowish-brown lateritic soils Laterite soils*
Intrazonal soils	1. Halomorphic (saline and alkali) soils of imperfectly drained arid regions and littoral deposits	Solonchak or Saline soils Solonetz soils Soloth soils
	2. Hydromorphic soils of marshes, swamps, seep areas, and flats	Humic-glei soils* (includes wiesenboden) Alpine meadow soils Bog soils Half-bog soils Low-humic-glei* soils Planosols Groundwater podzol soils Groundwater laterite soils
	3. Calcimorphic soils	Brown forest soils (braunerde) Rendzina soils
Azonal soils		Lithosols Regosols (includes dry sands) Alluvial soils

*New or recently modified great soil groups.

From "Higher Categories of Soil Classification: Order, Suborder, and Great Soil Groups," by James Thorp and Guy D. Smith, *Soil Science*, Vol. 67, January to June 1949, pages 117-126.

Soil Survey Manual, Handbook No. 18, U.S. Department of Agriculture, 1951.

Earl J. Felt, "Soil Series Names as a Basis for Interpretive Soil Classifications for Engineering Purposes," *Symposium on the Identification and Classification of Soils*, Special Technical Publication No. 113, American Society for Testing and Materials, Philadelphia, Pa.

Application to Soil-Cement Testing

The Department of Agriculture soil classification system has proved very helpful in soil-cement testing and construction work. It has been found that the cement requirement of a definite soil series and horizon is the same regardless of where it is encountered. Once the cement requirement has been determined by laboratory tests, no further soil-cement tests for that particular soil are needed when it is used on another project. Thus, by identifying the soil proposed for use by series and horizon, the need for conducting soil-cement tests can be sharply reduced or eliminated altogether for large areas. An increasing number of engineers are making use of this system of classification to reduce their soil-cement testing work.

Availability and Accuracy of Soil Maps

A large portion of the United States has been surveyed and mapped by the Soil Survey Division, Department of Agriculture. At the completion of a soil survey, which usually covers an area of one county, a soil map is made and a report written that describes the soil types occurring. These

339

Table 2. Percentage of Sand Sizes in Subclasses of Sand, Loamy Sand, and Sandy Loam
Basic Textural Classes as Defined by the U.S. Department of Agriculture

Basic soil class	Subclass	Soil separates				
		Very coarse sand, 2.0-1.0 mm.	Coarse sand, 1.0-0.5 mm.	Medium sand, 0.5-0.25 mm.	Fine sand, 0.25-0.1 mm.	Very fine sand, 0.1-0.05 mm.
Sands	Coarse sand	25% or more		Less than 50%	Less than 50%	Less than 50%
	Sand	25% or more			Less than 50%	Less than 50%
	Fine sand	Less than 25%		—or—	50% or more	Less than 50%
	Very fine sand					50% or more
Loamy sands	Loamy coarse sand	25% or more		Less than 50%	Less than 50%	Less than 50%
	Loamy sand	25% or more			Less than 50%	Less than 50%
	Loamy fine sand	Less than 25%		—or—	50% or more	Less than 50%
	Loamy very fine sand					50% or more
Sandy loams	Coarse sandy loam	25% or more		Less than 50%	Less than 50%	Less than 50%
	Sandy loam	30% or more —and—; Less than 25%			Less than 30%	Less than 30%
	Fine sandy loam	Between 15 and 30% —or—			30% or more	Less than 30%
	Very fine sandy loam	Less than 15% —or—			More than 40%*	30% or more

*Half of fine sand and very fine sand must be very fine sand.

reports and maps are available to the public and can be viewed at or obtained from the U.S. Department of Agriculture, county extension agents, colleges, universities, libraries, etc.

A tabulation of the counties in the United States that had published maps as of July 1, 1957, is given in Highway Research Bulletin No. 22-R, *Agriculture Soil Maps, Status, July 1957*. Highway Research Board Record No. 81, *Geophysical Methods and Statistical Soil Surveys in Highway Engineering*, 1965, contains a tabulation of the counties and other areas in each state for which soil surveys were published between July 1, 1957, and October 1, 1964. These tabulations also give the Department of Agriculture accuracy rating for each map available, inasmuch as the surveys were made over a period of many years and the

older reports and maps may not be as complete as the more recent ones. The ratings range from 1 to 5, according to each map's accuracy, completeness, and value as an engineering soil map. The more recent reports have included a section giving the engineering characteristics of each of the soil series occurring in the county.

The U.S. Department of Agriculture, Soil Conservation Service, has published a *List of Published Soil Surveys*, January 1972.

AASHO CLASSIFICATION SYSTEM

The American Association of State Highway Officials system of classifying soils is an engineering property classification based on field performance of highways. It was previ-

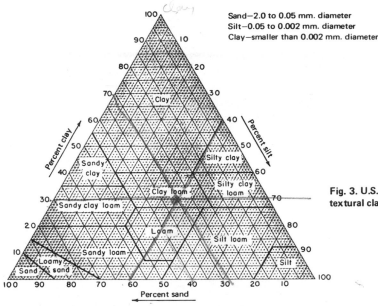

Sand—2.0 to 0.05 mm. diameter
Silt—0.05 to 0.002 mm. diameter
Clay—smaller than 0.002 mm. diameter

Fig. 3. U.S. Department of Agriculture textural classification chart.

ously referred to as the Public Roads Administration soil classification system since it was devised by that organization in 1931 (*Public Roads*, Vol. 12, No. 5, July 1931) and revised in 1942 (*Public Roads*, Vol. 22, No. 12, February 1942). The system was revised further by a subcommittee of the Highway Research Board in 1945 (Highway Research Board *Proceedings of the Twenty-fifth Annual Meeting*, Vol. 25, 1945, pages 375-392). In the same year, it became a standard of AASHO—AASHO M 145.

Grouping together soils of about the same general load-carrying capacity and service resulted in seven basic groups that were designated A-1 through A-7. The best soils for road subgrades are classified as A-1, the next best A-2, and so on, with the poorest soils classified as A-7.

Members of each group have similar broad characteristics. However, there is a wide range in the load-carrying capacity of each group as well as an overlapping of load-carrying capacity in the groups. For example, a borderline A-2 soil may contain materials with a greater load-carrying capacity than an A-1 soil, and under unusual conditions may be inferior to the best materials classified in the A-6 or A-7 soil group. Hence, if the AASHO soil group is the *only* fact known about a soil, only the broad limits of load-carrying capacity can be stated. As a result, the seven basic soil groups were divided into subgroups with a group index devised to approximate within-group evaluations. Group indexes ranged from 0 for the best subgrades to 20 for the poorest. Increasing values of the index within each basic soil group reflected (1) the reduction of the load-carrying capacity of subgrades and (2) the combined effect of increasing liquid limits and plasticity indexes and decreasing percentages of coarse material.

In 1966 the AASHO Recommended Practice was revised so that there is now no upper limit of group index value obtained by use of the formula. The adopted critical values of percentage passing the No. 200 sieve, liquid limit, and plasticity index are based on an evaluation of subgrade, subbase, and base course materials by several highway organizations that use the tests involved in the classification system.

Under average conditions of good drainage and thorough compaction, the supporting value of a material as a subgrade may be assumed as an inverse ratio to its group index, that is, a group index of 0 indicates a "good" subgrade material and group index of 20 or greater indicates a "very poor" subgrade material.

The charts and one of the tables used in AASHO M145-66, the Classification of Soils and Soil-Aggregate Mixtures for Highway Construction Purposes, are shown in Figs. 4 and 5 and Table 3. In addition to the charts and table given here, the AASHO Recommended Practice includes detailed descriptions of each classification group and the basis for the group index formula. Examples of the determination of the group index are also included.

Classification of materials in the various groups applies only to the fraction passing the 3-in. sieve. Therefore any specification regarding the use of A-1, A-2, and A-3 materials in construction should state whether boulders, retained on a 3-in. sieve, are permitted.

ASTM SOIL CLASSIFICATION SYSTEM
(Unified Classification)

The American Society for Testing and Materials soil classi-

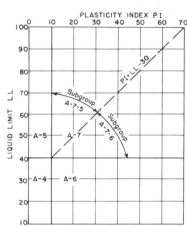

Fig. 4. Liquid limit and plasticity index ranges for A-4, A-5, A-6, and A-7 subgrade groups.

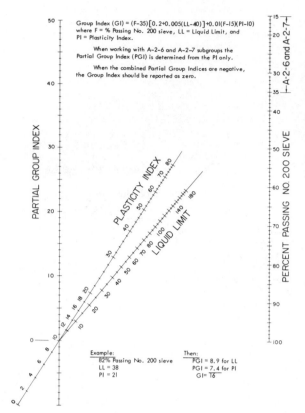

Group Index (GI) = (F-35)[0.2+0.005(LL-40)]+0.01(F-15)(PI-10)
where F = % Passing No. 200 sieve, LL = Liquid Limit, and
PI = Plasticity Index.

When working with A-2-6 and A-2-7 subgroups the
Partial Group Index (PGI) is determined from the PI only.

When the combined Partial Group Indices are negative,
the Group Index should be reported as zero.

Example:
82% Passing No. 200 sieve
LL = 38
PI = 21

Then:
PGI = 8.9 for LL
PGI = 7.4 for PI
GI = 16

Fig. 5. Group index charts.

Table 3. Classification of Highway Subgrade Materials (with Suggested Subgroups)

General classification	Granular materials (35% or less passing No. 200)							Silt-clay materials (More than 35% passing No. 200)			
Group classification	A-1		A-3	A-2				A-4	A-5	A-6	A-7
	A-1-a	A-1-b		A-2-4	A-2-5	A-2-6	A-2-7				A-7-5 A-7-6
Sieve analysis, percent passing:											
No. 10	50 max.	—	—	—	—	—	—	—	—	—	—
No. 40	30 max.	50 max.	51 min.	—	—	—	—	—	—	—	—
No. 200	15 max.	25 max.	10 max.	35 max.	35 max.	35 max.	35 max.	36 min.	36 min.	36 min.	36 min.
Characteristics of fraction passing No. 40:											
Liquid limit	—		—	40 max.	41 min.	40 max.	41 min.	40 max.	41 min.	40 max.	41 min.
Plasticity index	6 max.		NP	10 max.	10 max.	11 min.	11 min.	10 max.	10 max.	11 min.	11 min.*
Usual types of significant constituent materials	Stone fragments, gravel and sand		Fine sand	Silty or clayey gravel and sand				Silty soils		Clayey soils	
General rating as subgrade	Excellent to good							Fair to poor			

*Plasticity index of A-7-5 subgroup is equal to or less than LL minus 30. Plasticity index of A-7-6 subgroup is greater than LL minus 30.

fication system is based on the system developed by Dr. Arthur Casagrande of Harvard University for the Corps of Engineers during World War II. The original classification was expanded and revised in cooperation with the U.S. Bureau of Reclamation so that it now applies to embankments and foundations as well as to roads and airfields. This system became a standard of ASTM in 1969—ASTM D2487.

The ASTM soil classification system identifies soils according to their textural and plasticity qualities and their grouping with respect to their performances as engineering construction materials. The following properties form the basis of soil identification:

1. Percentages of gravel, sand, and fines (fraction passing the No. 200 sieve).
2. Shape of the grain-size distribution curve.
3. Plasticity characteristics.

The soil is given a descriptive name and a letter symbol indicating its principal characteristics.

Three soil fractions are recognized: gravel, sand, and fines (silt or clay).

The soils are divided as (1) coarse-grained soils, (2) fine-grained soils, and (3) highly organic soils. The coarse-grained soils contain more than 50 percent material retained on the No. 200 sieve, and fine-grained soils contain 50 percent or more passing the No. 200 sieve.

If the soil has a dark color and an organic odor when moist and warm, a second liquid limit should be performed on a test sample that has been oven-dried at 110 ± 5 deg. C. for 24 hours. The soil is classified as organic silt or clay (O) if the liquid limit after oven drying is less than three-fourths of the liquid limit of the original sample determined before drying (see ASTM D2217, Procedure B).

The coarse-grained soils are subdivided into gravels (G) and sands (S). The gravels have 50 percent or more of the coarse fraction (that portion retained on the No. 200 sieve) retained on the No. 4 sieve, and the sands have more than 50 percent of the coarse fraction passing the No. 4 sieve. The four secondary divisions of each group—GW, GP, GM, and GC (gravel); SW, SP, SM, and SC (sand)—depend on the amount and type of fines and the shape of the grain-size distribution curve. Representative soil types found in each of these secondary groups are shown in Table 4 under the heading "Typical names."

Fine-grained soils are subdivided into silts (M) and clays (C), depending on their liquid limit and plasticity index. Silts are those fine-grained soils with a liquid limit and plasticity index that plot below the *A* line in the diagram in Table 4, and clays are those that plot above the *A* line. The foregoing definition is not valid for organic clays since their liquid limit and plasticity index plot below the *A* line. The silt and clay groups have secondary divisions based on whether the soils have relatively low (L) or high (H) liquid limit (greater than 50).

The highly organic soils, usually very compressible and with undesirable construction characteristics, are classified into one group designated "Pt." Peat, humus, and swamp soils are typical examples.

FEDERAL AVIATION ADMINISTRATION CLASSIFICATION SYSTEM

The FAA has prepared a soil classification system based on the gradation analysis and the plasticity characteristics of soils.

The textural classification is based on a grain-size determination of the minus No. 10 material and the use of a soil chart, Fig. 6, that also includes definitions of sand, silt, and clay.

The mechanical analysis, liquid limit, and plasticity index data are referred to Table 5, and the appropriate soil group, ranging from E-1 to E-13 inclusive, is selected.

Two modifications of this procedure may be required. In one case, test results on fine-grained soils, groups E-6 through E-12, may place the soil in more than one group. When this occurs, the test results are referred to Fig. 7, where the appropriate soil group is determined.

The other modification is used when considerable material is retained on a No. 10 sieve since the classification is based on the material passing the No. 10 sieve. Upgrading the soil one to two classes is permitted when the percentage of the total sample retained on the No. 10 sieve exceeds 45 percent for soils of the E-1 to E-4 groups and 55 percent for the remaining groups, provided the coarse fraction consists of reasonably sound material. Further, it is necessary that the coarse fraction be fairly well graded from the maximum size down to the No. 10 sieve size. Stones or rock fragments scattered through a soil are not considered of sufficient benefit to warrant upgrading.

Table 4. ASTM Soil Classification System (Unified)

Major Divisions			Group Symbols	Typical Names	Classification Criteria
Coarse-Grained Soils (More than 50% retained on No. 200 sieve*)	**Gravels** (50% or more of coarse fraction retained on No. 4 sieve)	Clean Gravels	GW	Well-graded gravels and gravel-sand mixtures, little or no fines	$C_u = D_{60}/D_{10}$ Greater than 4; $C_z = \dfrac{(D_{30})^2}{D_{10} \times D_{60}}$ Between 1 and 3
			GP	Poorly graded gravels and gravel-sand mixtures, little or no fines	Not meeting both criteria for GW
		Gravels with Fines	GM	Silty gravels, gravel-sand-silt mixtures	Atterberg limits plot below "A" line or plasticity index less than 4
			GC	Clayey gravels, gravel-sand-clay mixtures	Atterberg limits plot above "A" line and plasticity index greater than 7
	Sands (More than 50% of coarse fraction passes No. 4 sieve)	Clean Sands	SW	Well-graded sands and gravelly sands, little or no fines	$C_u = D_{60}/D_{10}$ Greater than 6; $C_z = \dfrac{(D_{30})^2}{D_{10} \times D_{60}}$ Between 1 and 3
			SP	Poorly graded sands and gravelly sands, little or no fines	Not meeting both criteria for SW
		Sands with Fines	SM	Silty sands, sand-silt mixtures	Atterberg limits plot below "A" line or plasticity index less than 4
			SC	Clayey sands, sand-clay mixtures	Atterberg limits plot above "A" line and plasticity index greater than 7
Fine-Grained Soils (50% or more passes No. 200 sieve*)	**Silts and Clays** (Liquid limit 50% or less)		ML	Inorganic silts, very fine sands, rock flour, silty or clayey fine sands	
			CL	Inorganic clays of low to medium plasticity, gravelly clays, sandy clays, silty clays, lean clays	
			OL	Organic silts and organic silty clays of low plasticity	
	Silts and Clays (Liquid limit greater than 50%)		MH	Inorganic silts, micaceous or diatomaceous fine sands or silts, elastic silts	
			CH	Inorganic clays of high plasticity, fat clays	
			OH	Organic clays of medium to high plasticity	
Highly Organic Soils			PT	Peat, muck, and other highly organic soils	Visual-Manual Identification, see ASTM Designation D 2488.

Classification on basis of percentage of fines:
Less than 5% pass No. 200 sieve: GW, GP, SW, SP
More than 12% pass No. 200 sieve: GM, GC, SM, SC
5% to 12% pass No. 200 sieve: Borderline classification requiring use of dual symbols

PLASTICITY CHART

For classification of fine-grained soils and fine fraction of coarse-grained soils.

Atterberg Limits plotting in hatched area are borderline classifications requiring use of dual symbols.

Equation of A-line: PI = 0.73 (LL - 20)

*Based on the material passing the 3-in. (75-mm.) sieve.

344

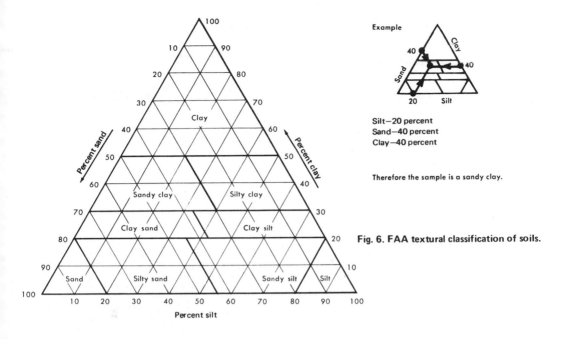

Fig. 6. FAA textural classification of soils.

Example

Silt—20 percent
Sand—40 percent
Clay—40 percent

Therefore the sample is a sandy clay.

Table 5. FAA Classification of Soils for Airport Construction

	Soil group	Mechanical analysis					LL	PI
		Retained on No. 10 sieve,* percent	Material finer than No. 10 sieve					
			Coarse sand passing No. 10, retained on No. 40, percent	Fine sand passing No. 40, retained on No. 200, percent	Combined silt and clay passing No. 200, percent			
Granular	E-1	0-45	40+	60 −	15 −		25 −	6 −
	E-2	0-45	15+	85 −	25 −		25 −	6 −
	E-3	0-45	−	−	25 −		25 −	6 −
	E-4	0-45	−	−	35 −		35 −	10 −
Fine grained	E-5	0-55	−	−	45 −		40 −	15 −
	E-6	0-55	−	−	45+		40 −	10 −
	E-7	0-55	−	−	45+		50 −	10-30
	E-8	0-55	−	−	45+		60 −	15-40
	E-9	0-55	−	−	45+		40+	30 −
	E-10	0-55	−	−	45+		70 −	20-50
	E-11	0-55	−	−	45+		80 −	30+
	E-12	0-55	−	−	45+		80+	−
	E-13	Muck and peat—field examination						

*If percentage of material retained on the No. 10 sieve exceeds that shown, the classification may be raised provided such material is sound and fairly well graded.

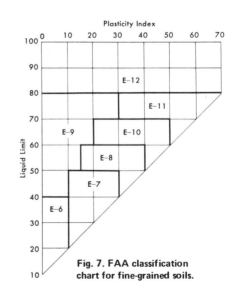

Fig. 7. FAA classification chart for fine-grained soils.

26

Reprinted from *Engr. J.* **35**:1199–1210 (1952)

Suggested Classification of Muskeg for the Engineer

Norman W. Radforth

McMaster University,
and
Royal Botanical Gardens, Hamilton, Ont.

Foreword

The Associate Committee on Soil and Snow Mechanics of the National Research Council has been concerned, since the start of its work, with studies of the terrain of Canada. Muskeg constitutes a large proportion of the total area of the country, although situated in the less accessible areas, its significance is not always appreciated by the city dwellers.

As the boundaries of national development steadily advance, the problems which muskeg can create when it is traversed, or when it interferes with engineering work, are becoming increasingly important. The Associate Committee is therefore pleased to be able to publish this first major report on a project which it has supported financially, and in which it has long been interested.

After an early investigation of the chaotic use of even the word "muskeg" in Canada, the Committee was glad to find in Dr. Radforth a scientist whose special field of interest was palaeobotany, through which discipline an approach to the scientific study of organic terrain alone seemed possible.

Working closely with the Defence Research Board, which is primarily interested in the interpretation of muskeg conditions from the air, the Committee has supported Dr. Radforth in field work at Churchill, Manitoba, in visits to European muskeg areas, and in his laboratory investigations at Hamilton, Ontario, at McMaster University, and the Royal Botanical Gardens.

Dr. Radforth has accumulated a vast amount of information, many samples of muskeg and an unusual library of photographs of organic terrain. Using this, and his wide field experience, he has managed to develop a rational system of classification for a material long regarded as "unclassifiable".

The suggested classification is admittedly a first approach only to a singularly difficult problem. Its imperfections will be shown only by its use. Its improvement and further development will come only after such field application. The Associate Committee and Dr. Radforth will therefore welcome comments and criticisms and, in particular, reports upon the use of the suggested classification in the field.

R. F. LEGGET
Chairman, Associate Committee on Soil and Snow Mechanics, National Research Council.

The interpretation of mineral soil aggregates has become a well-established study, and is now being approached from many directions in the scientific and engineering realms. Some branches of the field are relatively new. Their development is just beginning to broaden. The consideration of mechanical problems relating to mineral soils, for instance, is comparatively young.

By comparison with this, however, knowledge concerning the interpretation of organic soils is negligible. No organized reference system for organic soils is in existence. Until it is, the utilization, treatment and adjustment of organic soils for engineering purposes will remain a difficulty.

Under these circumstances, the comparison of results in problems of foundation engineering wherever an organic soil is involved will always be troublesome and at best, inadequate. This is particularly true when the soil medium in question is almost completely organic: the material called "muskeg".

Possibly because of its biological origin, muskeg is an extremely complex material. In degree of complexity, it compares favourably with that of any mineral soil and its variability is likewise well marked. Moreover, in assessing this material for engineering purposes, depth, form and other features that refer to the soil body as a whole, are just as significant for the organic as for the mineral soil.

Finally, in applying engineering techniques and interpretations, attendant factors such as water and physiographical phenomena are just as important for organic soils as for mineral soils as agents influencing soil character.

In the present study the writer has had in mind the need for adequate portrayal of muskeg features. It is thought, however, that this objective is secondary in importance to the development and presentation of a system whereby intelligent reference can be made to muskeg.

While it is true that this primary objective relates more to the need of the engineer interested in problems of trafficability, construction and foundation engineering, it is hoped that the information will be of use to foresters, geographers, biologists, pedologists, conservationists, and indeed to all those associated in one way or another with muskeg.

The Expression "Muskeg"

The writer has gone to considerable lengths to determine what is intended by the word "muskeg" by those who use the term. The results of the inquiry have been enlightening but bewildering. The range of meaning ascribed to muskeg is so broad that the word has questionable value when used as a precise reference term. Some would include swamp under the heading of "muskeg". Others would differ and claim that it signifies the shore of a bog-lake. It has been used as synonymous with tundra. It has been employed to designate imponded areas completely covered with vegetation, or the thickly carpeted floor

The author, Dr. Radforth (left) with R. F. Legget, M.E.I.C., chairman of the annual meeting session at which the paper was presented.

of open spruce woodland, and in many instances has been supplied as an acceptable reference term to signify "peat-bed".

When used in any of these ways, no special size limit is specified. Frequently it is applied in describing large areas of land more or less continuous for several thousands of square miles. This is especially so in the case of the Canadian North and Northwest. On the other hand, it may be utilized as referring to small patches of overburden, particularly in the eastern, western and southern parts of Canada and in some parts of the United States.

The flexibility and vagueness of meaning that "muskeg" implies leads to confusion, and provides serious difficulty when practical and intelligent use of the expression is attempted for purposes of discussion and report. This situation is further aggravated when the expression is used to label the field material itself. Because of this difficulty, it is tempting to eliminate "muskeg" as a reference term altogether. To discard it completely, however, would invite other problems.

Though the term is attended by such lack of agreement, it has, none the less, come into frequent use and doubtless will continue to be used. Also, though its meaning is not clear, those who have become accustomed to using it in connection with their work, regard it with some respect. They are content to ascribe to it their own particular set of values, no matter how this might be at variance with the views of others.

Literally interpreted, "muskeg" suggests simply "peat-bog" (Chippewan Indian, maskeg, meaning "grassy bog"). Though this definition is not adequate as it stands, it does furnish a fundamental meaning which can be widely appreciated, and which shows promise of acceptable expansion. Primarily, the idea of organic constitution is inferred. There is also in the literal definition the suggestion that a high percentage of water influences the character of the organic medium.

A third idea which may be inherent in the literal translation is spatial delimitation. This possibility coincides with the suggestion of some, that the term "bog" includes the saturated organic matter held in a more or less saucer-shaped depression, frequently two or three miles in diameter. This view does not support the impression that muskeg may be an extended sheet of organic material, with a depth generally uniform over great distances, in excess of say, twenty miles.

A workable definition of "muskeg" might thus be simply "organic terrain". The possible weakness may, however, be dismissed if it is agreed that the expression relates not only to materials, but also to states or conditions. Certainly, if any translation of "muskeg" is to exclude this proposal it will have to be regarded as inadequate at the outset.

The expression "organic terrain" seems to allow for the inclusion of the base of the organic matter and the top of the underlying mineral soil in considerations relating to "muskeg". It also allows any reference to topographic conditions and ground surface characteristics in general. The designation does not preclude the use of the expression "organic deposit". It does suggest, and in a sense specifies the nature of the deposit. Some organic deposits such as coal, lignite, oil and submerged peat beds would be excluded by the expression.

Organic Terrain Deposits

The kind of organic deposit included in the expression "organic terrain" has widespread distribution. In contemplating the limitation of "organic terrain" as a reference expression, the writer feels that geographic implications are important.

Organic terrain deposits occurring in Europe, Asia and North America all merit study. In Canada organic terrain predominates in areas having horizons separated by as much as a thousand miles. Such country may be found in the Yukon, in Saskatchewan, and Manitoba. Smaller areas, but still significantly large, occur in British Columbia, Ontario, Quebec, New Brunswick, Nova Scotia and Newfoundland.

Organic terrain also occurs in Washington, Oregon, North Dakota, Michigan, New York and the New England States. To a lesser degree it occurs in other regions within the United States. Perhaps the best known geographic regions in which organic terrain occurs are in the British Isles and in northern Europe. The largest continental areas containing deposits comparable with those of Canada are in Siberia.

The great bulk of the deposit, wherever located, arises as an accumulation of vegetable matter. The thickness of the material varies considerably. It may be one to a few inches in depth, or it may be between one and two hundred feet in depth. Whatever the depth, it is still significant as a potential problem, or as a predominant element providing character or limitation for large areas. For any given expanse of this kind of overburden the depth is more often than not variable. Constancy in depth is, however, a condition not uncommon, particularly in the far North.

Organization, The Basis for Classification

Any system of classification presupposes the existence of organization in whatever medium is to be classified. The higher the degree of orderliness in the medium under study, the better will be the chances for recognizing organization, and the more the promise of rendering a classification that will justify the widest acceptance. Whatever the form of classification, if it is to be successful, its procurement depends upon the manifestation of organization.

It is also important to realize that the procedure of revealing organization may depend on reference to natural relationships within the medium being classified, or on the other hand to artificial relationships arbitrarily revealed. It must also be expected that if the material to be classified is highly complex, the classification system ultimately derived may be also complex. It may include not one, but several, sets of classification systems. In cases of this kind, one system may parallel another, or one system may be subsidiary to another.

It would be unfortunate if, in establishing the basis for classification for a medium that has so far evaded interpretation, an attempt were not made to distinguish between identification and classification. In this account, classification will be used to segregate and define *properties* of materials and conditions. Identification will be used in referring to *examples* of materials and conditions. Thus, examples will be regarded as identifiable through classifiable properties. Thus, it seems proper to emphasize that the recognition of classifiable properties within the organic terrain is the fundamental task.

Survey of Conditions for Classification

Having set out the terms of reference for the basis of classification, it now remains to explore the material and conditions which confront the field observer. This exploration involves two realms: that which embraces the hidden or underground aspects, and that which involves surface features lying in full view.

Subsurface Constitution

In any sort of construction problem involving "muskeg", the field observer generally feels that the materials he is dealing with fall into the category of black, unstable muck. This is inadequate if the engineer desires as clear and complete a picture of organic soils as he insists upon for soils derived from a mineral source. It is therefore necessary that the constitution of organic terrain from the aspect of particle size, form and distribution should be studied and incorporated into the main framework of classification.

This will serve to parallel the increasingly important information existing with regard to the particulate constitution of mineral soils. In the latter field, not only is particle size an important factor for the engineer, but *kinds* of particles, their frequency of occurrence and their distribution, are also important. The relationship between one kind of particle and another in the physical, chemical and mechanical sense is also significant. It is not surprising, therefore, that the same point of view should be accepted and utilized with reference to an interpretation of organic terrain.

Size and Disposition of Constituent Particles

Interpretation of organic terrain depends upon reference to particles or units of two orders of size; one microscopic, the other observable with the naked eye (macroscopic) in the field. The attention of the reader is first directed to the latter.

On the assumption that the field engineer makes adequate preliminary investigation of the organic terrain materials, he will readily appreciate that dead plant remains, appearing as visible units of the terrain, differ widely as to size, quantity and spatial relationships. At one place in the organic blanket he is traversing, he may encounter "black muck" which, when dry, does not fall away but remains in "chunks" or "cakes". These have a fine granular texture, with larger units in negligible abundance.

Instead of this condition he may come across one in which the sample consists of non-woody (soft, resistant) fibres, short and interlacing to form a mesh. Or he may have selected instead a sample in which the fibres (again non-woody and quite numerous) are long, perhaps up to a foot in length, contrasting with an inch or two for the previous sample. These may be ribbon-like and oriented more or less in a vertical plane.

Or he may discover an example consisting mostly of woody, interlacing and thread-like fibres, forming a mat which offers considerable mechanical strength. These threads may be only a few inches in length, but could be several feet long and follow a tortuous course, conforming to no particular plane. The survey may reveal the possibility of the presence of examples in which the fibres are recognizable as tree roots or branch systems, which make it difficult to procure samples.

In some instances, this woody mesh may be made up of short "chunks", in others of extended "pole-like" units of different lengths. It is not the purpose of this article to discuss isolated mechanical properties or sets thereof which these various examples might furnish. It does, however, seem desirable to emphasize the structural contrast that exists between say, the first and last mentioned examples. Mechanical problems relative to these would also show obvious contrast.

Much could be written about variation in quantity of given particle types occurring within a unit volume of terrain. Shape, as well as size and content of the units, would figure prominently. Suffice it to state here that quantitive range with respect to frequency of given particle type must include levels near 0 per cent, as well as levels close to 100 per cent.

It is well to remember that before the observer there is a range of possible unit sizes which includes both particles of granular size just visible to the naked eye, and units which may be the size of fence posts or even larger. Regardless of orientation it is a natural assumption that some sort of three-dimensional mesh, net or web, will inevitably be present in the organic terrain. The interstices of the mesh may be very small or may be quite large, depending upon the size of the units making up the mesh.

Particularly where the interstices are large, it is not difficult to conceive that a mesh of a second order of size may exist in the interstices of the first order. Where the interstices are extremely small their contents are made up of particles microscopic in size, along with various organic derivatives which together form a material essentially colloidal in nature.

Cohesion and compactability of examples will obviously show up differently, depending upon the nature of the mesh, the size of the particles and the constitution of the interstices. Disintegration and tendency to flow or shift or shear, or to resist compressional forces, are other features that will vary with the physical characteristics of the materials expressed, and depending, too, on their orientation.

Spatial relationships suggest a consideration of the vertical dimension or depth of the organic terrain. The question may arise for instance, as to whether there is a change in range of particle size throughout the depth. This is frequently the case, and the situation may be readily revealed by exposing a profile through the organic blanket. It is important to recognize that this change is, in most cases an orderly one, conducive not only to inspection for quality, but also to measurement.

Where there is change in range of particle size and size of mesh there is noticeable change in the banding or layering that exists in the peaty material. The banding conforms or corresponds with the structural change, making structural correlations and comparisons possible in a given profile or volume of terrain. There are frequent examples, of course, in which the particulate constitution of the organic material is the same for all practical purposes throughout the depth of a section.

Finally, reference should be made to the horizontal dimensions which implies coverage over area. Where the peaty material is uniform in texture throughout its depth, examples are known of its covering areas four or five miles in extent. The deposit as a whole may have a shape which relates to physiographical characteristics of the area in question.

Shapes of deposits more often than not conform to recognizable features or conditions prevalent either in the landscape, or through the nature of the organic material under observation. Where the construction of the material is not the same throughout its depth, but is noticeably banded as revealed in profile, the frequency of change in peat construction over an area is comparatively high.

It can now be seen that reference to organic terrain on the basis of particle size and distribution cannot be considered adequate if use of expressions such as "black muck" are insisted upon. Intelligent appraisal of conditions existing in organic terrain will never be possible, as long as it is thought only necessary to refer to peaty material as constituted of leaves, twigs and other plant remains.

Water Content and Drainage

There must be acknowledgement of the presence of water either in liquid or solid form in relation to a survey of conditions for classification. The peaty material in the organic terrain forms a natural reservoir for water, and conditions are known where drainage is almost impossible. Regardless of particle size, shape and arrangement, the constitution of peat is such that it can retain large volumes of water. It can be shown that like volumes of peaty material differing in physical make-up will drain at different rates, will recharge at different rates and will have different potentialities for maximum water retention.

Structure and amount of organic terrain are also related to natural drainage channels, some of which are obvious to the eye because they exist as river-like systems; others of which are not so obvious, in that they travel within the peat or at the peat-mineral interface, and can only be recognized through imponding.

Before drainage in relation to organic terrain can be adequately assessed, seasonal climatic data have to be related to local field and organic terrain conditions. Enough has been surveyed to establish the claim that water content is not merely related to organic terrain. It enters into its constitution, and must be regarded as a part of "muskeg", along with the peaty material. Thus, water (liquid or solid) enters into the definition of organic terrain, and definitions are fundamental to classification.

Organic-inorganic Interface

In consideration of mineral soils, it often occurs that the engineer must direct his attention to regions within a soil body, where a soil identified as having one set of properties is related physically to a soil identified as having another set of properties. This region of interrelationship is often more important than either of the two parent bodies concerned in the relationship.

Sometimes the zone of contact is very narrow and mainly in one plane. Sometimes there are faults in it, and other structural deformities. Occasionally it may be a zone of some depth—a kind of transition bed which may vary in thickness throughout the zone. In the case of each of the parent bodies referred to, it must be recognized that the attributes of each include a consideration of the fact that the one soil body has the other as its neighbour.

A similar situation arises with reference to the peaty layer of the organic deposit and the mineral soil sub-layer in contact with it. The zone of contact is often quite thin and sharply marked. On the other hand, the writer has worked with an interface material which was mineral matter highly charged with organic material nearly a foot in depth, forming an interbed which varied in thickness and in structural composition.

It is not always the case, as is commonly supposed, that the mineral substrate is fine textured, falling into the category of silts and clays. On the contrary, organic overburden often lies directly upon a coarse grained substrate of a sandy, sometimes gravelly constitution. Indeed, it is not infrequent to find that the peaty layer may be in contact with coarse gravels containing quite large boulders. If the organic layer is only a foot or two in depth, the contour of the boulders or rocky outcrops is imparted to the peaty overburden, and becomes a topographic feature of the organic terrain.

Examination of the zone of interface often reveals wide differences as to relative amounts of water. Sometimes the interface is almost completely fluid. On the other hand, it may be just moist. Water by reason of its presence and abundance is an important component of the interface zone.

Interface problems may be just as great as those involved within the organic layer itself. This is not to suggest, however, that in specific problems, as for instance with highway subsidence involving failure of foundation materials, it is necessarily the interface that must be held responsible. Where a whole roadbed slides to one side, possibly the worst kind of failure found in "muskeg" country, studies suggest that failure is more often a function of structural relationships within the organic layers, which have not been accounted for in surveys that preceded construction.

Emphasis, so far, has been placed on sub-surface constitution. Composition of organic material, water relations and organic-inorganic interface have each been referred to separately. The ultimate system of classification may reveal a relationship linking the three subtopics. Even at this stage of "muskeg" interpretation, prediction of the constitution of the mineral sublayer is not ruled out once the constitution and extent of the organic layer has been determined.

Surface Coverage

For foundation, drainage, construction and trafficability problems, those concerned usually give first consideration to the soil body. With mineral soils, examination of exposures and interpretations from borings are always useful. The same should apply with organic soils. Profile exposures are none too frequent in this case, however, and while borings are always possible, the presence of other field marks can prove useful. In the latter connection, an observer traversing organic terrain is tempted to make use of the upper level of the organic layer. This comprises the aerial and sub-aerial parts of all the living plants affording coverage for the dead organic material in the rest of the organic overburden.

The Living Organic Layer

For purposes of this survey of conditions for classification, the most important matter is to emphasize that the living organic material is an essential part of the organic terrain. It is certainly included in the literal translation of "muskeg" referred to earlier. Recognition of this means a corresponding extension of the list of properties which have to be used to describe organic terrain.

The field investigator has the responsibility of recording field conditions, as he experiences them in studying organic terrain. He is always alert to the possibility of finding plant types which he may use as valid indicators for various purposes. More often than not, he attempts to find the common name for any plant which he may select. The writer is quite certain that this idea was not initiated by botanists, but by field workers of other professions, trades and vocations.

In an attempt to confirm the validity of this procedure and its degree of usefulness, the writer, particularly in his work in the Churchill area, made a fairly complete collection of plants typical of the organic terrain of those regions. Most of the striking, colourful types of plants have, however, proved troublesome when used as indicators. Perhaps this is partly because certain plant types of the nature referred to are not faithful as indicators for given terrain conditions.

This, however, does not eliminate plants of the living organic layer as analytical indices. Independent field investigators and observers have developed habits of referring to particular kinds of vegetative coverage. Though non-botanists, they apparently had little difficulty in establishing common denominators on which they could compare their

experiences, by reference to vegetation properties for which they provided their own descriptive terms. In discussions bearing on these matters such terms as trees, shrubs grasses and mosses would be used. Such reference methods are not new to engineering surveyors, as shown by old survey records. With crude reference terms where precise designations are required, the advice of a trained field botanist is always of value. These preliminary attempts at coverage recognition provide encouragement for the botanist (field ecologist).

The Topographic Factor

In its broadest meaning, topography suggests description of form outline of land areas. Physical features, relating to unevenness of contour, take a prominent position in terrain consideration. In practice, location of drainage basins is also important, but in organic terrain appraisals, expediency rules that attention be given to terrain unevenness. Where water bodies are concerned, the nature and stability of the organic foundation lining the water course must be observed, both beneath the water and at the banks. Whatever the point of view of the field surveyor responsible for the interpretation of field conditions, he is bound to concern himself with over-all form and form trends, if any, as he attempts to traverse organic terrain.

It has been suggested that topographic difference in organic overburden is sometimes due to irregularities in the mineral foundation of the organic blanket. On the other hand, much of the unevenness seen from the surface is due to topographic change within the organic material itself. In studies of mineral soils it is not always easy to demonstrate the reason for topographic change. The same difficulty arises for the organic soils. In the north, frost phenomena are prevalent, and these are the controlling agents in contour establishment and regulation. Other agents are water and wind.

There is evidence that these factors sometimes act together to produce some of the shapes of the organic terrain. Perhaps the most influential agent, however, is one that is a function of the vegetation (dead and living) which forms the organic overburden. This is recognized partly through change in character of the peaty material, and partly through growth habits of certain plants as they augment and accumulate organic debris at their growth sites.

Even if contour phenomena were eliminated as necessities in "muskeg" interpretation, field appraisers and those whose task it is to deal with "muskeg" would be among the first to insist that methods be devised to account for them, to assist in classification and in intelligent appraisal. Coverage and topography are important also in aerial interpretation of organic terrain. But whether the observer is in the air or on the ground, coverage and topography are the conspicuous factors of the organic blanket on which he is primarily dependent.

The Seasonal Aspect

An appraisal of the conditions for classification of organic terrain is not complete without reference to the time factor, as this associates itself through seasonal aspects. In most parts of Canada, "muskeg" is completely frozen throughout the winter. Possible exceptions to this may arise due to the lower depths of very deep deposits being protected from frost for at least a part of the winter by the insulating values of the upper layers of the deposit.

With the approach of high temperature conditions in the far north in later spring and early summer, melting of the active ice commences on and within the organic mass. Due chiefly to drainage lag, for which the peaty materials are partly responsible, the accumulation of water is very marked. This is notably true when rainfall in the early summer is high. If, on the other hand, rainfall is not excessive, and if winds are frequent and strong, excess water and the growth of ponds may not be too noticeable, as compared with average conditions.

The wasting of the active frost layer during the summer months is a slow process. This applies, of course, chiefly in the far north as for example, around Churchill and towards Le Pas in the south. The condition has its effect on the appearance of the coverage, imparting to it seasonal aspects which are characteristic for each time period during summer and early autumn months. The seasonal aspect will vary depending upon coverage and other properties of the organic deposit. Mechanical properties of the various kinds of peaty materials will also show relationship to this seasonal factor.

In field inspection, it is difficult to resolve the seasonal factor into something tangible which can be evaluated in terms other than qualitative. Some assistance is, however, afforded through the medium of colour. With the advance of season there are local and general background colour changes in the coverage. Though the task is not easy, it is possible to measure and record colour response in superficial segments of the coverage, as well as in background coverage. This is done with the aid of the Munsell Colour System.*

*Munsell Book of Color: Pocket Edition, Munsell Color Co. Inc., 1929-1942.

cipally stature, degree of woodiness, external texture and certain growth habits easily recognized by an inexperienced observer, especially with the help of photographs. Examples of plant material are given following each colour-type description.

Commenting briefly on method of application, it is important to note in the first place that, when an observer records an appropriate class designation by letter, that letter by reason of its relationships in the table signifies the properties suggested. Seldom do these properties defining the coverage exist alone. They occur rather in combination with properties for which other class letters are symbolic. In other words, one seldom finds pure classes existing by themselves.

For given areas requiring analysis, practice seems to decree that, if one set of coverage class properties is not present to the extent of twenty-five per cent, it lacks enough prestige to give it significant prominence in the composite cover description. The complete description may sometimes, therefore, be given, though not often, in terms of one, two or three letters.

The reader might expect an overwhelming number of combinations of classes. For selected areas in the Churchill region and far to the south, however, this is not the case. For instance, the descriptions involving the use of the letters F - E - I are quite frequent. In the formula, or combinations of letters,

The Classification System

It is now possible, against the foregoing background, to suggest the main lines of the classification system so far developed to meet the need for accurate identification of muskeg. This is a first presentation only, given in summary form on the basis of work extending throughout five field seasons and winter laboratory working periods.

Coverage Pattern and Typing

Classification depends upon the possibility of revealing organization and relationship. If this can be done, a major first step will have been accomplished. The only other fundamental prerequisite to classification is to plan for a level of adequacy appropriate to the degree of complexity of the materials and conditions being classified. For organic terrain, with its high degree of complexity, several subsidiary systems are required.

One of these systems should be based on coverage. For the purposes of this account, priority should be given to it, in that it is the first factor of the terrain which the observer notices when he encounters a "muskeg" expanse. If coverage classification is to be in terms of a subsidiary system, it would be of little value if its relationship to another subsidiary system dealing with another main factor of the terrain were not anticipated.

In segregating coverage character, therefore, the plan has been devised in accordance with the following rules:—

(1) Only those features of coverage most likely to show direct relationship with features of the dead organic material comprising the bulk of the "muskeg" deposit, will be utilized.

(2) The expressions used in classification will be those best suited to the recognition of organization in the coverage;

(3) The expressions and the level of organization ultimately depicted will be such that they can be used also for aerial inspection and interpretation. Travel over "muskeg" is predominantly by air.

(4) The terms employed will be such that they can be readily understood and utilized by non-botanists; and

(5) Types of coverage that the terms express must be photographed, and authentic photographs must be available for wide use.

As may well be expected, meeting the challenge of this grouping called for much checking and rechecking from the air and from the ground, both in field observation and in laboratory analysis of results. An added complication arose through the need for anticipating trafficability problems and mechanical problems in general, sometimes relating to industrial, sometimes to engineering and sometimes to agricultural, geological, forestry or military requirements.

The accompanying table (Table I) presents the descriptive information for nine coverage class types. Photographs depicting these types appear in Figs. 8-16. An examination of the schedule given in the table will show that the properties mentioned are not referable to species of plants. They relate instead to qualities of vegetation. These are prin-

Table I—Summary of Properties Designating Nine Pure Coverage Classes

Coverage Type (Class)	Woodiness vs. Non-woodiness	Stature (approx. height)	Texture (where required)	Growth Habit	Example
A	woody	15 ft. or over	—	tree form	Spruce Larch
B	woody	5 to 15 ft.	—	young or dwarfed tree or bush	Spruce Larch Willow Birch
C	non-woody	2 to 5 ft.	—	tall grass-like	Grasses
D	woody	2 to 5 ft.	—	tall shrub or very dwarfed tree	Willow Birch Labrador tea
E	woody	0 to 2 ft.	—	low shrub	Blueberry Laurel
F	non-woody	0 to 2 ft.	—	mats, clumps or patches, sometimes touching	Sedges Grasses
G	non-woody	0 to 2 ft.	—	singly or loose association	Orchid Pitcher plant
H	non-woody	0 to 4 in.	leathery to crisp	mostly continuous mats	Lichens
I	non-woody	0 to 4 in.	soft or velvety	often continuous mats, sometimes in hummocks	Mosses

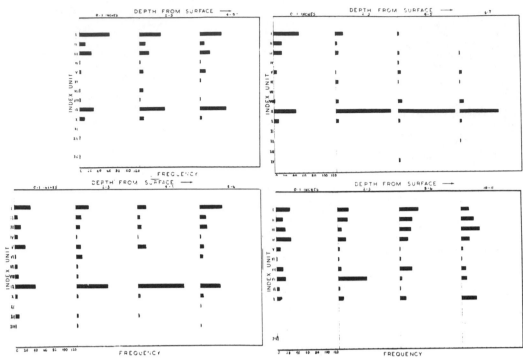

Top left to lower right: Fig. 1, Fig. 2, Fig. 3, Fig. 4.

that letter which represents the most prominent set of properties is placed first. If other letters are involved, they follow in the formula in order of prominence.

Thus, the formula gives a description of coverage composition in terms of properties and lays emphasis on the set of values most typical for the area. It is instructive to select at random areas of, say, ten square miles in extent and to analyse them by application of the methods suggested. Within a given area so selected, the frequency and degree of influence that each set of class properties shows is easily derived from the sets of formulae obtained. The formula having greatest frequency will also be revealed.

Finally, if the formulae are mapped, directional trends are exhibited. If several areas are compared with reference to properties, the degree of importance with reference to these factors can be derived, and the pattern exhibiting coverage properties becomes manifest. In short, coverage organization is revealed.

Referring again to the guiding principles, there is good reason to suggest that points 2 to 5 have been satisfied. The aspect of aerial interpretation is being tested on the basis of the method prescribed here,

and evidence has been suggested to show that it can be applied satisfactorily. Improvements and adjustments of a secondary nature will doubtless be forthcoming, but fundamentally, main requirements are met. Discussion concerning the application of the method and system in relation to rule (1) will be left to follow the discussion of the next subsidiary system.

Indexing for Subsurface Character

Turning now to the dead organic material in the organic terrain, a different approach is needed to derive a workable classification system. In making the survey of the materials available to which classification is to be prescribed, reference has been made so far to macroscopic features only. Attention of the reader has been called to information suggesting the presence of a mesh structure and related interstices. There is some evidence suggesting that differentiation of peat samples might be accomplished through an understanding of mesh size, construction and prevalence.

This alone is not enough, because during the formation of the organic body the mesh characters might well remain relatively constant, particularly with the larger sizes of

mesh, while characters of the peaty material might change. The nature of a secondary mesh, for instance, could easily change in constitution without noticeable change in the structural relationships of the primary mesh. The material in the interstices might also change with time, in amounts sufficient to alter the physical properties of the peaty deposit as a whole.

Consequently, if some method of classification could be devised to give a record of the succession of vegetation, built into the deposit as a whole in the course of time, this would provide a more certain basis. It would also permit a better understanding of the macroscopic ingredients from the point of view of both classification of properties and the appreciation of mechanical problems.

To accomplish this task, it is necessary to look to a study of microscopic constituents. As the components of past vegetation matured and contributed their remains to peat formation, their presence was recorded by the numerous pollen grains, which are normally preserved as they fall into the peat. Many of these are doubtless carried some distance before they alight. It is, however, a reasonable

352

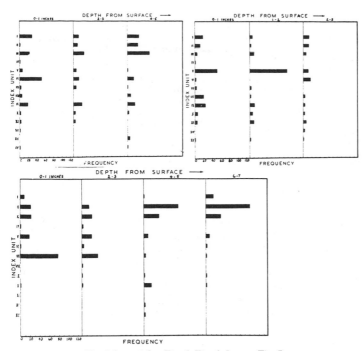

Top left to right: Fig. 5, Fig. 6. Lower: Fig. 7.

assumption that large numbers, if not the majority, fall in the area where they are produced and provide a continuous record of the vegetation.

Should it occur that, during this recording, climatic conditions at intervals are such that the surface of the deposit is dessicated, preservation would be interrupted and the continuity of the record broken. This, however, is not a serious matter if these fragments (microfossils) are to be used to symbolize that vegetation which had been preserved, and whose record had not been destroyed by climatological conditions. It is only the existing portion of the deposit that is before the observer for interpretation. It is to these microfossils that the writer turned for help in classification problems relative to sub-surface conditions.

In the initial part of the investigations there was much hesitancy concerning this approach. This was partly because much of the field work was being carried out in the north, sometimes at latitudes further north than Fort Churchill, Manitoba.

Most of the work done in connection with microfossils has, however, been related to forest history and the study of past climates. For this the chief kind of microfossil utilized is

that derived from tree pollens, and in the north there are vast areas which are treeless. For the present work the kind of microfossil which would carry most influence would be that which has been produced by plants growing on the site under examination. The question arose whether there were enough of these relicts throughout the peaty deposit to provide material that would stand statistical analysis. It took two summers to convince the writer that this was a reasonable possibility, and detailed analyses were then pursued on a broad and intensive scale.

This approach for recent organic deposits is entirely new, particularly when it is associated with the idea of revealing properties for classification purposes. It is not new, however, applied to coal structure, stratigraphy, and utilization problems. The difficulty of readapting techniques of analysis and providing new ones caused some complications. Planning and carrying out the field work on a sufficiently broad scale was time-consuming.

The materials studied for microfossils were obtained from the field, either as cylindrical cores taken with a peat borer (Hiller Model), or as channel samples from an exposed peat profile. The ultimate analyses were made from portions one cubic

inch in size, removed at intervals from different levels of each core. Usually every second inch was examined. The microfossils were released from the rest of the peaty matrix with the use of chemical reagents (ten per cent potassium hydroxide is most frequently used by palaeobotanists for this purpose). Following washing and centrifuging, a few drops of the sediment were transferred to microscope slides for permanent record.

It was from each of these sets of preparations that microfossil identifications and counts were made. For this, the usual practice adopted by palaeobotanists is to count two hundred microfossils with the aid of a microscope fitted with a mechanical stage, then determine from this the proportion of each kind or category of microfossil. Sometimes the total number counted may be less than two hundred where statistical comparisons seem to warrant it. This was the method adopted for the investigations reported here.

When it was reasonably certain that enough samples had been analysed to reveal a complete range of microfossil types, attention was centred on variation in frequency for each kind of microfossil, at different depths in the peat and in many locations over a wide area.

The fundamental aim was to utilize the microfossils as identifiable reference units, which would point to state, organization, and change, if any, for all manner of peaty profiles and examples which showed even slight structural differences. For the non-botanist, recognition need only be given to the fact that in the process of analysis these microfossils are related to the plants from which they are derived. If this is appreciated, it is then only necessary to call each microfossil category an index unit and to give it a number. The strictly botanical aspects can be dealt with elsewhere.

One more brief point requires explanation before the inspection of results should be made. This has reference to microfossil frequency. Not only is the actual presence of particular microfossils significant; numbers at given levels in the peaty deposit provide data equally, if not more significant, in evaluating results. It must be emphasized also that wherever numbers of index units are mentioned, *relative numbers* are inferred.

Attention may now be directed to Fig. 1. Each of the three sections of the diagram records the occurrence and relative frequency of the microscopic index units for three depth intervals within a core sample. The core was analysed according to the method described earlier.

There should be no difficulty in understanding the terminology expressed at the co-ordinates of the graph in the figure. Each configuration in successive histograms designates both presence and frequency for an index unit. Because these configurations differ as to length, collectively they provide a histogram pattern useful in classifying the condition with respect to the microfossil features for the appropriate depth. For the three sets of histograms indicated, observe there is close similarity in appearance and therefore in classification value. Examination of the detail within each histogram reveals important information. It is of primary importance to recognize the more general feature, namely that which demonstrates relationship in histogram pattern.

Similarity, in the form of faithful repetition suggests stability in the constitution of the peaty deposit throughout its depth. This is in addition to the more fundamental fact that these methods provide the means for a practical comparison —one which is confirmable and which lends itself to quantitive measurement.

It is encouraging to the investigator, and to all those who desire to interpret organic terrain conditions, to discover that these first data assist in establishing confidence in the matter of classification of the dead organic material, not only in principle, but also at the level of practical necessity.

Referring now to Figs. 2, 3 and 4, similarity in histogram pattern will also be observed. It is true that there are slight differences. These are of a secondary nature, however, and can be rationalized when matters of detail are under consideration. This lends support to the view that the analytical methods have validity, and that a reliable basis for classification is established.

The next significant feature of the results is that for a given set of histograms, the set of patterns is peculiar to that example. For instance, a comparison of Figs. 1 to 4 shows that marked contrast exists among the four examples, in spite of the high degree of similarity in each. Thus peats, stable in constitution throughout their depth, may differ in constitution one from the other. This then is another major and encouraging demonstrable fact revealed through application of these methods. It is further proof of

organization, it affords classifiable data, and it refers directly and indirectly to constitutional features utilized as factors in the interpretation of peaty materials in the muskeg. Deliberate attempts have been made to avoid the use of botanical expressions, in order that field reference may be made in terms understandable to all.

There is yet a third general organization feature discernible. For this, reference should be made to Figs. 5, 6, and 7. A comparison of these histogram sets shows:—

(1) Dissimilarity of histogram pattern for each set;
(2) Each set differs in pattern from the others;
(3) Organization still reflected in spite of variation as suggested in points 1 and 2;
(4) Constitutional trends (not stability) which have arisen with growth in depth of the peaty deposit; and
(5) Each item (1 to 4) can be assessed on a quantitative measureable basis.

The data used in these graphs are derived from only seven bore samples from the Churchill area. Hundreds of such samples have now been taken, and many of them examined for index unit relationships. Samples are also available from northern Sweden, the Orkney and Shetland Islands, as well as from several places in Canada. When enough histogram patterns are available for wide comparison, relative prevelance of pattern, and hence relative prevalence for organization trends, may well be discernable.

In the meantime a workable and adaptable system for identifying peats is now available. That it is workable has been demonstrated. That it is adaptable has been only partly demonstrated. In connection with the latter, the question arises, how can the system be utilized in demonstrating organization from the standpoint of structural features, significant from the aspect of mechanical problems? The answer depends in part on a consideration of other subsidiary systems of classification, two of which have not yet been reviewed. One of these which deals with topography will be presented next.

Descriptive Terminology for Topographic Features

In the section dealing with the relationships and importance of the topographic factor to problems of surface coverage classification, an account of the precise terms of classification for topography was omitted, for inclusion under this main chapter on classification. As was the case in selecting type categories for vegetation coverage, and in selecting type index units for subsurface appraisal, so with topography type expressions are necessary. It is important in making the selection that they prescribe to all contour phenomena, that they conform if possible to expressions already in use, and that they be as few in number as circumstances permit.

Table II—Terminology and Properties for Topographic Classification

Contour Type	Formation	
a	Hummock	includes Tussock and Nigger-head, has tufted top, usually vertical sides, occurring in patches, several to numerous.
b	Mound	rounded top, often elliptic or crescent shaped in plane view.
c	Ridge	similar to Mound, but extended, often irregular, and numerous; vegetation often coarser on one side.
d	Rock gravel plain	extensive exposed areas.
e	Gravel bar	eskers and old beaches.
f	Rock enclosure	grouped boulders overgrown with organic deposit.
g	Exposed boulder	visible boulders interrupting organic deposit.
h	Hidden boulder	single boulder overgrown with organic deposit.
i	Peat plateau (even)	usually extensive and involving sudden elevation.
j	Peat plateau (irregular)	often wooded, localized, and much contorted.
k	Closed pond	filled with organic debris, often with living coverage.
l	Open pond	water rises above organic debris.
m	Pond or lake margin (abrupt)	
n	Pond or lake margin (sloped)	
o	Free polygon	forming a rimmed depression.
p	Joined polygon	formed by a system of banked clefts in the organic deposit.

Through field experience it has developed that the accompanying table of expressions is adequate in classifying topographic properties (Table II). Many of the features listed in the table occur, as far as this writer is aware, only in regions north of, say, Le Pas, and more frequently as one proceeds northward. The system is therefore somewhat detailed for prevailing conditions in the south or in northern Europe. All the criteria mentioned are useful, however, especially in aerial inspection and in enhancing applications of the classification system devised for vegetative coverage of the organic terrain. Reference to this aspect will be made later.

hoped that a future report will be prepared to cover this phase of the work. Pocket editions of the Munsell System are obtainable for field use, and range of colour formulae can be worked out on the spot for various coverages at different times.

Little can be said at this stage about classification with respect to organic and mineral soil interface. Temporary assistance, however, can be attained for this purpose through a discovery of interface conditions wherever sampling is desirable. Reference has already been made to the fact that various kinds of substrates can be expected when exploring at the base of the organic layer, whether this be a few inches or many feet in depth.

Other Physical Phenomena

The importance of drainage in muskeg problems has already been indicated. For purposes of classification, drainage becomes a secondary problem and its features can best be understood by a more thorough knowledge and application of the subsidiary classification systems mentioned in preceding sections. It would seem that this same view still holds, even if we regard the water factor not from the point of view of drainage, but rather from the aspect of presence or absence of water in given terrain examples, both fluid and solid. No subsidiary classification system seems necessary, therefore, to deal with the water factor.

As has been suggested, the water factor is largely a function of seasonal and climatological influence. The latter seems to be secondary from the point of view of classification for organic terrain; the former, though directly pertinent, is highly intangible and difficult. The utilization of colour,

however, to display seasonal trends, and visible indices in the terrain is not impracticable. Much is being done to exploit it for classification purposes, as well as for aerial interpretation of organic terrain.

If an attempt is made to assess colour in terms of absolute values rather than range, it will meet with certain failure. When, however, classification is based on range, and is applied to the contrasting vegetation groupings represented in coverage formulae, there is much to be gained. By the application of the Munsell Colour System the seasonal factor in the organic terrain can be intelligently appraised. It is

Colour in terms of hue, degree of greyness, and intensity in the subject material, along with interface character expressed in qualitative terms, can readily be added to information derived through the application of the other subsidiary systems, to give the over-all picture of the inter-relationship of properties for the organic terrain.

Application of the System and its Potential Value

Whatever the final form may be for a suggested system of classification of "muskeg", it is clear that the integration of several subsidiary systems is required. This idea is directed towards the need for differentiating muskegs, whereas rules

of classification normally demand that an adequate definition be given first of the general material to be classified. "Muskeg" has become the term designating organic terrain, the physical condition of which is governed by the structure of the peat it contains, and its related mineral sub-layer, considered in relation to topographic features and the surface vegetation with which the peat co-exists.

The next logical step in "muskeg" classification is to apply the subsidiary systems to given examples of "muskeg". Part of this is best done in the field, as in the case of mineral soils. The remainder can be done in the laboratory, again as is done with mineral soils after appropriate bore samples have been procured and analysed in systematic order. On the record sheet attached to each sample will appear, "coverage description" (Table I). Following this will be the terminology pertinent to topography classification (Table II).

In both of these cases care should be taken to note the position of the sample or samples, relative to the limits of the coverage formula location, and the proximity to topographic change. This factor may be recorded in terms of linear measurement. Third on the list are the Munsell reference formulae appropriate to the position of the sample

---

Data Sheet I

Sample . . . . Bore 173.
Location . . . Fort Churchill, Manitoba. Area P$_5$, A run at intersection with B in the airphoto. (Sonne x26, x38, x201, x202, x210).
Date . . . . . 30 August, 1948.

INTERPRETIVE DATA (FIELD)

	Direct Method	Photo Method (Sonne)	
Coverage description	H E (Table I)	H E	
Topographic classification . . .	i (Table II)	i	
Munsell formula	$10.0Y \dfrac{5\text{-}6\text{-}78}{2\text{-}4}$	$2.5Y \dfrac{8}{6\text{-}8}$	$7.5Y \dfrac{8}{8}$
	$2.5GY \dfrac{6\text{-}7\text{-}8}{2\text{-}4}$	$5.0Y \dfrac{8}{6\text{-}8}$	
Proximity to topographic change	50 ft. approx.	50 ft. approx.	
Field photos	x29 Nos. 13 to 16		

Data Sheet II

INTERPRETIVE DATA (LABORATORY)

Slide Nos. . . . . 1040, 1046, 1127.
Analysis Sheet . . . . . No. 14.
Intervals Examined . . . 0-1″, 1-2″, 2-3″.
Index Units . . . . Histogram pattern deposited in laboratory Royal Botanical (Occurrence & Frequency)   Gardens, Hamilton. (See Fig. 3 and remarks below).
Macroscopic Features . . . Not sharply banded, fine mesh, granular to non-woody at base changing to: granular to woody fibrous at top resting on sand/gravel base. Near to highly decomposed less stable non-woody fibrous derivative.
Remarks . . . . Histogram pattern reveals predominance of index unit IX at all depths but decreasing at the surface. Units V and IV and VI are prominent at the surface.

Top row: Fig. 8. Coverage Class A.
Centre row: Fig. 11. Coverage Class D.
Bottom row: Fig. 14. Coverage Class G.

Fig. 9. Coverage Class B.
Fig. 12. Coverage Class E.
Fig. 15. Coverage Class H.

Fig. 10. Coverage Class C.
Fig. 13. Coverage Class F.
Fig. 16. Coverage Class I.

or samples in the field, and based on coverage colour in the region where the samples were taken. Information concerning date and local prevailing climatological conditions should be appended to this record.

This information should then be sent to the laboratory, along with the sample or samples, in order that further data can be added following analysis. This will appear in the form of series of histogram patterns showing index unit relationships. These will be few or several, depending upon the number of bore samples procured and the extent and detail of information desired. Appended to this should be notes concerning the macroscopic or gross structural features of the peaty material, as they appear in the sample or samples. Field notes concerning the proximity of other structural types in neighbouring locations would be useful in the appendix. This would be particularly the case if bore samples were few in number and the area concerned was one which appeared to contain a

variety of structural types of organic terrain conditions. An example of a reference sheet accompanying a field sample is shown in the accompanying Data Sheets.

The analysis suggests fine textured wood-fibrous organic matrix will predominate, but will be interrupted at intervals of twenty-five to a hundred feet with discontinuous areas of coarser, mechanically resistant woody matrix. This general type of terrain will thin out in some areas, giving place to relatively low-lying non-woody fibrous peat. The latter kind of matrix will probably be fairly continuous as a common base for all the organic coverage.

Topography will be generally moderately irregular, with abrupt and frequent amplitude of change averaging about a foot. Ponding will not be so great as to render traversability unpractical. The terrain will be moderately well drained. Sub-surface ice contour will be irregular with isolated well covered prominences until towards the end of August. Tree coverage is generally

sparse for this kind of peaty coverage.

When the classification picture thus derived is complete, interpretation of the organic terrain conditions is then possible. To those who have not worked with "muskeg" it may seem unfortunate that so much by way of classification data is required. Those who have worked with mineral soils however, will perhaps not be surprised because the degree of complexity involved in this procedure does not exceed that which the mineral soils demand.

In any event, it would appear that with organic soils, this apparently high level of complexity is further reduced through the possible existence of relationship between sub-surface and surface conditions in the organic terrain, and which are shown in the classification record.

Thus, where coverage class F predominates for a given set of examples, and index unit IX in the histogram configurations is consistently prominent throughout the depth of the organic matter, it is highly probable that the macroscopic or visible component of the peat that will predominate is of the fibrous ribbon-like structure, and that this condition will hold for the entire depth of the muskeg at that place.

It is a simple matter to check on this with histogram configurations available, but it is most convenient, especially for those anxious for quick field appraisal methods, to discover that by scanning the coverage and making sure of the appropriate formula, they can assess the structure of that which lies beneath. This would obtain for the situation as represented in Fig. 1. It would also hold for the situation as shown in Fig. 2, and for the same reasons.

It would also apply for the case indicated in Fig. 3, but the situation which brings index units IV, V, VI, into relatively greater prominence would also be reflected in the surface coverage formula. Predominance of woodiness, and greater strength of fibre would be noted in the peaty material—factors which again could have been predicted. Shift in character of histogram pattern as exemplified in Figs. 6 and 7, also relates to surface properties as depicted in coverage formulae.

For the peaty material represented in Fig. 7, fine woody fibrous mesh at the top with coarser and stronger mesh at the base would be expected. The writer believes most of this detail could be predicted from knowledge of the vegetation coverage formula. In this case as with the others, laboratory analysis provides the best means for confirmation, particularly if the knowledge is to be applicable for areas extending a good distance beyond the zone of sampling.

The index unit composition (Fig. 4), suggests a gross structure predominantly non-fibrous, and non-woody, changing near the surface, however, to a slightly fibrous-woody constitution. In order to predict this from surface observations, however, more formulae depicting coverage conditions would have to be secured than usual. Also without laboratory analysis the change in peaty constitution might not be detected. Certainly its extent could not readily be estimated.

For a superficial examination, especially for purposes of aerial interpretation, surface and subsurface relationship holds to a workable and encouraging degree. There is no doubt but that the microfossil analysis on the index unit basis suggests structural charac-ter, particularly with regard to occurrence and possible extent of secondary mesh and presence or absence of constitutional stability.

Referring more generally to potential value of the suggested classification system, it should be noted that it is the combined or integrated (master) result, rather than a single subsidiary system, which lends itself to adequate interpretation and possible solving of mechanical problems. Such problems range from trafficability to drainage; from insulation and utilization problems to questions involving possibility of support for forest coverage in developmental and control programs.

The correlating of organic terrain features is most important for Canada. Like subsurface geology, it is a subject which depends upon intelligent prediction. Complete success in this can follow only from proper application of field and laboratory classification records. As these accrue, not only will the possibility for better correlating benefit, but the classification system will improve. Certainty can be obtained only by the study of many samples. To this end the writer asks the help of all and invites everyone to have "muskeg" samples sent to him from any area in Canada.

The system is the first ever devised to assist in identifying and distinguishing "muskeg" samples; to reveal and to record the organization in "muskeg"; and as a basis of appraisal for "muskeg" structure and its mechanical properties. It will be used also for opening the way to intelligent aerial interpretation. The attempt to establish this to the exclusion of botanical expressions has been deliberate, in order that all who work with "muskeg" and who are interested in it may use the system and make suggestions in the anticipation of refinement and improvement. There is much to be contributed to the new and specialized field of applied botany on which "muskeg" interpretation depends—a field which for want of a better expression has been designated as palaeovegetography!

Acknowledgements

The writer wishes to acknowledge the interest of the National Research Council and the Defence Research Board, which have provided financial assistance towards this and related investigations. Special thanks are due to Robert F. Legget, Chairman of the Associate Committee on Soil and Snow Mechanics, National Research Council, and Director of the Division of Building Research of the National Research Council of Canada. He has been a source of sustained inspiration and encouragement.

To colleagues, assistants and students whose help in the field and in the laboratory and whose interest in "muskeg" continues, the writer is grateful. Appreciation is due also to Colonel Graham W. Rowley, Arctic Research Section, Defence Research Board, and to members of the "Sub-Committee Muskeg" of the Associate Committee on Soil and Snow Mechanics, N.R.C., for their encouragement, criticism and suggestions.   √

Part VI

SOIL CORRELATION

Editor's Comments
on Paper 27

27 **SIMONSON**
Soil Correlation and the New Classification System

Soil scientists as a group use more than one soil classification system at any given time. There are many different approaches to the subject, as is evident from the papers reprinted in this volume, as well as holdovers from older systems that have been revised or replaced. If such multiplicity of use were not controlled to some degree, chaos would reign supreme in the realm of soil classification. In a sense, soil correlation serves such a purpose by ensuring that kinds of soils are adequately defined, mapped, and named. The process is applicable to detailed surveys, national soil-survey programs, and to comprehensive international mapping efforts. The successes of soil correlation are, however, dependent on current classification systems, because they determine the kinds of information used in descriptions and definitions of soil units. The results of such efforts, no matter how well intentioned, are thus no better than the taxonomic systems themselves. Simonson's discussion of soil correlation in the United States and of the potential problems pending formal introduction of a new soil classification system such as *Soil Taxonomy* is germane to even broader aspects of the subject. To this important activity, which relates soil bodies represented on maps to taxonomic classes, the last word is given to Roy Simonson without further comment.

27

SOIL CORRELATION AND THE NEW CLASSIFICATION SYSTEM

R. W. Simonson
U. S. Dept. of Agriculture

Whenever a system of soil classification is replaced or modified appreciably, the correlation of soils is affected for a long time to come. Some of the effects follow promptly, others appear much later. Because of the importance of the soil classification system in use to soil correlation, an effort is made in this paper to illustrate the impacts of past systems and to outline probable consequences of the new system being developed. Since not all the effects of a new system can be foreseen, the outline cannot be complete, but it is believed that an effort to appraise the probable effects of the new system (11) may be helpful. The nature and functions of soil correlation must be kept in mind, however, for an understanding of the probable impacts of the new system. Soil correlation is, therefore, reviewed before the effects of classification systems are considered. It should also be pointed out that this discussion is centered primarily on soil classification and correlation as they have been carried forward in the United States.

GENERAL NATURE AND PURPOSE OF SOIL CORRELATION

In a narrow sense, soil correlation is concerned with the definition, mapping, naming, and classifying of the kinds of soils in specific survey areas. In a wider sense, soil correlation is concerned as well with the improvement of standards and techniques for describing soils and with the application and development of soil classification. (5).

Relating the soil bodies represented on maps to taxonomic classes at some level in a classification system is accomplished through soil correlation. The process of correlation, as used in soil surveys, thus requires scrutiny and testing of the concepts of individual soil series and of the series category as a whole. The use of soil series in the correlation process is a major application of

[1] Soil Survey, Soil Conservation Service, Washington 25, D. C.

classes in that category, and such use is also an important application of a system of soil classification.

The data obtained and recorded in the study of soils, both in the field and in the laboratory, form the evidence upon which correlations rest. The validity of these data depends upon the available standards and techniques for description and characterization of soils. These standards and techniques, described earlier (10, 12), thus form part of the foundation for soil correlation.

Any system of soil classification affects the kinds of observations made in the study of soils, whether these studies are outdoors or in the laboratory. The system in current use also governs in large measure the selection and weighting of characteristics as criteria for defining and differentiating soil series. Consequently, classification systems also form part of the foundation for the correlation process. Moreover, classification systems themselves are modified as a result of their application and testing in completing the correlation of soils in specific survey areas. In the past, several systems of soil classification have been used in this country (1, 4, 9), and some of the effects of these systems will be considered in a subsequent part of this paper.

The ultimate purposes of correlation are to ensure that kinds of soils are adequately defined, accurately mapped, and uniformly named in all soil surveys. These are large objectives. If they are to be achieved they demand concerted effort on the part of every soil scientist concerned with every soil survey. The work required for satisfactory correlation of soils in a survey area begins with the onset of preliminary studies for construction of the initial legend and continues until a final legend is approved for the published soil survey report. The quality of soil correlation in each survey area thus depends upon the caliber of work done at every stage of the survey, beginning with the construction of the initial mapping legend and continuing through the preparation of the final correlation memorandum. More importantly, this also governs the usefulness of the results of the soil survey for any purpose.

ELEMENTS OF SOIL CORRELATION

The nature of soil correlation can be indicated most easily by a discussion of the main elements for individual soil survey areas. For any one soil survey area, the three main elements are (a) construction of the mapping legend, (b) review of the legend and mapping, and (c) preparation of the formal correlation memoranda. Each of these main elements could be subdivided further with discussion of each subdivision (5). Because of limitations of space, however, no more than the main elements are considered.

The construction of the mapping legend is the first step in the soil survey of an area. As part of this step, the soils of the area are examined and descriptions prepared for a number of profiles, according to approved standards. Observations are also made on soil features other than profiles. On the basis of the profile descriptions and other notes, concepts are developed for possible kinds of soil bodies to be represented on the maps. The data collected on the morphology and other characteristics of soils are further synthesized into tentative definitions of soil types and series. These tentative definitions are compared with those of already recognized series to test the spans being allowed in such classes and to find out if the kinds of soils in the survey area have been encountered elsewhere and have already been classified.

What is done in the early stages of each soil survey depends upon available knowledge about the soils of the region. If soil surveys have been made in the region within recent years, series concepts and definitions will be available for many of the kinds of soils occurring in the survey area. Conversely, if no mapping has been done in a region within recent years, concepts and definitions for tentative series will need to be developed. More work, therefore, is required in regions of the latter kind than in those of the former.

The legend constructed at the outset of each soil survey has three principal parts: a list of symbols for each kind or combination of kinds of soils to be represented on the field sheets; names for the various kinds, insofar as names can be assigned; and descriptions of map units, or of the kinds of soil to be identified by symbol and boundary on the field sheets, based primarily on field notes but possibly partly on laboratory data.

The initial legend in a survey area can be no better than the available knowledge about the soils. At the outset of a survey, such knowledge must be incomplete on several counts. An initial legend is thus a first approximation which must be tested and then modified as may be necessary while the survey goes forward.

The nature of the mapping legend for a given survey is of great importance to the future usefulness of the survey data. Ideally, a legend should provide for every separation significant to soil genesis and behavior within current understanding. In other words, every separation should be made that has significance to soil genesis or behavior, and no additional ones should appear on the maps. Furthermore, *the legend should record what is shown on the maps and why the separations are made.* The actual value of the survey results is governed to a large extent by the quality of the legend constructed as a basis for mapping.

The second main element in the survey of a specific area consists of periodic reviews of work in progress. Such reviews comprise one or more activities. In the course of these reviews, help is provided the field party on problems of describing, classifying, and mapping soils. The accuracy of boundary placement is examined, usually on a sample basis. The ranges allowed in the major component soils within map units are examined and compared with those permitted in established or proposed series. Attention is also given to proportions of inclusions other than the major component soils in map units. Last but not least, the significance of the separations on the field sheets to the genesis or to the behavior of soils in agriculture, forestry, and engineering must be considered. These review activities are meant to ensure that the findings obtained will meet the objectives of the survey.

Although the assignment of names to map units begins with preparation of the initial legend, it is given special emphasis in the completion of the formal correlation memoranda, the third main element in soil correlation for a survey area. Map units are named in order to relate the delineated kinds of soils to taxonomic classes at some level in a classification system. The prime objective is to provide a basis for transferring the results of experience and research on soils at one location to like soils elsewhere.

The preparation of formal correlation memoranda after the field mapping is done completes the process of assigning names to map units. The bulk of the map units recognized in detailed surveys are named as phases of soil types or as soil types; some may be named as complexes or undifferentiated groups. Some map units may lack soil or consist of soils that are not classified and named; such map units are termed miscellaneous land types. In less-detailed surveys, map units may be named as soil associations. All of these kinds of map units are described in the *Soil Survey Manual* (10).

Because map units include pedons from two or more series, conventions have been worked out for permissible proportions of inclusions in a map unit named as a phase or as a soil type (10). A map unit named as one phase of a soil type is expected to consist dominantly of pedons falling within the range of that soil type. Approximately 85 per cent of the pedons are expected to be within the range of the soil type, the name of which is assigned the map unit. In other words, roughly 15 per cent may fall outside the limits of the type. Some weight is given to the degree of contrast between the dominant kind of soil and the inclusions within a map unit. If the contrast is great, a proportion as low as 10 per cent may need to be recognized by naming the map unit as a complex or undifferentiated group, depending upon pattern of occurrence of component kinds of soils. If the contrast is small, the inclusions may comprise as much as 20 per cent without being identified in the name. As indicated earlier, few if any map entities consist exclusively of pedons that would lie within the defined limits of a single soil type or soil series (8).

An important part of the work done in preparation of correlation memoranda over and above that of assigning names to map units is the critical examination of the bases for setting apart map units and of the validity of separations in terms of soil genesis and soil behavior. Beyond that, the preparation of formal correlation memoranda also includes review of concepts and definitions of series and of the placement of the series into classes of higher categories. These last two activities comprise an application of the system of classification in use. Furthermore, they also affect the structure of the system, because series must be grouped into progressively broader classes in a comprehensive scheme.

DESCRIPTIONS OF PROFILES, MAP UNITS, AND SERIES

Continued emphasis has been placed in the preceding discussion of legends on descriptions of the soil bodies shown as delineations on maps. These descriptions provide a record that is basic not only to soil correlation but also to subsequent interpretation and use of survey data.

Descriptions which may be prepared in the course of the soil survey of a specific area are of three kinds. The initial descriptions are for individual soil profiles. Subsequently, descriptions are prepared for the map units or the segments of the soil mantle represented as delineations on maps. Last of all, descriptions may be prepared as definitions of soil series. Since the descriptions of individual profiles are often used as parts of the descriptions of mapping units and of series, the three kinds are not clearly differentiated at all times.

Each delineation on the map represents a small segment of the soil mantle identified by a symbol and set apart from its neighbors by a boundary. This small segment is a bundle of contiguous pedons (8). Efforts are made to place boundaries so that as many as possible of the included pedons will be within the range of a single-class of low categoric rank, for example a soil series. Despite such efforts, few, if any, of the bodies of soil represented by delineations consist of pedons classifiable in a single series. Most, if not all, consist of pedons which are classifiable into two or more series. In most map units in detailed surveys, the bulk of the pedons in one map entity are within the range of a single series, but some are not. The pedons other than those of the dominant series within a map unit are called inclusions if the total proportion is small. Conventions have been worked out for the proportions to be permitted, and these were considered in the immediately preceding section.

The descriptions of map units are a record of the nature of a set of like segments of the soil mantle. They must, therefore, record the characteristics of the dominant pedons and of any others within the map unit. The descriptions should also provide information on the patterns of occurrence in the map unit; on features such as stoniness, slope, and degree of erosion; and on relationships to other map units.

A part of the description of many map units is a description of a soil profile which represents a vertical slice down through one pedon among the bundle comprising a map entity. Consequently, a profile description cannot record ranges in characteristics among pedons within a map unit; for this purpose, additional information is necessary. Profile descriptions are an important part but not the whole of the record of what map units may be.

Descriptions of soil series must also be broader than the descriptions of individual profiles. Each series description does include the description of a typical profile which represents a pedon selected as the norm for that series. Each series, however, consists of a group of pedons considered a single class because of similarities in characteristics. Thus, the basis for defining series differs from that for subdividing the soil mantle in the making of maps. In mapping, the boundaries must be drawn around a bundle of contiguous pedons. A series, however, comprises a group of pedons similar in selected characteristics. Limits are not placed upon the definition of a series by the actual occurrence of pedons within the soil continuum. Pedons need not be contiguous, as must be true for those comprising a single map entity. Because the series is a group of pedons, information beyond that provided by the description of the typical profile is needed to define each series.

Descriptions of individual profiles are commonly parts of descriptions of map units and are always parts of series descriptions. Thus, the description of a map unit and the description of a series must be more comprehensive than the description of a profile. Moreover, the emphasis appropriate in descriptions of series would not be appropriate in descriptions of map units. In the case of series, attention is focused on a group of pedons because of their similarities, without regard to distribution, whereas, in the case of map units, attention is focused on groups of pedons which occur together in the soil continuum. All three kinds of descriptions are closely related but each has a distinctive character as well.

FUNCTIONS OF SOIL CORRELATION

The earlier section on the main elements of soil correlation for a given survey area in itself indicates a number of the functions. Major emphasis has been given to the definition and description of the soil bodies represented on maps, and to the process of naming these soil bodies so as to show their relationships to taxonomic classes in a comprehensive system. Soil correlation has other functions as well.

The map units recognized in every survey area must be tested, as must concepts and definitions of series. The testing of the validity of map units is necessary to see that they can be distinguished consistently and that they have meaning to either soil genesis or soil behavior, or to both. The testing of validity is based on evidence in the form of descriptive legends, profile notes, series descriptions, results of laboratory analyses, interpretive groupings, ratings of various kinds, and placement of series in a classification system. The testing begins with the initiation of field work and continues through the preparation of the final correlation memorandum.

During the preparation of formal correlation memoranda, the available information on the nature of soils in the survey area and on their relationships to soils elsewhere are examined. The findings of the soil survey are thus given critical scrutiny. The process of completing the correlation memoranda furnishes the last opportunity for soil scientists to test their own findings in a survey area before those findings are made available to the public, and the findings must meet ultimate tests of application and use. It is a basic premise of all scientific work that the findings can withstand critical scrutiny and can be supported by evidence acceptable to other competent scientists. Hence it is one function of the correlation process to provide for the testing of survey findings before the findings are offered to the public at large.

As was pointed out in an earlier section, one purpose of soil correlation is to ensure that any one kind of soil is given the same name wherever it occurs. To put this in another way, one purpose is to identify the same kind of soil by the same name in all places. This is only possible if there is a degree of uniformity of approach in describing, mapping, and classifying soils. The approach followed in each survey area is reflected in the mapping legend and in the supporting evidence for the map units and the nomenclature. Although attention is given to these throughout the course of a survey, in the preparation of formal correlation memoranda they are reviewed with special care. Comparisons are made of the spans allowed in map units, the definitions of these units, the definitions of series, and the classification of series for a given area with reference standards and with information from other areas. This serves to identify differences in approach and the problems that follow.

Deficiencies in concepts and definitions of soil series, or in the standards for describing soils, may come to light in the correlation of soils of individual areas. Such deficiencies must be identified before they can be corrected. Soil correlation thus has a role, as was mentioned earlier, in the improvement of standards and techniques for describing soils and in the improvement of the classification system. This illustrates once again the reciprocal relationships among the description, mapping, correlation, and classification of soils.

GENERAL EFFECTS OF CLASSIFICATION SYSTEMS

Soil correlation is dependent upon the classification system in current use as well as upon the available data on morphology and composition of soils. The general impacts of systems of soil classification on soil correlation are therefore discussed here before the effects of earlier systems are considered.

A system of classification focuses attention on the characteristics selected as differentiating for classes in different categories. This can be illustrated by one example. The proposal by Marbut (6) to distinguish two orders of soils focused attention on concentrations in the profile of calcium carbonate *versus* concentrations of aluminum and iron. Pedocals were defined as soils in which calcium carbonate accumulated, whereas Pedalfers were defined as soils in which iron and aluminum accumulated. These characteristics were subsequently studied by many investigators.

While it is in use, any classification system calls attention to those characteristics of soils used to differentiate classes at various levels. The system thereby affects the kinds of observations made by soil scientists in soil surveys. The records of those observations later become part of the data upon which soil correlation must rest.

A system of classification has an important bearing on the selection and weighting of soil characteristics as series criteria. Every kind of soil has a host of characteristics, relatively few of which are used to differentiate one series from another. Some characteristics are selected as differentiae, whereas others are not. Furthermore, several characteristics must be considered simultaneously and assigned relative weights or importance. This assignment of relative weights must be done without benefit of any numerical common denominator for the characteristics under consideration. The choice of characteristics to differentiate series and the relative weight given to any one characteristic in a given combination are both dependent upon the system of soil classification in current use. To return to the earlier example of the Pedocals and Pedalfers, much weight was given to the presence or absence of a horizon of carbonate accumulation in the profile, though some weight was also given to an interpretation of how the horizon was formed. A horizon of carbonate accumulation was to be a major criterion, provided it was a reflection of soil formation. A general boundary was drawn between the regions of Pedocals and Pedalfers, and the boundary was a basis for setting apart, as series, soils which were alike in their morphology.

The classification system in use has an important bearing on the kinds of information that go into series descriptions and definitions. The definitions of series tend to emphasize characteristics used to differentiate soils in classes in higher categories. During the first decade of this century (7), major emphasis was given to physiographic provinces and underlying rock in the classification of soils. The descriptions of soil series prepared during that time stressed the occurrence of the series in a given physiographic province and also gave much attention to the nature of the underlying rock. As classification systems have been modified or replaced, there have been corresponding changes in the kinds of information in the descriptions of series. Within the last decade and a half, emphasis has been placed on the classification of soils into great soil groups. As a consequence, the descriptions of series tend to stress features which are characteristic of great soil groups. Any system of soil classification in use invariably affects the descriptions and definitions of series as well as series concepts.

EFFECTS OF PRIOR CLASSIFICATION SYSTEMS

Several systems have been used in the classification of soils in this country in the past, each reflecting the state of knowledge in the soil science of its day (9). In addition to their direct effects on soil correlation while they were in use, these systems continue to exert an influence long after they have been superseded.

Soils must be correlated within some framework, which includes a system of classification as well as the concept of soil itself. Correlation decisions for individual survey areas in the first part of the century were made within some framework, as is done now. The nature of the earliest framework used in soil correlation is indicated in the instructions to field parties published near the beginning of the century (2) and also in descriptions of soil series in published soil survey reports.

The earliest framework for soil correlation was a combination of physiographic provinces or divisions, underlying rock, and soil texture. This framework did emphasize one soil characteristic, texture. Soil series, comprising a category first introduced in 1903, were to be recognized in physiographic provinces or divisions for sets of soils formed from sediments deposited at the same time but differing in texture. Each series was to be restricted to one physiographic province or division.

Changes were made in the framework for soil correlation soon after 1903. Study of the descriptions of soil series in the reports of field operations shows that thickness of the dark-colored surface layer was used to differentiate series as early as 1907. The Miami and Marshall series, then defined as a pair of soils underlain by glacial drift or loess and occurring in the same physiographic province, were initially set apart on this basis. Additional changes in the framework soon followed. These were due in part to changes in the classification system and in part to the growth in the understanding of soils.

A long-term effect of each classification system on soil correlation follows from the establishment of soil series. Each series is first recognized within the prevailing classification system. Each is defined in terms of characteristics emphasized by that system. The range of characteristics permitted within series depends upon the approach in soil classification at the time that the series is first recognized. As indicated earlier, a number of series in the current list were established within classification systems of the past. The concept of each of these series reflects the system of classification under which the series was established, and this is not necessarily consistent with the present-day understanding of soils. Long after a classification system has been superseded, its effects remain in the definitions of soil series established within that classification. Series must, therefore, be redefined in terms of characteristics important to a new or modified system, or some other action must be taken.

Numerous examples are available to illustrate the effects of classification systems in the establishment of series. In many instances in the past, two series have been recognized for soils that are identical, or nearly so, in morphology and composition. At one time this was done on the basis of physiographic province boundaries. Later it was done on the basis of the boundary drawn between the regions of Pedocals and Pedalfers. It was also done on the basis of the boundary drawn between regions of Gray-Brown Podzolic and Red-Yellow Podzolic soils. Each classification system that is constructed has a long-time impact on soil correlation through the series that are then established. The replacement or modification of a classification system entails a continuing review of series concepts and definitions. Furthermore, changes in concepts and definitions of certain series become necessary.

As a group, soil scientists make use of more than one classification system at any one time. Many soil scientists now at work originally learned a classification system that has been greatly modified or replaced. What is first learned is not readily abandoned. As John Dewey (3) wrote, "It is not easy to break away from established interpretations and classifications of the world." If "soils" were substituted for "the world," the statement would lose none of its validity. Even when an effort is made to substitute a new or modified classification system for one learned in the past, a person may revert to the old one from time to time. Some scientists believe that an older system is better than the new, and they will continue to use the older system as part of their frameworks for soil correlation. All this means that more than one framework for soil correlation is in use concurrently.

Systems of classification used in the past thus affect present efforts in soil correlation in several ways. Some residue of earlier systems remains among the concepts of series and in their definitions. Another and more important residue of earlier systems remains in the minds of men and affects the outlook and approach of such men in soil correlation. This last effect is the more important.

EFFECTS OF NEW CLASSIFICATION SYSTEM

It is possible now to foresee some of the effects of soil correlation on the new classification system. Five such effects are discussed in the following paragraphs. As was indicated earlier, not all effects can be foreseen and there are likely to be consequences which are not expected.

When a new system of classification is adopted, the use of old ones is not discontinued everywhere by all scientists concerned with soil correlation. Thus, the adoption of a new system does not make the old ones disappear, but, rather, adds another framework to those already in existence. For a number of the scientists concerned with soil surveys, earlier classification systems will continue to be parts of frameworks for soil correlation. The long-term impacts of classification systems, discussed earlier, can be expected to continue to have the same kinds of impacts on soil correlation each time a new one is introduced. Problems that follow in soil correlation will be of the same kind as those that have followed introduction of new classification systems in the past. There will be conflict between the framework for soil correlation that includes the new classification system and the older frameworks that included earlier systems.

After a new classification system is adopted, the basic problems of subdividing the soil continuum, both in mapping and in classification, remain unchanged. Furthermore, the basic problems in classification hold for all classes in all categories of a system and are not peculiar to the definition and differentiation of soil series. This discussion is restricted to the series, however, because their definition and differentiation is carried forward largely through the correlation process.

The quantitative limits between classes in higher categories of the new system will in themselves generate some problems in soil correlation. For example, it can be expected that efforts will be made to set apart two series of soils where the range in a given characteristic is narrow but happens to stretch across the limit between a pair of adjacent classes in one of the higher categories. Arguments will be offered for recognizing two series for soils with a limited range in carbon-nitrogen ratio because the range straddles the limits between a pair of great groups. Ranges in other properties of the soils in question could be narrow as well. Similar arguments will be related to other characteristics such as base saturation, conductivity, and clay mineralogy. Problems in soil correlation will follow from attempts to split narrow ranges in one or more characteristics of soils at the limit specified for a class in a higher category.

Once the new system is in use, there will be a great temptation to set apart series on the basis of a limit in a single characteristic for a great group, a suborder, or an order. It is far easier to accept limits already given and then apply them as differentiae without reflection than it is to weigh each characteristic in the given combination for each kind of soil. This practice avoids the major problems in soil classification, namely, the selecting and weighting of characteristics as class criteria and the defining of classes. Yet there seems to be no *a priori* reason to expect that mechanical use of limits for classes in the higher categories will provide the most useful concepts and definitions of soil series. There will continue to be need to examine and weigh whole sets of characteristics of given kinds of soils to arrive at appropriate concepts for soil series and to relate map units to taxonomic classes. Mechanical application of class limits of higher categories for the differentiation of series does not promise to be the most useful approach, inasmuch as this would assume perfection of the system.

Adoption of a new system will eventually require redefinition of some series and subdivision of others. It will also require the combining of some series established in the past. These kinds of changes among series in current use have been necessary in the past and will continue to be necessary in the future. Such changes are essential if new knowledge is to be incorporated into the concepts and definitions of series. Furthermore, there has also been a gradual narrowing of permissible ranges within soil series over the years, and this trend will be reinforced by the new classification system.

Benefits as well as problems in soil correlation can be expected to follow from use of the new system. As a matter of fact, work done in the development of the system thus far has resulted in the sharpening of concepts of a number of series. Furthermore, available data on morphology and composition are better known because they have been studied and re-studied in the search for the most useful class criteria.

The effort to define classes in higher categories in quantitative terms as fully as possible will also encourage efforts to define soil series and to describe map units more completely and more accurately. Thus, the trend toward more comprehensive and exact descriptions of series and map units will be reinforced. This has been the trend over the years, as is evident from comparisons of series descriptions prepared since the first standard ones were written 25 years ago.

Completion of the classification scheme by the grouping of all series into classes in higher categories will facilitate comparisons of series among areas, as is required in the correlation process. Such grouping of all series has not been completed in any system used in the past. When the new system is completed, it will therefore help field scientists to identify possible competing series when a new series is being proposed. It will also serve to identify some of the conflicts or overlapping among series established in different parts of the country.

The greater emphasis placed on soil morphology and composition in the new system as compared with older systems will encourage the same practice for concepts of series and definitions of map units. It is evident now that classification systems constructed in the past have been based on one or more of the factors of soil formation, underlying rock, physiographic position, or other features related to but not characteristic of soil (9). It is also evident that with the passing of time, progressively more emphasis has been given to soil morphology and composition in descriptions of series and map units. This can be demonstrated by comparisons of descriptions of soil series prepared at different times since the first standard ones of 25 years ago. It is true that standards and techniques for describing soils have been improved greatly during this same period. These improvements have also contributed to the

increased emphasis on morphology and composition. The total effort, however, has meant more attention to the characteristics of soil itself. Prospects are that such attention will be enhanced by the emphasis given to soil morphology and composition in the class definitions in the new system.

REFERENCES

(1) BALDWIN, M., KELLOGG, C. E., AND THORP, J. 1938 Soil classification. [In *Soils and Men* (U. S. Dep. Agr. Yearbook), pp. 979–1001.]

(2) BUREAU OF SOILS 1903 *Instructions to Field Parties and Descriptions of Soil Types.* U. S. Dep. Agr.

(3) DEWEY, J. 1958 *Experience and Nature*, p. 219. Dover Publications, New York.

(4) KELLOGG, C. E., *et al.* 1949 Soil classification. *Soil Sci.* 67: 77–191.

(5) KELLOGG, C. E. 1959 *Soil Classification and Correlation in Soil Survey.* U. S. Dep. Agr., Soil Conservation Service, processed.

(6) MARBUT, C. F. 1927 A scheme of soil classification. *Proc. Intern. Congr. Soil Sci. 1st Congr.* 4: 1–31.

(7) MARBUT, C. F., *et al.* 1913 *Soils of the United States.* U. S. Dep. Agr. Bur. Soils Bull. 96.

(8) SIMONSON, R. W., AND GARDNER, D. R. 1960 Concept and functions of the pedon. *Trans. Intern. Congr. Soil Sci., 7th Congr. (Madison)* 4: 127–131.

(9) SIMONSON, R. W. 1962 Soil classification in the United States. *Science* 137: 1027–1034.

(10) SOIL SURVEY STAFF 1951 Soil survey manual. *U. S. Dep. Agr. Handbook No. 18.*

(11) SOIL SURVEY STAFF 1960 *Soil Classification—A Comprehensive System. 7th Approximation.* U. S. Dep. Agr., Soil Conservation Service.

(12) SOIL SURVEY STAFF 1962 *Identification and Nomenclature of Soil Horizons.* Supplement to the U. S. Dep. Agr. Handbook No. 18.

BIBLIOGRAPHY

Aandahl, A. R., S. W. Buol, D. E. Hill, and H. H. Bailey, eds., 1974, Histosols: Their Characteristics, Classification, and Use, *Soil Sci. Soc. America Spec. Pub. No. 6*, Madison, Wis., 136 p.

Ahn, P. M., 1961, *Soils of the Lower Tano Basin, South-Western Ghana*, Ghana Ministry of Food and Agriculture, Kumasi, 265 p.

Albareda Henera, J. M., 1943, Clasificaciones y tipos de seulos, *Inst. Edafol. Madrid Anales*, pp. 151–192, 272–407.

Albareda Henera, J. M., 1944, Clasificaciones y tipos de seulos, *Inst. Edafol. Madrid Anales*, pp. 142–155.

Arany, S., 1960, Classification of the Hungarian Salt-affected Soils, *A. Thaer. Arch.* **4**:23–36 (English summary).

Arkley, R. J., 1968, Statistical Methods in Soil Classification, *9th Intern. Congr. Soil Sci. Trans.* **4**:187–192.

Aubert, G., 1955, Sur quelques problèms de pédogenase et de classification des sols-abordés à Leopoldville, *Assoc. Francaise Étude Sol Bull. No. 63.*

Aubert, G., 1964, La classification des sols utilisée par les pédologues francais en zone tropicale au aride, *Sols Africains* **9**:92–105.

Aubert, G., 1965a, Classification des sols Tableaux de classes, sous-classes, groupes et sous-groupes utilises par la section de pédologie de l'O.R.S.T.O.M., *Cahier O.R.S.T.O.M. ser. Pédologie* **3**:269–288.

Aubert, G., 1965b, La classification Pédologie utilisée en France, *Pedologie*, Spec. Ser. **3**:25–56.

Avery, B. W., 1965, Soil Classification in Britain, *Pedologie*, Spec. Ser. **3**:75–90.

Avery, B. W., 1968, General Soil Classification: Hierarchial and Coordinate, *9th Intern. Congr. Soil Sci. Trans.* **4**:169–175.

Barrat, B. C., 1969, A Revised Classification and Nomenclature of Microscopic Soil Materials with Particular Reference to Organic Compounds, *Geoderma* **2**:257–271.

Basu, J. K., and S. S. Sirur, 1938, Soils of the Deccan Canals, I: Genetic Soil Survey and Classification, *Indian Jour. Agric. Sci.* **8**:6–37.

Bennett, H. H., and R. V. Allison, 1928, *The Soils of Cuba*, Tropical Plant Research Foundation, Washington, D. C., 409 p.

Bidwell, O. W., L. F. Marcus, and P. K. Sarkar, 1964, Numerical Classification of Soils by Electric Computer, *8th Intern. Congr. Soil Sci. Trans.* **5**:933–941.

371

Bibliography

Blackburn, G., 1962, The Uses of Soils Classification and Mapping in Australia. *Intern. Soc. Soil Sci. Trans.*, Comm. IV and V, (New Zealand), pp. 284–290.

Blasco, M. L., 1968, Informacion preliminar de los suelos de Amazonas Colombiano, *Anales Edafol, Agrobiol.* **24**:47–55.

Buol, S. W., F. D. Hole, and R. J. McCracken, 1980, *Soil Genesis and Classification*, Iowa State University Press, Ames, Ia., 404 p.

Buringh, P., 1979, *Introduction to the Study of Soils in Tropical and Subtropical Regions*, Centre for Agricultural Publishing and Documentation, Wageningen, The Netherlands, 146 p.

Campbell, I. B., and G. G. C. Claridge, 1969, A Classification of Frigic Soils—the Zonal Soils of the Antarctic Continent, *Soil Sci.* **107**:75–85.

Campbell, N. A., M. J. Mulcahy, and W. M. McArthur, 1970, Numerical Classification of Soil Profiles on the Basis of Field Morphological Properties, *Australian Jour. Soil Res.* **8**:43–58.

Cardoso, J. C., 1966, Classification of the Soils of Southern Portugal According to the 7th Approximation, *Conf. Mediterranean Soils Trans.* (Madrid), pp. 395–399.

Charter, C. F., 1954, Colloquium on Soil Classification, *5th Intern. Congr. Soil Sci. Trans.* **4**:497–501.

Cline, M. G., 1963, Logic of the New System of Soil Classification, *Soil Sci.* **96**:17–22.

Cline, M. G., 1977, Historical Highlights in Soil Genesis, Morphology, Classification, *Soil Sci. Soc. Am. J.* **41**:250–254.

Coffey, G. N., Chmn., 1914, Progress Report of the Committee on Soil Classification and Mapping, *J. Am. Soc. Agron.* **6**:284–287.

Crocker, R. L., 1946, The Simpson Desert Expedition, 1939. Scientific reports: No. 8, the Soils and Vegetation of the Simpson Desert and Its Borders, *Royal Soc. South Australia Trans.* **70**:235–258.

Crowther, E. M., 1932, Climate, Clay Composition and Soil Type, *2nd Intern. Congr. Soil Sci. Trans.* **5**:15–23.

de Bakker, H., 1973, Hydromorphic Soils in the System of Soil Classification for The Netherlands, in *Pseudogley and Gley*, E. Schlichting and U. Schwertmann, eds., Verlag Chemie, Weinheim, W. Germany, pp. 405–412.

Dement, J. A., and L. J. Bartelli, 1969, The Role of Vertic Subgroups in the Comprehensive Soil Classification System, *Soil Soi. Soc. Am. Proc.* **33**:129–131.

Del Villar, E. H., 1939, A New Contribution to a Universal Objective Classification of Soils, *Bodenkd. Forschungen* **6**:221–226.

Didio, V., 1964, La classification des sols selon leus valurs pour l'irrigation, *8th Intern. Congr. Soil Sci. Trans.* **5**:891–896.

Dokuchaev, V. V., 1967, *Russian Chernozem*, Israel Program for Scientific Translations, Jeruslaem, 419 p. (Translated by N. Kaner from the original Russian publication "russki chernozem" of 1883.)

Duchaufour, P., 1963, Soil Classification: A Comparison of the American and the French Systems, *J. Soil Sci.* **14**:149–155.

Duchaufour, P., 1970, *Précis de Pédologie*, Maason, Paris, 481 p.

Duchaufour, P., 1977, *Pédologie: 1. Pedogenese et classification*, Masson, Paris, 477 p.

Duchaufour, P., and B. Souchier, 1966, Note sur les criteres de classification des sols lessives, *Conf. Mediterranean Soils Trans.* (Madrid), pp. 401–405.

Dudal, R., ed., 1965, Dark Clay Soils of Tropical and Subtropical Regions, *FAO Development Paper No. 83*, 147 p.

Dudal, R., and M. Soepratohardjo, 1957, Soil Classification in Indonesia, *Contr. Gen. Agric. Res. Station Bogor, Indonesia* **148**:1–16.

Dregne, H. E., 1976, *Soils of Arid Regions,* Elsevier, Amsterdam, 237 p. (especially Chapter 2: Soil Classification).

Drew, J. V., ed., 1967, Selected Papers in Soil Formation and Classification, *Soil Sci. Soc. Am. Spec. Pub. No. 1*, 428 p.

Edelman, C. H., 1950, *Soils of the Netherlands,* North Holland, Amsterdam, 177 p.

Ehwald, E., 1965, Die neue amerikanische Bodeniklassification, *Deutsch. Akad. Landwirt. Berlin Sitzungsber.* **14**:1–108.

Ehwald, E., I. Lieberoth, and W. Schanecke, 1966, Zur Systematik der Boden der Deutsche Demokratischen Republik besonders in Hinblick auf die Bodenkartierung, *Deutsch. Akad. Landwirt. Berlin Sitzungsber.* **15**:1–95.

Fallou, F. A., 1962, *Pedologie oder allgemeine uder besondere Bodenkunde,* G. Shoenfeld, Dresden, Germany, 488 p.

Farnham, R. S., and H. R. Finney, 1965, Classification and Properties of Organic Soils, *Ad. Agron.* **17**:115–162.

Feld, J., 1948, Early History and Bibliography of Soil Mechanics, *2nd Intern. Conf. Soil Mech. Found. Eng. Proc.* (Rotterdam) **1**:1–7.

Fields, M., 1968, Constitutional Classification of Soils, *9th Intern. Congr. Soil Sci. Trans.* **5**:37–40.

Filipovaki, G., V. Neugebauer, M. Ćisiś, A. Skoric, and M. Zivković, 1964, Soil Classification in Yugoslavia, *8th Intern. Congr. Soil Sci. Trans.* **4**:177–185.

Finkl, C. W., Jnr., 1967, Geographers, Pedolgists, and the New System of Soil Classification: A Commentary for Geographers, *Prof. Geographer* **19**:239–243.

FitzPatrick, E. A., 1967, Soil Nomenclature and Classification, *Geoderma* **1**:91–105.

FitzPatrick, E. A., 1971, *Pedology,* Oliver and Boyd, Edinburgh, 306 p. (especially Chapter 6: Soil Classification—A Review).

Fölster, J., 1971, Ferrallitische Böden aus sauren metamorphen Gestein in den feuchten und wechselfeuchten Tropen Africas, *Göttingen Bodenkdi. Berichte* **20**:

Forest Soils Division Staff, 1975, Classification of Forest Soils in Japan, *Gov. Forest Exp. Station Bull.* (Tokyo) **280**:21–28 (English summary).

Ganssen, R., 1957, *Bodengeographie,* Koehler, Stuttgart, 219 p.

Gaucher, G., 1973, Can the Geochemistry of Pedological Processes Be Used Eventually as a Fundamental Principle in Soil Classification?, *Compt. Rend. Seances Acad. Agric. France* **59**:284–292.

Gedroits, K. K., 1929, Der absorbierende Bodenkomplex und die adsorbierten Bodenkationen als Grundlage der genetische Bodenklassifikation, *Kolloidchem. Beihefte* **30**:1–112. (Translated by the Israel Program for Scientific Translations, Jerusalem, 1966, as Genetic Soil Classification Based on the Absorbed Soil Cations.)

Gerasimov, I. P., and E. N. Ivanova, 1959, Comparison of Three Scientific

Trends in Resolving General Questions of Soil Classification, *Soviet Soil Sci.* pp. 1190–1205. (Translated from *Pochvovedenie* **11**:1–18, 1960.)

Geze, B., 1942, Observation sur les sols du Cameroun occidentale, *Ann. Agron.* **12**:103–131.

Gibbons, F. R., 1968, Limitations to the Usefulness of Soil Classification, *9th Intern. Congr. Soil Sci. Trans.* **4**:159–167.

Glangeaud, L., 1956, Intervention sur la classification des sols, *6th Intern. Congr. Soil Sci. Trans.* **1**:176–179.

Greene, H., 1945, Classification and Use of Tropical Soils, *Soil Sci. Soc. Am. Proc.* **10**:392–396.

Gruijter, J. J. de, 1977, Numerical Classification of Soils and Its Application in Survey, *Agric. Res. Report* **855**:1–117 (English summary.)

Hai, L. T., 1962, *Soils of Taiwan*, Society of Soil Scientists and Fertilizer Technologists of Taiwan, Taipei.

Hardy, F., 1949, Soil Classification in the Caribbean Region. *Commonwealth Bur. Soil Sci. Tech. Commun.* **46**:64–75.

Hole, F. D., and M. Hironaka, 1960, An Experiment in Ordination of Some Soil Profiles. *Soil Sci. Soc. Am. Proc.* **24**:309–312.

Hubrich, H., 1973, The Classification of Soil Types According to Hydrological Characteristics. *Archiv. Acker-und Pflanzenbau Bodenkunde* **17**:795–805.

International Society of Soil Science, Committee Report, 1967, Proposal for a Uniform System of Soil Horizon Designations, *Intern. Soc. Soil Sci. Bull.* **31**:4–7.

Jansen, I. J., and R. W. Arnold, 1976, Defining Ranges of Soil Characteristics, *Soil Sci. Soc. Am. J.* **40**:89–92.

Jenny, H., 1930, Hochgebirgaboden, *Blanck's Handbuch der Bödenlehre* **3**:96–118.

Jones, T. A., 1959, Soil Classification—a Destructive Criticism, *J. Soil Sci.* **10**:196–200.

Kellogg, C. E., 1936, Development and Significance of the Great Soil Groups of the United States, *U.S. Dept. Agric. Misc. Pub. 229.*

Kellogg, C. E., 1938, Recent Trends in Soil Classification, *Soil Sci. Soc. Am. Proc.* **3**:253–259.

Kellogg, C. E., 1949, Preliminary Suggestions for the Classification and Nomenclature of Great Soil Groups of Tropical and Equatorial Regions, *Commonwealth Bur. Soil Sci. Tech. Commun.* **46**:78–85.

Kellogg, C. E., 1963, Why a New System of Soil Classification, *Soil Sci.* **76** (Special issue):1–5.

Kovda, V. A., and E. V. Lobova, eds., 1968, *Geography and Classification of the Soils of Asia*, Israel Program for Scientific Translations, Jeruslaem, 267 p.

Kowalinski, S., 1966, Attempt at a New Classification of the Soils of Europe, *Soviet Soil Sci.* pp. 76–79. (Translated from *Pochvovedenie* **3**:37–41, 1966.)

Kubiëna, W. L., 1953, *The Soils of Europe*, Thomas Murby, London, 317 p.

Kyuma, K., 1972, Numerical Methods in Soil Classification, *Pedologist* **16**:49–60.

Le Roux, J., and D. M. Scotney, 1970, *A Key to the Soils of Natal.* Department of Agricultural Technical Services, Pretoria, 96 p.

Liverosky, Y. A., 1969, Some Unresolved Problems in Classification and Systematization of USSR Soils, *Soviet Soil Sci.* **1**:106–116. (Translated from *Pochvovedenie* **2**:119–130.)

McCraw, J. D., 1960, Soils of the Ross Dependency, Antarctica, *New Zealand Soc. Soil Sci. Proc.* **4**:30–35.

MacFarlane, I. C., and G. P. Williams, 1974, Some Engineering Aspects of Peat Soils, in *Histosoils, Their Characteristics, Classification, and Use,* Soil Science Society of America Spec. Pub. No. 6, pp. 79–93.

MacVicar, C. N., and J. de Villiers, eds., 1977, *Soil Classification: A Binomial System for South Africa,* Department of Agricultural Technical Services Sci. Bull. 390, Pretoria.

MacVicar, C. N., 1980, Advances in Soil Classification and Genesis in Southern Africa, *8th Congr. Soil Sci. Soc. South Africa Proc.,* South Africa Dept. Agric. Tech. Services Tech. Comm. No. 165, **165**:22–40.

Manil, G., 1956, Repport general sur le probleme de la classification des sols, *6th Intern. Congr. Soil Sci. Trans.* **1**:166–174.

Manil, G., 1959, Aspects pedologiques du problem de la classification des sols forestiers, *Pedologie* **9**:214–226.

Marbut, C. F., 1923, Soils of the Great Plains, *Assoc. Am. Geog. Ann.* **13**:41–66.

Marbut, C. F., 1926, The Classification of Arid Soils, *4th Intern. Conf. Pedology Actes* (Rome) **3**:362–375.

Marbut, C. F., 1951, Soils, Their Genesis and Classification, in *Soil Science Society of America Memorial Volume: The Life and Work of C. F. Marbut,* Soil Science Society of America, Madison, Wis., 134 p.

Marbut, C. F., H. H. Bennett, J. E. Lapham, and M. H. Lapham, 1913, Soils of the United States, *U.S. Bur. Soils Bull.* 96, 791 p.

Martini, J. A., 1967, Principales grupos de suelos de America Central y Mexico, *Fititecnia Latinoamericanan* **4**:57–79.

Matsui, T., 1966, A Proposal on a New Classification System of Paddy Soils in Japan, *Pedologist* **10**:68–87.

Matsui, T., 1968, General Characteristics of the Soil Geography of Japan, *Pedologist* **12**:25–35.

Middleburg, H. A., 1950, Tentative Scheme for Classification of Tropical and Subtropical Soils, *4th Intern. Congr. Soil Sci. Trans.* **4**:139–142.

Milne, G., 1935, Some Suggested Units of Classification Particularly for E. African Soils, *Soil Res.* **4**:183–198.

Mohr, E. C. J., F. A. van Baren, and J. van Schuylenborgh, 1972. *Tropical Soils: A Comprehensive Study of Their Genesis,* Mouton-Ichtiar Baru-van Hoeve, The Hague, 481 p.

Moore, A. W., and J. S. Russell, 1966, Potential Use of Numerical Analysis and Adansonian Concepts in Soil Science, *Australian J. Sci.* **29**:141–142.

Moorman, F. R., 1961, The Soils of the Republic of Vietnam, Ministry of Agriculture, Saigon, 61 p.

Moorman, F. R., and C. R. Panabokke, 1961, Soils of Ceylon—A New Approach to the Identification and Classification of the Most Important Soil Groups of Ceylon, *Trop. Agricst.* **117**:1–73.

Mückenhausen, E., 1962, *Entstehung, Eigenschaften und Systematik der Boden der Bundesrepublik Deutschland,* DLG-Verlag, Frankfurt am. Main, 148 p.

Mückenhausen, E., 1970, Fortschritte in der Systematik der Böden der

Bundesrepublik Deutschland, *Mitteil. Deutsch. Bodenkund. Gesell. Bd.* **10**.

Muckenhirn, R. J., E. P. Whiteside, E. H. Templin, R. F. Chandler, Jr., and L. T. Alexander, 1949, Soil Classification and the Genetic Factors of Soil Formation, *Soil Sci.* **67**:93–106.

Nikiforoff, C. C., 1931, History of A, B, C Horizons, *Am. Soil Surv. Assoc. Bull.* **12**:67–70.

Norris, J. M., and M. B. Dale, 1971, Transition Matrix Approach to Numerical Classification of Soil Profiles, *Soil Sci. Soc. Am. Proc.* **35**:487–491.

Northcote, K. H., 1965, A Factual Key for the Recognition of Australian Soils, *C.S.I.R.O. Div. Report 2/65* (Australia).

Odell, R. T., J. C. Dijkerman, W. van Vuure, S. W. Melsted, A. H. Beavers, P. M. Sutton, L. T. Kurtz, and A. Miedema, 1974, Characteristics, Classification, and Adaptation of Soils in Selected Areas in Sierra Leone, West Africa, *Univ. Ill. Agric. Exp. Station Bull. No. 748.*

Orvedal, W. R., and M. J. Edwards, 1941. General Principles of Technical Grouping of Soils, *Soil Sci. Soc. Am. Proc.* **6**:386–391.

Owen, G., 1951, A Provisional Classification of Malayan Soils, *J. Soil Sci.* **2**:20–42.

Oyama, M., 1962, A Classification System of Paddy Rice Field Soils Based on Their Diagnostic Horizons, *Nat. Inst. Agric. Sci. Bull.* **B12**:303–372.

Pendleton, R. L., 1949, Classification and Mapping of Tropical Soils, *Commonwealth Bur. Soils Sci. Tech. Commun. 46* (Harpenden), pp. 93–97.

Pelisek, J., 1965, Genetische klassifikation und characteristik der Böden in der Tschechoslowakischen Sozialistischen Republik, *Pedologie* Spec. Issue **3**:185–201.

Pomerening, J. A., and E. G. Knox, 1962, A Test for Natural Soil Groups Within the Willamette Catena Population, *Soil Sci. Soc. Am. Proc.* **26**:282–287.

Powell, J. C., and M. E. Springer, 1965, Composition and Precision of Classification of Several Mapping Units of the Appling, Cecil, and Lloyd Series in Walton County, Georgia, *Soil Sci. Soc. Am. Proc.* **29**:454–458.

Prescott, J. A., and C. S. Piper, 1932, The Soils of the South Australian Mallee, *Royal Soc. South Australia Trans.* **56**:118–147.

Prescott, J. A., and R. L. Pendleton, 1952, Laterite and Lateritic Soils, *Commonwealth Bur. Soil Sci. Tech. Commun. No. 47.*

Ragg, J. M., and B. Clayden, 1973, The Classification of Some British Soils According to the Comprehensive System of the United States, *Soil Survey England Wales Tech. Monogr. No. 3*, 227 p.

Raychaudhuri, S. P., 1954–1955, Classification and Nomenclature of Indian Soils, *J. Soil Water Conserv.* (India) **3**:92–96.

Raychaudhuri, S. P., et al., 1964, *Soils of India*, Indian Council of Agricultural Research, New Delhi, 496 p.

Raychaudhuri, S. P., and S. V. Govindarajan, 1971, Soil Genesis and Soil Classification, in *Review of Soil Research in India*, Indian Council of Agricultural Research, New Delhi.

Riecken, F. F., 1963, Some Aspects of Soil Classification in Farming, *Soil Sci.* **76**:49–61.

Rieger, S., 1974, Arctic Soils, in *Arctic and Alpine Environments*, J. D. Ives and R. G. Barry, eds., Methuen, London, pp. 749–769.

Rode, A. A., 1962, *Soil Science,* Israel Program for Scientific Translations, Jerusalem, 517 p. (Translated by A. Gourevitch.)

Rozov, N. N., N. A. Karavayeva, and A. A. Rode, 1960, Second plenum of the Committee on the Nomenclature, Systematics, and Classification of Soils—for the Academy of Sciences of the USSR, *Soviet Soil Sci.* **pp.** 1036–1041. (Translated from *Pochvovedenie* 9:109–115, 1958.)

Rozov, N. N., and E. N. Ivanova, 1968, Soil Classification and Nomenclature Used in Soviet Pedology, Agriculture and Forestry, *FAO World Soil Resour. Report No. 32,* pp. 53–77.

Ruhe, R. V., and R. B. Daniels, 1958, Soils, Paleosols, and Soil Horizon Nomenclature, *Soil Sci. Soc. Am. Proc.* **22:**66–69.

Sarker, P. K., O. W. Bidwell, and L. F. Marcus, 1966, Selection of Characteristics for numerical Classification of Soils, *Soil Sci. Soc. Am. Proc.* **30:**269–272.

Seki, T., 1926, Soil Classification and Mapping System for Japan, *3rd Pan-Pacific Sci. Congr. Proc.* (Tokyo).

Selby, M. J., and H. S. Gibbs, 1971, A New World Soil Map and International Soil Classification System Proposed by FAO/UNESCO, *6th New Zealand Geogr. Conf.* (Christchurch), pp. 111–118.

Shaw, C. F., 1927, A Uniform International System of Soil Nomenclature, *1st Intern. Congr. Soil Sci. Trans.* **4:**32–37.

Shaw, C. F., 1930, The Soils of China: A Preliminary Survey, *China Geol. Surv. Soil Bull.* **1:**1–38.

Sigmond, A. A. de, 1932–1933, Principles and Scheme of a General Soil System, *Soil Research* (Berlin) **3:**103–126.

Simonson, R. W., 1954, Morphology and Classification of the "Regur Soils" of India, *J. Soil Sci.* **5:**275–288.

Simonson, R. W., 1964, The Soil Series as Used in the USA, *8th Intern. Congr. Soil Sci. Trans.* **5:**17–24.

Simonson, R. W., 1980, Soil Survey and Soil Classification in the United States, *8th Congr. Soil Sci. Soc. South Africa Proc.* **165:**10–21.

Smith, G. D., 1973, Soil Moisture Regimes and Their Use in Soil Taxonomies, *Soil Sci. Soc. Am. Spec. Pub.* **5:**1–7.

Sneath, P. H. A., and R. R. Sokal, 1962, Numerical Taxonomy, *Nature,* **193:**855–860.

Stace, H. C. T., G. D. Hubble, R. Brewer, K. H. Northcote, J. R. Sleeman, M. J. Mulcahy, and E. G. Hallsworth, eds., 1968, *A Handbook of Australian Soils,* Rellim, Glenside, S. Austrialia, 435 p.

Standards Association of Australia, 1967, Australian Standard Methods of Testing Soils for Engineering Purposes, *Australian Standard A89–1966,* 121 p.

Stephens, C. G., 1953, *A Manual of Australian Soils,* C.S.I.R.O., Melbourne, 48 p.

Stephens, C. G., 1954, The Classification of Australian Soils, *5th Intern. Congr. Soil Sci. Trans.* **4:**155–160.

Stevens, J. G., 1968, The Soils of the Trucial States: Classification and Capability, *9th Intern. Congr. Soil Sci. Trans.* **4:**253–260.

Sys, C., 1959a, Cartographie et classification regionale des sols au Congo belge, *3rd Interafricain Sols Conf.* **1:**291–302.

Sys, C., 1959b, La classification des sols congolais, *3rd Interafricain Sols Conf.* **1:**303–312.

Sys, C., 1968, Suggestions for the Classification of Tropical Soils with Lateritic Materials in the American Classification, *Pedologie* **18**:189–198.

Tache, B., 1930, Die Humusböden der gemüssigten Breiten, *Blanck's Handbuch Bodenlehre* **4**:124–183.

Tavernier, R., and R. Marechal, 1962, Soil Survey and Soil Classification in Belgium, *7th Intern. Congr. Soil Sci. Trans.* Comm. IV and V, pp. 3–12.

Tavernier, R., and C. Sys, 1965, Classification of the Soils of the Republic of Congo (Kinshasa), *Pedologie* Spec. Ser. **3**:91–136.

Tedrow, J. C. F., and J. R. Cantlon, 1958, Concepts of Soil Formation and Classification in Arctic Regions, *Arctic* **11**:166–179.

Tedrow, J. C. F., and F. C. Ugolini, 1966, Antarctic Soils, in Antarctic Soils and Soil Forming Processes, J. C. F. Tedrow, ed., *Am Geophys. Union, Antarctic Res. Ser.* **8**: 161–177.

Thorp, J., 1936, *Geography of the Soils of China*, National Geologic Survey, Nanking, China, 552 p.

Thorp, J., and T. Tschau, 1936, Notes on Shantung Soils: A Reconnaissance Soil Survey of Shantung, *China Geol. Surv. Bull.* *14*, 132 p.

Thorp, J., and M. Baldwin, 1938, New Nomenclature of the Higher Categories of Soil Classification as Used in the Department of Agriculture, *Soil Sci. Soc. Am. Proc.* **3**:260–268.

Tiurin, I. V., 1930, Genesis and Classification of Forest Soils, *Pedologie* Spec. Ser. **25**:104–141.

Tiurin, I. V., 1965, The System of Soil Classification in the U.S.S.R., *Pedologie* Spec. Ser. **3**:7–24.

Tomaszewski, J., 1964, A System of World Soil Classification, *8th Intern. Congr. Soil Sci. Trans.* **5**:59–67.

Tsyganov, M. S., 1955, Fundamental Principles of Genetic Classification and Nomenclature of Soils, *Pedologie* **12**:52–63.

Van Wambeke, A. R., 1961, Les sols du Rwanda-Burundi, *Pedologie* **11**: 289–353.

Van Wambeke, A. R., 1962, Criteria for Classifying Soils by Age, *J. Soil Sci.* **13**:124–132.

Van Wambeke, A. R., 1967, Recent Developments in the Classification of the Soils of the Tropics, *Soil Sci.* **104**:309–313.

Veatch, J. O., 1927, The Classification of Organic Soils, *1st Intern. Congr. Soil Sci. Trans.* **4**:123–126.

Verster, E., and J. M. de Villiers, 1968, Criteria for the Recognition of Oxisols—a Preliminary Examination, *South Africa J. Agric. Sci.* **11**: 1–8.

Vilensky, D., 1925, The Classification of Soils on the Basis of Analogous Series in Soil Formation, *Intern. Soc. Soil Sci. Proc.*, New Series **5**: 224–241.

Vine, H., 1949, Nigerian Soils in Relation to Parent Materials, *Commonwealth Soil Sci. Tech. Bull.* **46**:22–29.

Webster, R., 1968, Fundamental Objections to the 7th Approximation, *J. Soil Sci.* **19**:354–365.

Westin, F. C., J. Avilon, and A. Buatomente, 1968, Characteristics of Some Venezuelan Soils, *Soil Sci.* **105**:92–102.

Wilde, S. A., 1940, Classification of Gley Soils for the Purpose of Forest Management and Reforestation, *Ecology* **21**:24–44.

Bibliography

Wilde, S. A., 1949, Glinka's Later Ideas on Soil Classification, *Soil Sci.* **67**: 411–413.

Wilde, S. A., 1958, *Forest Soils*, Ronald, New York, 537 p.

Winters, E. 1949, Interpretative Soil Classification: Genetic Groupings, *Soil Sci.* **67**:131–140.

Yaalon, D. H., 1959, Classification and Nomenclature of Soils in Israel, *Res. Council Israel Bull.* **8**:91–118.

Yaalon, D. H., 1960, Some Implications of Fundamental Concepts of Pedology in Soil Classification, *7th Intern. Congr. Soil. Sci. Trans.* **4**:119–123.

AUTHOR CITATION INDEX

382

SUBJECT INDEX

A-B-C horizon designations, 2
Accidental characteristics, philosophy of
    classification, 80
Accumulating Differentia, Principle of, 87–
    88
Afansiev, J. N., classification system, 95
A horizon
    anthropogenic, 253
    definition of, 56, 222–223
Alfisols, 199
Alkali soils, 32, 50
All-Union Conference of Soil Scientists,
    169
Alluvial soils, 22, 157, 235
Al$_2$O$_3$/Fe$_2$O$_3$ ratio, 248
Alpine Meadow soils, 311
Alpine Turf soils, 311
American Association of State Highway
    Officials (AASHO), classification
    system, 340–341
American Society for Testing and Materials
    (ASTM), classification system 341–
    343
Andosols, 272
    gleyed, 274
    wet, 274
Arctic soils, 318
    brown, 311
Argillic horizon, 199, 202, 233
Aridosols, 199
Arid soils, unleached, 49
Artificial soils, 16
Australia, classification system, 295–301.
    See also Soils Atlas; Factual key
Automorphic soils, 173
Azonal soils, 13, 24, 241

B horizon
    definition of, 3, 56
    latosolic, 247
    podzolic, 233
    textural, 253
    weathered (Bw), 234

Biogenic soils, 101–102, 183, 306
Biohalogenic soils, 101–102, 183
Biohydrogenic soils, 183
Biolithogenic soils, 101–102, 183
Black earth, 34, 156
Bleisand, 19, 30
Bog soils, 285
Borowina, 34
Braunerde, 157
Braunlehm, 293
Brazil, classification system, 247–258
Brown soils, 42
Burozem, 185

Calcareous soils, 43, 235
Canada, classification system, 240–246. See
    also National Soil Survey Committee
    of Canada; Saskatchewan Soil Survey
Categorical rank, 81
Catena, 155, 241
Ceiling of independence, 84–85
Chemical soil properties
    composition, 43
    use in classification, 23, 43
Chernozemic horizon, 257
Chernozem soils, 13, 18, 19, 25–26, 156
Chestnut soils, 13, 25
C horizon, definition of, 56, 149, 222–223
Classification criteria
    Marbuts's list, 86–87, 148
    Whiteside's list, 90–91
Clay triangle, 261
Cluster analysis, 333
Composition, mechanical, 41
Concretions, 30
Corn soils, 44
Cosmopolitan soils, 95
Cotton soils, 44
Crude soils, 21
Cryoborolls, 311
Cryosols, 311–320
Cryoturbation, 317
Cryumbrepts, 311

About the Editor

CHARLES W. FINKL, JNR., an Associate Professor at Nova University, is Director of the Institute of Coastal Studies in the Ocean Sciences Center at Port Everglades, Florida. Previous academic appointments were at Oregon State University, University of Western Australia, and Florida International University. He received the B.Sc. and M. Sc. degrees in Natural Resources and Soil Science from Oregon State University and the Ph.D. in Soil Science from the University of Western Australia. He served as staff geochemist for International Nickel Australia Limited from 1971 to 1974.

Dr. Finkl's research interests focus on the role of deep chemical weathering in landscape development, soil stratigraphy, coastal soils, and the geomorphology of cratonic regions. He has written many scientific papers and is editor of the The Encyclopedia of Soil Science, Parts I and II, in the Encyclopedia of Earth Science Series.

Dr. Finkl is a member of numerous professional societies including the Association of Southern Agricultural Scientists, Australian Society of Soil Science, British Society of Soil Science, Deutschen Bodenkundlichen Gesellschaft, International Society of Soil Science, Soil Science Society of America, and is also listed with the American Registry of Certified Professionals in Agronomy, Crops, and Soils as a Certified Professional Soil Scientist. He also serves as a corresponding member of the International Geographical Union, Commission on Geomorphological Survey and Mapping, and is a full member to the Subcommission on River and Coastal Plains.